T0325399

# The Future of Bluefin Tunas

# The Future of Bluefin Tunas

## Ecology, Fisheries Management, and Conservation

Edited by Barbara A. Block

Johns Hopkins University Press
Baltimore

Printed in Canada on acid-free paper
9 8 7 6 5 4 3 2 1

Johns Hopkins University Press
2715 North Charles Street
Baltimore, Maryland 21218-4363
www.press.jhu.edu

Library of Congress Cataloging-in-Publication Data

Names: Block, Barbara A. (Barbara Ann), 1958–, editor.
Title: The future of bluefin tunas : ecology, fisheries management, and conservation / edited by Barbara A. Block.
Description: Baltimore : Johns Hopkins University Press, 2019. | Includes bibliographical references and index.
Identifiers: LCCN 2018041203 | ISBN 9781421429632 (hardcover : alk. paper) | ISBN 1421429632 (hardcover : alk. paper) | ISBN 9781421429649 (electronic) | ISBN 1421429640 (electronic)
Subjects: LCSH: Bluefin tuna fisheries. | Bluefin tuna.
Classification: LCC SH351.B56 F88 2019 | DDC 639.2/758—dc23
LC record available at https://lccn.loc.gov/2018041203

A catalog record for this book is available from the British Library.

*Special discounts are available for bulk purchases of this book. For more information, please contact Special Sales at 410-516-6936 or specialsales@press.jhu.edu.*

Johns Hopkins University Press uses environmentally friendly book materials, including recycled text paper that is composed of at least 30 percent post-consumer waste, whenever possible.

# Contents

Preface vii
Contributors xi

ATLANTIC

1. The Journey from Overfishing to Sustainability for Atlantic Bluefin Tuna, *Thunnus thynnus* 3
Clay E. Porch, Sylvain Bonhommeau, Guillermo A. Diaz, Haritz Arrizabalaga, and Gary Melvin

2. Otolith Microchemistry: Migration and Ecology of Atlantic Bluefin Tuna 45
Jay R. Rooker and David H. Secor

3. Life History and Migrations of Mediterranean Bluefin Tuna 67
Haritz Arrizabalaga, Igor Arregui, Antonio Medina, Naiara Rodríguez-Ezpeleta, Jean-Marc Fromentin, and Igaratza Fraile

4. Use of Electronic Tags to Reveal Migrations of Atlantic Bluefin Tunas 94
Barbara A. Block

5. Spatial Mixing Models for Atlantic Bluefin Tuna 117
Nathan Taylor

PACIFIC

6. Life History of Pacific Bluefin Tuna, *Thunnus orientalis* 131
Tamaki Shimose

7. Migrations of Pacific Bluefin Tuna Tagged in the Western Pacific Ocean 147
Takashi Kitagawa, Ko Fujioka, and Nobuaki Suzuki

8. Transpacific Migration of Pacific Bluefin Tuna with Chemical Tracers 165
Daniel J. Madigan

9. Tagging to Reveal Foraging, Migrations, and Mortality of Pacific Bluefin Tuna 184
Rebecca E. Whitlock, Murdoch K. McAllister, and Barbara A. Block

SOUTHERN

10. Keys to Advancing the Management of Southern Bluefin Tuna, *Thunnus maccoyii* 211
Jessica H. Farley, Ann L. Preece, Mark V. Bravington, J. Paige Eveson, Campbell R. Davies, Karen Evans, Toby A. Patterson, Naomi P. Clear, Peter M. Grewe, Jason R. Hartog, Richard M. Hillary, Alistair J. Hobday, Matthew J. Lansdell, and Craig H. Proctor

11. Rebuilding Southern Bluefin Tuna: Past, Present, and Future 248
Richard M. Hillary, Ann L. Preece, and Campbell R. Davies

12. Bluefin Tunas in a Changing Ocean 270
Alistair J. Hobday, Barbara A. Muhling, Elliott L. Hazen, Haritz Arrizabalaga, J. Paige Eveson, Mitchell A. Roffer, and Jason R. Hartog

AQUACULTURE

13. Spain's Atlantic Bluefin Tuna Aquaculture 299
Aurelio Ortega and Fernando de la Gándara

14. The Resource and Environmental Intensity of Bluefin Tuna Aquaculture 312
Dane H. Klinger and Nicolas Mendoza

Index 335

# Preface

Ocean waters at least 3 km deep cover more than half of Earth's surface. One can travel by boat thousands of kilometers from Portugal or Japan and see, at first glance, only various shades of blue stretching toward the horizon. But for those of us lucky enough to have studied the oceans' top predators, the explosion of water created by a school of bluefin tuna in pursuit of their prey is a striking hint of the life below.

Bluefin tunas are extraordinary predators that ply our planet's open oceans and coastal waters. Three species of bluefin tunas—Atlantic (*Thunnus thunnus*), Pacific (*Thunnus orientalis*), and Southern bluefin (*Thunnus maccoyii*)—swim our subpolar, temperate, and tropical seas. These tunas can live for 40 years, dive more than 2,000 meters, and reach a body size of 650 kg. Bluefins are rare among fish for their endothermic physiology, unique stiff-bodied thunniform swimming, hydraulic fins, and powerful lunate tails. They have a sleek, fusiform shape packed in a thinly scaled skin that flashes metallic silver from below and neon blue, bronze, and yellow from above. The power afforded by their warm bodies coupled with their hydrodynamic efficiency enables remarkable migrations of tens of thousands of kilometers across our planet's oceans. They are the pinnacle of bony fish evolution, kings of their realm.

Our relationship to bluefins as a resource stretches back to antiquity, when these fish fueled the growth of western civilization around the

Second-century coin depicting two Atlantic bluefin tuna and Hercules.

Mediterranean Sea. For more than 2,000 years, we fished them sustainably, but bluefin tunas are currently among the species most sought after in global wildlife trade. Their ruby red flesh has a rich taste and unique texture prized for sashimi and sushi across the globe. In fewer than 50 years, the onslaught of industrial fishing has put bluefin tuna populations at risk across the planet. It is clear that we have the capacity to fish these magnificent animals to extinction.

Despite the overexploitation, careful management practices built on international collaborations and newly acquired scientific knowledge are now enabling recovery of some populations. The commercial trade and the removal of bluefin tunas from the ocean must now be carefully monitored and controlled to ensure that future generations can experience the power of these spectacular fish. This book is about the present and future of bluefin tunas.

I have spent most of my academic life studying northern bluefin tunas, first exploring their physiology, pondering their endothermic traits, measuring their metabolism with my students and postdocs, and studying their extraordinary cardiac dynamics. More recently, our lab in Monterey, California, at the Tuna Research and Conservation Center, which is a unique partnership between Stanford University and the Monterey Bay Aquarium, has been focused on the physiology and genomics of bluefin tuna maturation, the evolution of endothermy, the biomechanics of locomotion and maneuverability, the sequencing of the genome of Atlantic and Pacific bluefins, and the impacts of the Deepwater Horizon oil spill on the Gulf of Mexico population of tuna. Inspired first by my innovative and insightful mentors Dr. Francis G. Carey and Knut Schmidt Nielsen, I have strived for most of my career to understand the unique physiology of bluefin tuna in the context of their open ocean ecology. I learned the art of tracking a fish with my Woods Hole Oceanographic colleagues, Drs. Carey and John Kanwisher, and in our Stanford lab, together with engineers from multiple countries, we've pioneered the modern biologging technologies that enable us to track transoceanic tuna migrations and to probe their deepest dives—beyond where our pressure sensors can go. The information we've obtained on the bluefin's travels has been extraordinary. However, our curiosity has always been interrupted by the intensity of the fisheries that seek these fish for profit.

I asked my colleagues from around the globe to contribute to *The Future of Bluefin Tunas* because I am concerned about the status of the planet's bluefin tuna populations. Removal of their biomass globally would inevitably contract bluefin genomic biodiversity. It remains up to our generation to preserve it—to ensure that bluefins have a future in our rapidly changing seas. There is an urgent need for a new generation of pelagic fisheries models that unite the flood of new types of biological knowledge obtained with modern techniques (e.g., biologging, genomics, genetics, and isotopic ecology) with the quantitative and technological possibilities of modern fisheries science to allow for proper spatial management and dynamic protection of spawning populations as well as better estimates of population biomass.

Our collective goal in compiling this knowledge into a book is to inspire conservation of all three species, to work together to translate advances made on one bluefin species in one ocean basin to a different species in another—and to generate a collaborative atmosphere between biologists, oceanographers, fisheries scientists, policymakers, conservationists, and industry.

This volume, the e-book, and the accompanying short video are products that demonstrate our collaborative potential among nations to address the issues confronting proper management of these incredible species. Bluefin tunas have inspired mankind for millennia. They solved the longitude problem before we did. And as they reveal their secrets, they inspire biomimetic designs such as those for underwater vehicles. They deserve a future that ensures that they, and the biodiversity to which they contribute, remain on this planet for those who follow us and for all of time.

I thank Stanford University and the Monterey Bay Aquarium, particularly Ms. Julie Packard, for providing the funds that allowed us to accomplish this project. Additional support was provided by the TAG-A-Giant fund of the Ocean Foundation. I am indebted to the contributors to the volume, the referees, and Ms. Sayzie Koldys and Ms. Shana Miller for reviewing, editing, and production efforts that enabled the completion of the volume.

# Contributors

**Igor Arregui** is Senior Researcher, AZTI–Tecnalia, Marine Research Division.

**Haritz Arrizabalaga** is Principal Researcher, AZTI–Tecnalia, Marine Research Division.

**Barbara A. Block** is the Charles and Elizabeth Prothro Professor in Marine Sciences, Hopkins Marine Station, Stanford University.

**Sylvain Bonhommeau** is Marine Ecology Researcher, IFREMER DOI.

**Mark V. Bravington** is Principal Research Scientist, CSIRO, Data61.

**Naomi P. Clear** is Experimental Scientist, CSIRO, Oceans and Atmosphere.

**Campbell R. Davies** is Senior Principal Research Scientist, CSIRO, Oceans and Atmosphere.

**Fernando de la Gándara** is Senior Researcher, Spanish Institute of Oceanography, Planta de Cultivos Marinos.

**Guillermo A. Diaz** is Research Fishery Biologist, US National Marine Fisheries Service, Southeast Fisheries Science Center.

**Karen Evans** is Senior Research Scientist, CSIRO, Oceans and Atmosphere.

**J. Paige Eveson** is Senior Experimental Scientist, CSIRO, Oceans and Atmosphere.

**Jessica H. Farley** is Senior Experimental Scientist, CSIRO, Oceans and Atmosphere.

**Igaratza Fraile** is Researcher, AZTI–Tecnalia, Marine Research Division.

**Jean-Marc Fromentin** is Director of Research, IFREMER, UMR MARBEC—Marine Biodiversity, Exploitation and Conservation.

**Ko Fujioka** is Researcher, National Research Institute of Far Seas Fisheries, Japan Fisheries Research and Education Agency.

**Peter M. Grewe** is Senior Research Scientist, CSIRO, Oceans and Atmosphere.

**Jason R. Hartog** is Senior Experimental Scientist, CSIRO, Oceans and Atmosphere.

**Elliott L. Hazen** is Research Ecologist, Environmental Research Division, Southwest Fisheries Science Center, National Marine Fisheries Service, National Oceanic and Atmospheric Administration; and Associate Researcher, Department of Ecology and Evolutionary Biology, University of California–Santa Cruz.

**Richard M. Hillary** is Principal Research Scientist, CSIRO, Oceans and Atmosphere.

**Alistair J. Hobday** is Research Director, CSIRO, Oceans and Atmosphere.

**Takashi Kitagawa** is Associate Professor, Atmosphere and Ocean Research Institute, University of Tokyo.

**Dane H. Klinger** is Aquaculture Innovation Fellow, Conservation International, and Affiliated Scholar, Center on Food Security and the Environment, Stanford University.

**Matthew J. Lansdell** is Research Technician, CSIRO, Oceans and Atmosphere.

**Daniel J. Madigan** is HUCE Environmental Fellow, Harvard University Center for the Environment.

**Murdoch K. McAllister** is Associate Professor in the Quantitative Modeling Group and Canada Research Chair in Fisheries Assessment, Institute for the Oceans and Fisheries, Aquatic Ecosystems Research Laboratory, University of British Columbia–Vancouver.

**Antonio Medina** is Facultad de Ciencias del Mar y Ambientales, Departamento de Biología, Universidad de Cádiz, Campus de Excelencia Internacional del Mar.

**Gary Melvin** is Research Scientist, Department of Fisheries and Oceans, Population Ecology Division, St Andrews Biological Station.

**Nicolas Mendoza** is CEO and Founder, OneForNeptune.

**Barbara A. Muhling** is Associate Research Scholar, Princeton University Program in Atmospheric and Oceanic Sciences and NOAA Geophysical Fluid Dynamics Laboratory.

**Aurelio Ortega** is Head of Aquaculture, Spanish Institute of Oceanography, Planta de Cultivos Marinos.

**Toby A. Patterson** is Senior Research Scientist, CSIRO, Oceans and Atmosphere.

**Clay E. Porch** is Director, US National Marine Fisheries Service, Southeast Fisheries Science Center.

**Ann L. Preece** is Senior Experimental Scientist, CSIRO, Oceans and Atmosphere.

**Craig H. Proctor** is Senior Experimental Scientist, CSIRO, Oceans and Atmosphere.

**Naiara Rodríguez-Ezpeleta** is Senior Researcher, AZTI, Herrera Kaia.

**Mitchell A. Roffer** is Founder, Roffer's Ocean Fishing Forecasting Service, Inc.

**Jay R. Rooker** is Regents Professor, Department of Marine Biology and Department of Wildlife and Fisheries Sciences, and McDaniel Chair of Marine Fisheries, Department of Marine Biology, Texas A&M University.

**David H. Secor** is USM Regents Professor, Chesapeake Biological Laboratory, University of Maryland Center for Environmental Science.

**Tamaki Shimose** is Researcher, Research Center for Subtropical Fisheries, Seikai National Fisheries Research Institute, Japan Fisheries Research and Education Agency.

**Nobuaki Suzuki** is Researcher, National Research Institute of Far Seas Fisheries, Japan Fisheries Research and Education Agency.

**Nathan Taylor** is Bycatch Coordinator, International Commission for the Conservation of Atlantic Tunas.

**Rebecca E. Whitlock** is Researcher, Department of Aquatic Resources, Swedish University of Agricultural Sciences.

ATLANTIC

CHAPTER 1

# The Journey from Overfishing to Sustainability for Atlantic Bluefin Tuna, *Thunnus thynnus*

Clay E. Porch, Sylvain Bonhommeau, Guillermo A. Diaz,
Haritz Arrizabalaga, and Gary Melvin

## Introduction

The Atlantic bluefin tuna (ABT), *Thunnus thynnus* (Linnaeus, 1758), is the largest of the tunas and among the largest of all bony fish, reaching to 3.3 m and 725 kg (Cort et al. 2013). The species is highly migratory and is broadly distributed through most of the Atlantic Ocean and its adjacent seas (Figure 1.1), thanks in large measure to a highly developed thermoregulatory system that allows it to thrive in waters as cold as 3°C (Carey and Lawson 1973, Block et al. 2001). Its great size and power have captivated fishermen and scientists alike since ancient times. Aristotle, Pliny the Elder, and Oppian wrote of tunas 2,000 years ago, and their bones have been excavated from prehistoric sites dating back to the Stone Age (Oppianus 177 BCE; Plinius 65 CE; Aristotelis 1635; Ravier and Fromentin 2001; Di Natale 2012, 2014; Puncher et al. 2016).

The fascination with Atlantic bluefin tuna has only grown in modern times. The demand for bluefin tuna for the sashimi market in Japan fuels a lucrative commercial fishery where a single fish can be worth tens of thousands of dollars. Researchers passionately pursue investigations of ABT, writing dozens of scientific papers every year. Public interest in this charismatic species, fanned by warnings of overfishing, has risen to a level usually reserved for whales (Porch 2005). The story of ABT has been told in compelling documentaries and popular books such as Safina's (1998) *Song for the Blue Ocean*, Maggio's (2000a,b) *Mattanza*, and Ellis's (2008) *Tuna: A Love Story*. The ABT even has its own reality show in National Geographic's *Wicked Tuna*. As Ellis puts it, ABT just may be "the world's best-loved fish" (Ellis 2008).

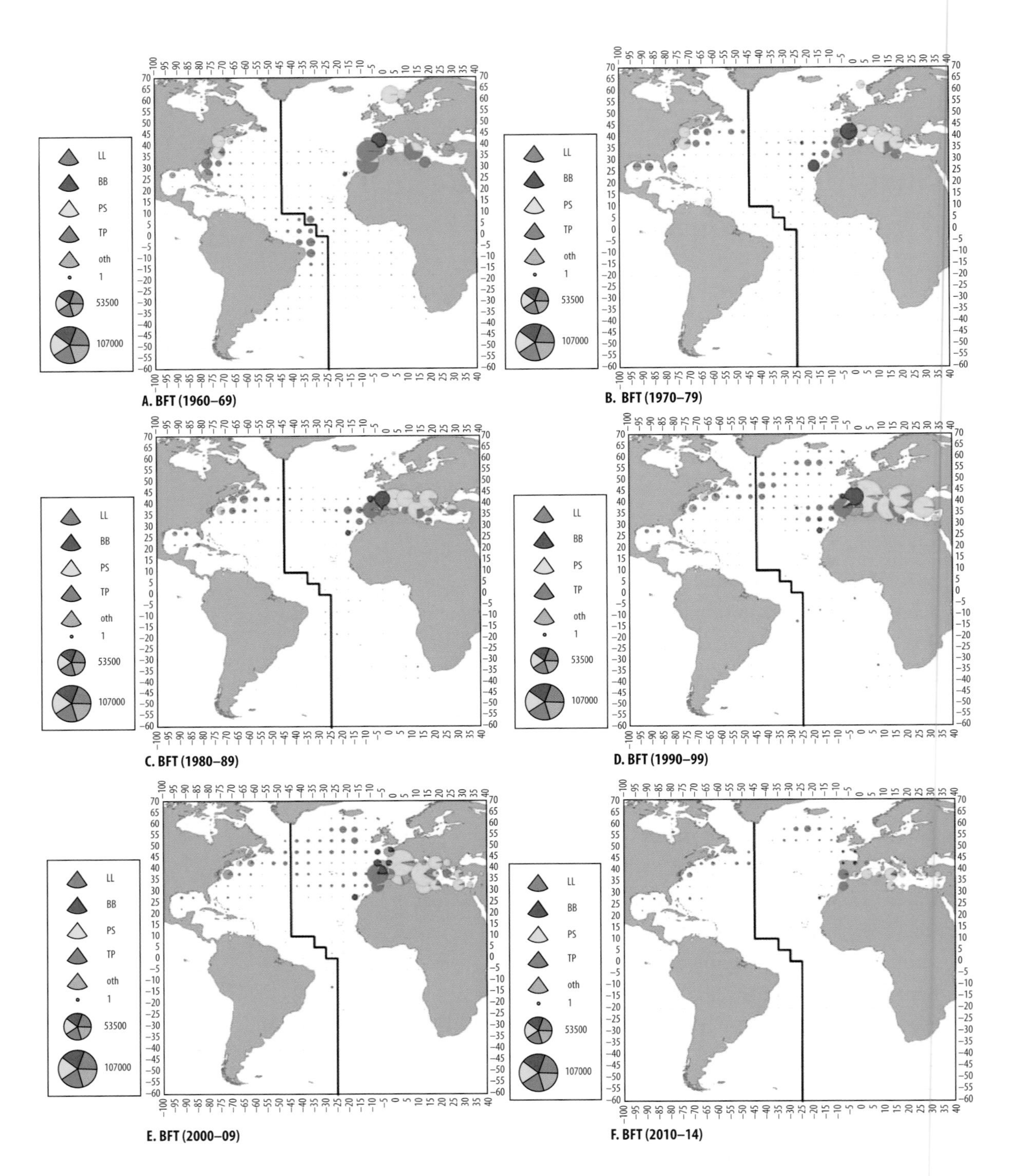


**Figure 1.1.** Geographic distribution of Atlantic bluefin tuna (BFT) catches per 5° × 5° and per main gears for each decade (*a–f*) from 1960 to 2014. The vertical black line beginning at 45°W in the North Atlantic represents the delineation of the two Atlantic bluefin tuna stocks as defined by the International Commission for the Conservation of Atlantic Tunas.

dation [Rec.] 02-07) would result in stock declines of approximately 3% per year, whereas a TAC of 2,100 t would allow the stock to increase at approximately 1.5% per year, and a TAC of 2,300 t would maintain the stock at the 2006 level (Anonymous 2007). In response, ICCAT adopted Recommendation 06-06, which reduced the TAC from 2,700 t to 2,100 t for 2007 and 2008 and maintained other fishing limitations. The 2007 SCRS acknowledged that this reduction in TAC was a step in the right direction but warned that the management regulations in place might be insufficient to rebuild the stock to maximum sustainable yield (MSY) levels (Anonymous 2008b).

A new stock assessment was conducted in 2008, and on the basis of the results, the SCRS strongly advised against an increase in TAC (Anonymous 2009b). The SCRS expressed concern that the rebuilding plan had been unsuccessful inasmuch as the biomass halfway through the rebuilding period was still below the level when the plan was initiated. As such, the SCRS advised ICCAT "to adopt more conservative catch levels that will result in a higher probability (for example, a 75% chance) that $B_{MSY}$ is achieved by the beginning of 2019" (Anonymous 2008a, p. 78). The SCRS indicated that, for example, under the more optimistic low-recruitment scenario, this target might be achieved with a TAC of 2,000 t. However, it also pointed out that the appraisal was much less optimistic when the index for the small Gulf of St. Lawrence fishery was excluded from the assessment, suggesting that a TAC of 1,500 t or lower might be necessary. Moreover, under the high-recruitment scenario, rebuilding could not have been achieved during the expected time frame even with zero catches. Responding to the SCRS, ICCAT reduced the TAC for years 2009 and 2010 to 1,900 t and 1,800 t (Rec. 08-04). The following year, the SCRS commented that the TACs adopted by ICCAT in Recommendation 08-04 had less than a 50% chance of meeting the target under the high-recruitment scenario (Anonymous 2009b).

*Eastern Atlantic and Mediterranean*

The 2006 stock assessment estimated a substantial decline in the biomass of the eastern Atlantic and Mediterranean stock (Anonymous 2007), prompting the SCRS to recommend substantial reductions in fishing mortality. In particular, the SCRS advised ICCAT to close the Mediterranean bluefin tuna fishery during the spawning season and to raise the minimum size, estimating that a full implementation of these actions would result in a catch of ~15,000 t. The SCRS also warned that the fishing capacity of the eastern Atlantic and Mediterranean bluefin tuna fleet was excessive and that actions had to be taken to reduce the impact of overcapacity as well as illegal fishing. Although the reported catches for 2004 (the last year considered in the 2006 assessment) were 32,567 t, preliminary analyses of fishing capacity suggested the total catch was closer to 50,000 t.

In response to the SCRS advice, ICCAT adopted Recommendation 06-05, which established a 15-year recovery plan for the eastern Atlantic and Mediterranean bluefin tuna stock. The main objective of the plan was to ensure the sustainability of the fishery by 2022 with a probability of 60%. The TAC was reduced to 29,500 t in 2007 and stepped down to 25,500 t by 2010; this was significantly lower than in 2006 (32,000 t) but much higher than the 15,000 t TAC advised by the SCRS. In addition, ICCAT established a number of other fishing restrictions, including closed fishing seasons, larger minimum size limits, and prohibiting the use of airplanes and helicopters.

The 2007 SCRS report reiterated the advice it provided in 2006, including a 15,000 t TAC (Anonymous 2008b). Nevertheless, in 2007 ICCAT did not adopt any new management measures for eastern Atlantic and Mediterranean bluefin tuna; the TAC remained unchanged. The 2008 stock assessment confirmed the results from the 2006 assessment, estimating that the current fishing mortality rate was three times as high as the level that would permit the stock to stabilize at the MSY level (Anonymous 2009b). The SCRS acknowledged the positive steps taken previously by ICCAT but once again advised a TAC of ~15,000 t. Perhaps more importantly, the SCRS formally confirmed widespread suspicions of illegal, unreported, and unregulated fishing. Catches from the mid-1990s through 2007 were estimated to be in the order of 50,000 to 61,000 t, nearly double what had been officially reported (Anonymous 2009b). In response, ICCAT extended the seasonal closures, froze fishing capacity, and imposed measures to adjust farming capacity. The TACs were also reduced (to 22,000 t, 19,950 t, and 18,500 t, for 2009, 2010, and 2011, respectively), although not to the level recommended by the SCRS.

Two significant events occurred in 2009 that raised the stakes for ICCAT: first, a proposal to list Atlantic bluefin tuna under Appendix I of the Convention on International Trade in Endangered Species of Wild Fauna and Flora (CITES) would have precluded any international trade for this species, and second, a report from an independent panel of experts contracted by ICCAT to review its performance concluded that the management of eastern Atlantic and Mediterranean bluefin tuna was "widely regarded as an international disgrace" (Anonymous 2009c). These two factors, combined with mounting pressure from many stakeholders, nongovernmental organizations, and advice from the SCRS warning of potential fishery collapse, spurred ICCAT to reduce the TAC to 13,500 t (which for the first time was commensurate with the scientific advice). Moreover, closed seasons were further extended, fishing capacity was again reduced, and monitoring and control measures were strengthened considerably.

Despite the conclusion of the Food and Agriculture Organization (FAO) of the United Nations panel of experts that bluefin tuna stocks met the biological criteria to be listed under Appendix I of CITES, the proposal was ultimately defeated at the CITES meeting in March 2010. However, there is little doubt that the threat of listing bluefin tuna under CITES pressured ICCAT to finally adopt meaningful management measures for eastern Atlantic and Mediterranean bluefin tuna. Arguably, the adoption of these measures by ICCAT in November 2009 was a major reason for the rejection of the CITES listing.

## The Current Situation

### The Two-Stock Hypothesis and the Potential Importance of Mixing

Stock assessments of ABT have relied heavily on virtual population analyses (VPAs), although a number of other models have been considered. The first VPA, conducted in 1974, treated the entire Atlantic population as a single stock. However, even at that time, there was evidence from tagging data, larval surveys, and fishery patterns that there might be at least two intermixing stocks. The overall sentiment of the SCRS at the time was that, while the evidence was weak, it favored the hypothesis of separate eastern and western stocks with a small and variable interchange (Anonymous 1982). To protect the apparently flagging western stock, ICCAT formally recommended (Rec. 81-01) that CPCs take steps to prohibit the capture of bluefin tuna in the western Atlantic Ocean as defined by the equal-distance line shown in Figure 1.1 (Anonymous 1982, 88). The SCRS has used this equal-distance line to delineate the eastern and western stocks ever since, but always with the caution that a failure to adequately manage either stock could have serious repercussions on the other (Anonymous 1980).

The assessments in 1978 and 1981 analyzed the two putative stocks separately as well as combined; the end result was a recommendation to substantially reduce western catches (Anonymous 1980, 1982). Over the next decade, subsequent assessments benefited from better data and a longer time series but essentially employed the same methods; they drew considerable criticism because bluefin tuna had often been observed to move from one side of the Atlantic to the other. However, as Brown and Parrack (1985) pointed out, the question is not whether there is interchange, but whether there is enough interchange to invalidate the two-stock designation for management purposes.

The SCRS began quantifying the possible implications of various interchange hypotheses in 1993 by use of a two-area virtual population analysis model (Anonymous 1994, Butterworth and Punt 1994, Punt and Butterworth 1995). These early analyses showed that interchange rates of even a few percent could have a substantial impact on appraisals of the western stock status. Shortly thereafter, the US National Research Council (NRC) independently estimated the rates of interchange from tag-recapture data and showed that estimates of the abundance of mature fish in the western stock were higher with those interchange rates than when no interchange was assumed (NRC 1994). Subsequently, ICCAT resolved that the SCRS should develop recovery options for bluefin tuna that take into account the possible effects of mixing.

The two-area model used by the SCRS and the NRC during the early 1990s assumed that fish movement depended on where the fish were located, essentially implying that fish moving from one side of the ocean to the other "forget" where they came from (the so-called "diffusion" or "no-memory" model). However, the SCRS expressed concern that such a model is an unlikely characterization of bluefin tuna migration and that it may be more likely that bluefin tuna return to the area where they were born (Anonymous 1995, 108–10). Cooke and Lankester (1996) suggested a possible alternative in which the ranges of the eastern and western populations are assumed merely to overlap, which would imply that fish "remember" which side of the ocean they came from and how to return. They found that this "overlap" model was at least as consistent with the tagging data as the diffusion model. Several fairly extensive investigations have since been conducted on the performance of the diffusion and overlap models when the interchange rates are estimated within the VPA from tagging and stock composition data (Porch et al. 1995, 1998, 2001; Anonymous 2009d). These studies have generally found that the estimates of abundance were similar to those obtained conducting two separate VPAs, except in simulations where the mixing rates were large and of a diffusive nature. Otherwise, the results tended to be more sensitive to parameters such as the natural mortality or the type of data used to "tune" the model (e.g., whether tagging data were used or not) than to the specific nature of the mixing model.

The most recent investigation of the effect of mixing on the VPA results was conducted during the 2008 SCRS stock assessment (Anonymous 2009d). When stock composition data (derived from otolith microconstituent analyses; Rooker et al. 2008, Fraile et al. 2015) were used to estimate the degree of overlap between the two stocks, the estimates of stock status for both the eastern and western populations were very similar to the estimates obtained from the corresponding VPAs that ignored overlap. However, when conventional tagging data were used to estimate the overlap, the estimates of recent stock status were more optimistic for both the east and west. In effect, the stock composition data, while a more direct measure of the effect of interchange between the two stocks, contains little additional information that would affect other parameters estimated by the VPA (such as the fishing mortality rates on the oldest age class). Tagging data, however, potentially contain a great deal of information on fishing mortality rates and other parameters, and therefore have a greater impact on the VPA results.

Other models have been developed that use electronic tagging data to estimate interchange and stock status. For example, Taylor et al. (2011) introduced the multistock age-structured tag-integrated stock-assessment model (MAST) of ABT, which used extensive tagging data as well as stock composition derived from otolith microchemistry. Their appraisals of the western stock indicated that only strong catch limitations for western and eastern stock could enable the stock rebuilding. Although the SCRS did not consider these models to be ready for use as a source of management advice, it did note that they were useful tools to help identify research priorities for future assessments.

The SCRS noted also that the stock composition and tagging data available at the time were incomplete and not necessarily representative of the entire population, as most fish have been tagged in the western Atlantic Ocean. They concluded that the analyses of the effects of mixing were not yet reliable enough to use as the basis for ICCAT's rebuilding plans for the eastern and western ABT. Accordingly, subsequent assessments continued to focus on conducting separate assessments of the two stocks, with the perennial caveat that "management actions taken in the eastern Atlantic and Mediterranean are likely to influence the recovery in the western Atlantic, because even small rates of mixing from East to West can have considerable effects on the West due to the fact that eastern plus Mediterranean resource is much larger than that of the West" (Anonymous 2015a).

It should be noted that a great deal of progress has been made in recent years, both in the information that is available about mixing and in devising models that are flexible enough to utilize the diverse types of data now available (conventional tagging, electronic tagging, otolith microchemistry, and genetics; Carruthers et al. 2016). Evidence has increased in support of the notion that there are two main stocks that overlap in time and space, and it was expected that these new developments would contribute heavily to the 2017 assessment.

### A Brief Review of Current Methods: Virtual Population Analysis with Two Independent Stocks

The most recent assessment of ABT was conducted in 2014. The projections were updated in 2016 using the specifications of the 2014 assessment and the realized catch in 2014 and 2015. The primary analytical tools used for management advice were tuned single-stock VPAs, which were implemented using the program VPA-2BOX, and forecasts, which were implemented using the program PRO-BOX (Porch 2002). The VPA approach used here is fundamentally based on a series of extensions to the method of Parrack (1986), widely referred to as the ADAPT framework (Powers and Restrepo 1992, Porch 2002). The method assumes that the catch history of any given year class is known with little error, permitting the historical abundance and fishing mortality rates to be computed deterministically from an estimate of the fishing mortality rate on the oldest (terminal) age of the year class. The estimates are then refined in an iterative process until the best fit is achieved to the time series of relative abundance (a process often referred to as tuning). The key parameters of the VPA implemented in the VPA-2BOX software include the fishing mortality rate on the oldest age class (expressed as a multiple of the fishing mortality rate on the next younger age, the F-ratio), the fishing mortality rate on each age class in the previous year, the natural mortality rate at age, the scaling coefficients relating indices of relative abundance to absolute abundance (catchability coefficients), and the standard error of each abundance index. The specifications for the

2014 assessments of the two ABT stocks are summarized below, and a more detailed description is available in the original SCRS assessment document (Anonymous 2015b).

*Western Atlantic Stock: Data and Methods*

The primary catch statistics (Figure 1.2) come from four major fishing nations: Japan, the United States, Canada, and Mexico. Catch-at-size data were particularly scarce prior to 1970; therefore, the 2014 assessment was based on data from 1970 through 2013. Unfortunately, age-length keys were not available for bluefin tuna until very recently (2016), therefore the catch-at-size data had to be converted to catch-at-age data by using a version of cohort slicing (a method that uses a growth curve to convert size to age, which is generally regarded as less accurate than an age-length key). The low number of samples and use of cohort slicing suggest that the resulting catch-at-age matrix may be relatively error prone, in which case a statistical catch-at-age model may be more appropriate. However, over the years, numerous authors have applied various forms of statistical catch-at-age and catch-at-size models to ABT (e.g., Porch et al. 1994, Porch 1996, Taylor et al. 2011, Butterworth and Rademayer 2015), and in many cases the results have been less sensitive to the choice of statistical models than to other key drivers such as the natural mortality rate.

Twelve indices of abundance covering different life history stages were used in the assessment, including two fishery-independent indices (see Anonymous 2015b, table 9). Information for some of the indices no longer exists, but a number of them have been updated through 2015 (Figure 1.4). The catch rates of juvenile bluefin tuna (ages 2–3) in the US rod and reel fishery suggest strong year classes in 2002 or 2003 but none since. The catch rates of adults in the US rod and reel fishery decreased between 2011 and 2013 but increased somewhat in the two years after the assessment. Catch rates of the Japanese longline fishery fluctuated substantially over time, peaking in 2012 and declining thereafter (although they remain higher than the average in the 1990s and early 2000s). The catch-rate series from the US Gulf of Mexico longline fishery shows no clear trend prior to 1991 but a generally increasing trend since the early 1990s, with a peak in 2012 followed by a small decline. The indices for the Gulf of St. Lawrence show the most dramatic rise; the SCRS questioned whether such an increase was biologically plausible for the stock as a whole, noting that other factors may not have been accounted for (changes in stock distribution, management

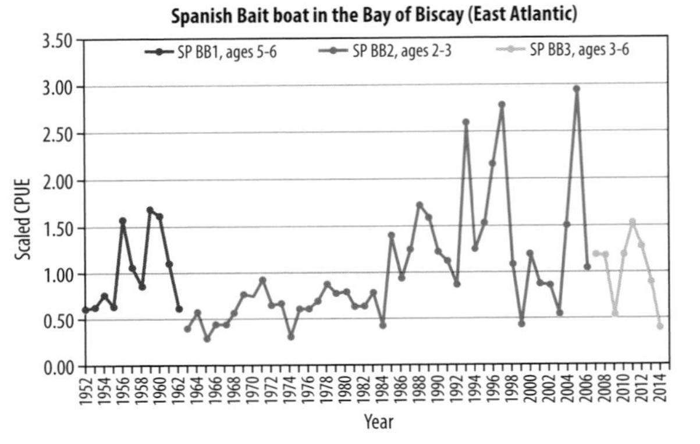
Spanish Bait boat in the Bay of Biscay (East Atlantic)
SP BB1, ages 5-6
SP BB2, ages 2-3
SP BB3, ages 3-6
Scaled CPUE
Year

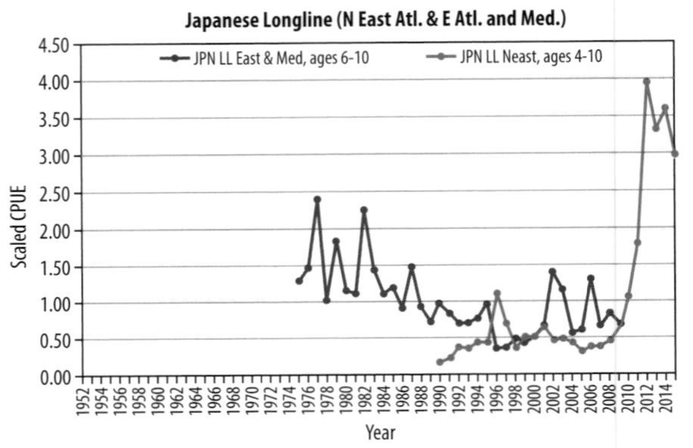
Japanese Longline (N East Atl. & E Atl. and Med.)
JPN LL East & Med, ages 6-10
JPN LL Neast, ages 4-10
Scaled CPUE
Year

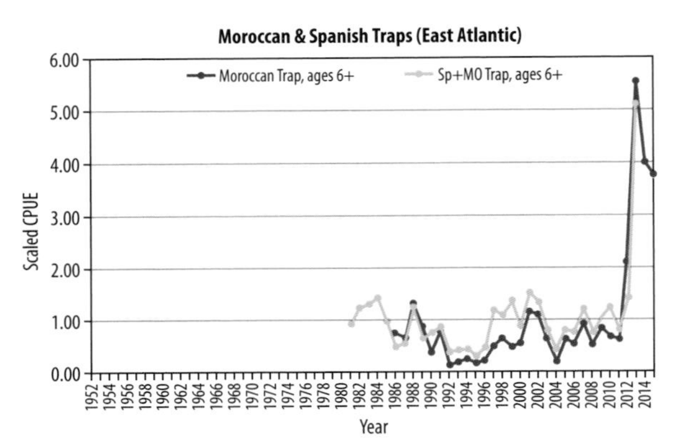
Moroccan & Spanish Traps (East Atlantic)
Moroccan Trap, ages 6+
Sp+MO Trap, ages 6+
Scaled CPUE
Year

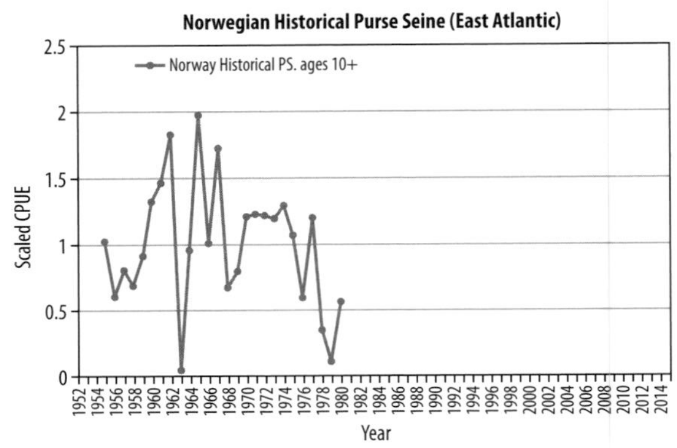
Norwegian Historical Purse Seine (East Atlantic)
Norway Historical PS. ages 10+
Scaled CPUE
Year

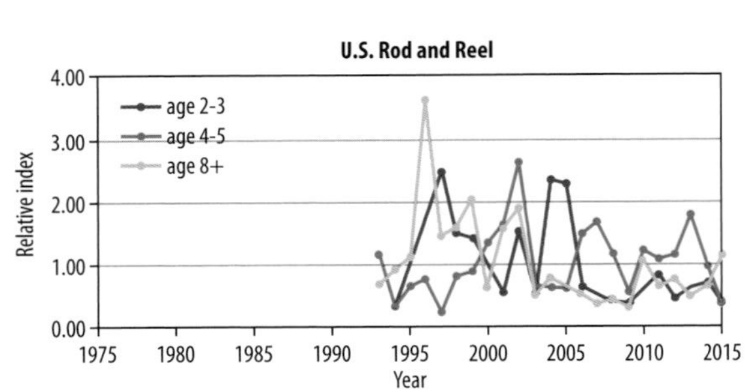
U.S. Rod and Reel
age 2-3
age 4-5
age 8+
Relative index
Year

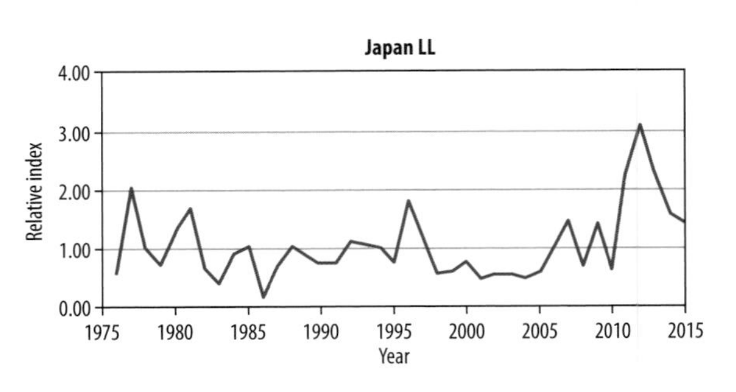
Japan LL
Relative index
Year

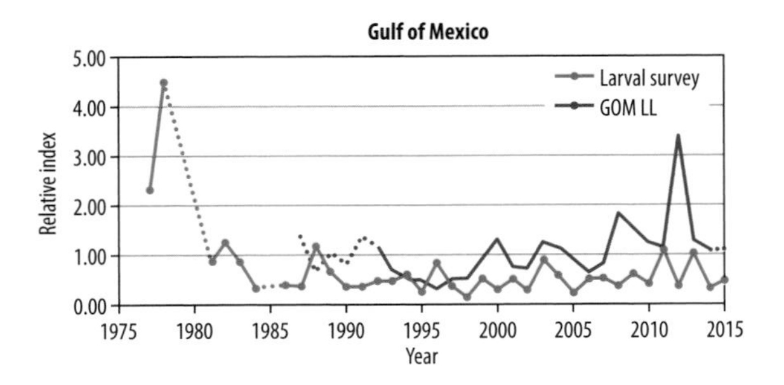
Gulf of Mexico
Larval survey
GOM LL
Relative index
Year

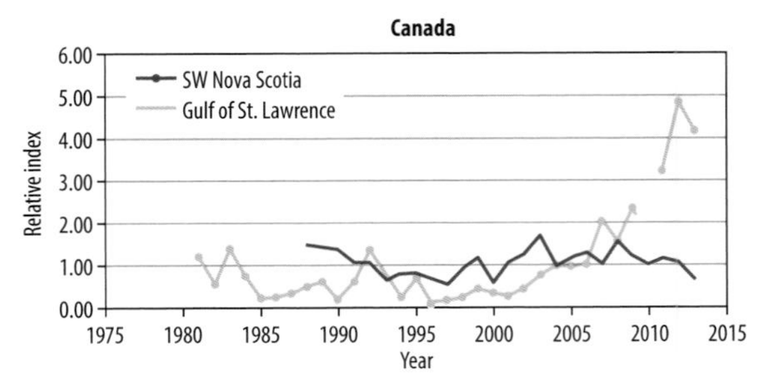
Canada
SW Nova Scotia
Gulf of St. Lawrence
Relative index
Year

regulations, fishing behavior, and environment). The Gulf of Mexico larval survey (the only active fishery-independent indicator) has fluctuated around relatively low levels since the 1980s, although 2011 and 2013 were relatively high.

The tuned VPA tracked the abundance and mortality of each age class from 1970 through 2013, where the oldest age class was a plus-group (ages 16 and older). The natural mortality rate was assumed age-independent at 0.14 $yr^{-1}$, and the fishing mortality rate on the plus-group was set equal to that at age 15 for the entire period (it was determined that selectivity was unlikely to differ on fish age 15 and older, because growth is relatively slow at this age and all animals are likely mature). The fishing mortality rates for each age in the last year were estimated with a mild constraint restricting the amount of change in the vulnerability pattern during the most recent three years. Projections (forecasts) of the abundance of the western stock under a variety of possible catch scenarios were based on the bootstrap replicates of the fishing mortality-at-age and numbers-at-age matrices produced by the VPA-2BOX software. Projections and benchmarks were computed for two scenarios of potential future recruitment (Figure 1.5): a low-recruitment scenario (two-line model) that assumes average recruitment cannot reach the high levels from the early 1970s (ostensibly owing to a change in the environment), and a high-recruitment scenario that assumes the number of recruits is a Beverton and Holt function of the spawning biomass in the previous year. Medium-term projections were conducted to cover the time of the rebuilding plan (2019). Projected levels of spawning stock biomass (SSB) were expressed relative to the SSB associated with

**Figure 1.4.** ***(opposite)*** The top four panels correspond to the updated catch-per-unit-effort (CPUE) time series fishery indicators for the East Atlantic and the Mediterranean bluefin tuna stock. All CPUE series are standardized series except the nominal Norway purse seine (PS) index. The Spanish (SP) bait boat (BB) series (*top left panel*) was split into three series to account for changes in selectivity patterns, and the latest series in 2014 was updated using French BB data owing to the sale of the quota by the Spanish fleet. The Japanese longline (LL) CPUE for the Northeast Atlantic has been updated until 2015. The Moroccan and Spanish (SP+MO) trap CPUE was not updated. The Moroccan CPUE up to 2013 was used only for the sensitivity analysis in the 2014 stock assessment and has been updated to 2015. The bottom four panels correspond to the updated indices of abundance for western bluefin tuna. The dashed portions of the larval survey, US Gulf of Mexico indices, and Canada Gulf of St. Lawrence indices bridge the gaps when data were missing or otherwise considered unreliable by the Standing Committee on Research and Statistics. The two Canadian indices have not been updated since 2014.

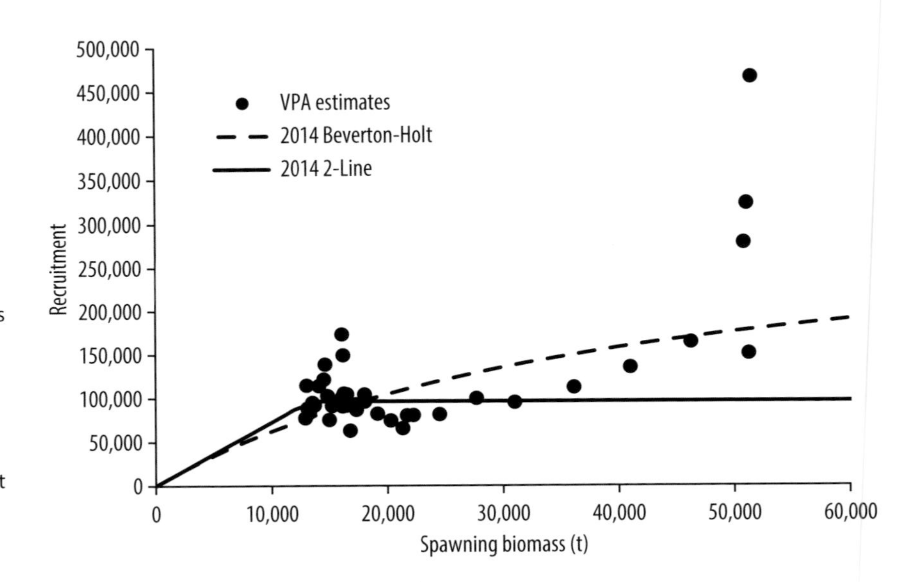


**Figure 1.5.** Recruitment scenarios for western bluefin tuna derived from the 2014 stock assessment. The low-recruitment-potential scenario (2-line) implies future recruitment will remain near present levels even if stock size increases. The high-recruitment-potential scenario (Beverton-Holt) implies future recruitment increases with stock size and has the potential to achieve levels that occurred in the early 1970s. Points represent the estimates from the 2014 base assessment. VPA = virtual population analysis.

the maximum sustainable yield ($SSB_{MSY}$, the current target of the ICCAT rebuilding plan) corresponding to each of the two recruitment scenarios.

*Eastern Atlantic and Mediterranean Stock: Data and Methods*

The catch statistics (Figure 1.2) for the eastern Atlantic and Mediterranean represent a variety of fisheries from nearly 50 countries. Since the 1980s, the bulk of the catch has been taken by purse seiners, mostly for the farming industry (especially France, Italy, Spain, Croatia, and Turkey), although trap, bait boat, and longline fisheries have remained important. Unlike the catch statistics for the west, those for the east are believed to be reasonably good from 1950 through the early 1970s. However, it became increasingly difficult to obtain representative size samples from the purse seine fleets in the Mediterranean as more and more of the catch was destined for farming operations. Much of the catch-at-size information has therefore been constructed using a variety of substitutions based on sparse samples borrowed from other fleets or, in the case of purse seiners, on observations of mean individual weights. Moreover, as mentioned previously, the SCRS estimated that the catches by purse seiners and other fleets were substantially greater than had been declared to ICCAT. Indeed, the situation was of such concern to the SCRS that no VPA was conducted during 2006; analyses focused instead on the pieces of data that were thought to be most reliable (Fromentin et al. 2007). It is worth noting that the situation has improved in recent years, especially since 2014, when ICCAT mandated the use of stereoscopic cameras to estimate the size frequency at caging (with errors of less than 5%).

The SCRS went back to using VPA during the 2008 assessment, owing in part to ICCAT's directive to provide Kobe matrices quantifying the probability that the stock would rebuild under various constant catch levels and owing in part to the need for a common platform for stock mixing analyses. The 2014 assessment was conducted using data from 1950 through 2013, but it considered two catch scenarios, one based on the declared catch and the second based on independent estimates made by the SCRS (i.e., the inflated catch scenario—50,000 t from 1998 to 2006 and 61,000 t in 2007, and reported catches thereafter). Age-length keys were not available, and the catch-at-size data had to be converted to catch-at-age data using cohort slicing (as with the western Atlantic stock, and with the same concerns).

Six indices of abundance were used to tune the base VPA in 2014 (Figure 1.4; see also Anonymous 2015a, table 6): catch-per-unit-effort (CPUE) data for Norwegian purse seiners in the North Sea (ages 10+), Spanish and Moroccan traps in the Mediterranean (ages 6+), Moroccan traps alone (ages 6+), Japanese longliners in the Northeast Atlantic (ages 4+), Japanese longliners in the East Atlantic and the Mediterranean Sea (ages 6+), and Spanish bait boats in the Bay of Biscay (ages 2–6, depending on the year). The Norwegian purse seine fishery no longer exists, and the remaining indices have been affected by recent regulation-induced changes in the operational patterns of the fishery or the size of fish being targeted. The changes were particularly profound for the Spanish bait boat index of juvenile fish, which had to be treated as a new index after the adoption of the recovery plan and then was updated using French bait boat data from 2011 to 2014 because the Spanish bait boat fishery sold most of its bluefin tuna quota. The bait boat index was effectively discontinued after 2014, as the number of French bait boats operating was too low to ensure the reliability of this index (the quota transfer ended in 2016, and this index may hence be continued). The most influential series in the assessment were those derived from the Japanese longline fishery in the northeast Atlantic and the Moroccan and Spanish traps, which suggested sharp increases in the abundance of large fish in recent years. Interestingly, these two indices have shown significant downturns since the 2014 assessment (Figure 1.3).

The VPA for the eastern stock set the plus-group to age 10, and the natural mortality rate $M$ varied with age according to estimates derived from tagging data for southern bluefin tuna (0.490, 0.240, 0.240, 0.240, 0.240,

0.200, 0.175, 0.150, 0.125, 0.100 $yr^{-1}$ for ages 1 to 10+, respectively; Hampton 1991). The F-ratios were fixed to values defined during the 2010 assessment based on independent analyses of changes in the size composition of large fish: 0.7 for 1950–1969, 1.0 for 1970–1984, 0.6 for 1985–1994, and 1.2 from 1995 onward. Projections and benchmarks were computed for the two sets of historical catch estimates (reported and inflated), three scenarios of potential future recruitment (the high average recruitments estimated for 1990–2000, the medium average for 1955–2006, and the low average during 1970–1980), and two selectivity patterns (geometric mean of the relative fishing mortality rate estimates over 2007–2009 or 2009–2011). Medium-term projections were conducted to cover the time of the rebuilding plan (2022). Projected levels of SSB were expressed relative to the SSB associated with the level corresponding to F0.1 (the proxy selected for estimating MSY).

## Estimated Stock Status and Outlook

### *Western Atlantic Stock*

The base VPA model from the 2014 assessment estimated that SSB of the western Atlantic stock had decreased rapidly from 1970 to 1992 but then began to stabilize, fluctuating without trend through the turn of the century. The estimates for more recent years indicate a gradually increasing trend through 2013, when SSB was estimated to be about 55% of the level estimated for 1970, thanks in part to a strong 2002–3 year class. The perceived status of the stock relative to the level that corresponds to MSY depends on the assumption about future recruitment (Figure 1.6). Under the low-recruitment scenario, the stock appears to be neither overfished nor undergoing overfishing; recent levels of F were ~36% of $F_{MSY}$, and SSB was about twice $SSB_{MSY}$. Under the high-recruitment scenario, the stock also did not appear to be undergoing overfishing ($F_{(2010-2012)} = 88\%$ of $F_{MSY}$) but was severely overfished ($SSB_{2013} = 48\%$ of $SSB_{MSY}$).

The projections of stock status suggest that if the low-recruitment scenario is correct, catches of 2,500 t or lower would maintain the stock above the $SSB_{MSY}$ level, and constant catches of 2,250 t would cause the stock to decline somewhat initially but allow a recovery to the 2013 level by the end of the rebuilding period (2019). If the high-recruitment scenario is correct, then the western stock would be highly unlikely to rebuild by 2019 even with no fishing, although catches less than 2,500 t are predicted to prevent overfishing (Table 1.1).

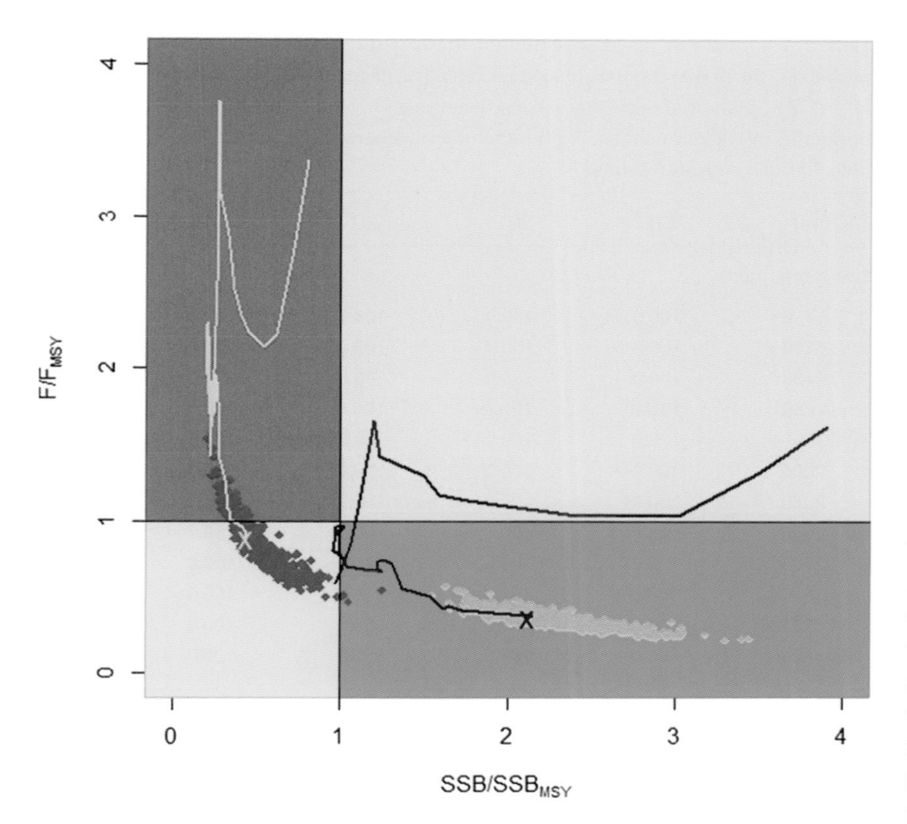


**Figure 1.6.** Estimated status of western stock relative to the objectives of the International Convention for the Conservation of Atlantic Tunas (maximum sustainable yield [MSY]) by year (1973 to 2013) and recruitment scenario based on the 2014 stock assessment (light blue = high recruitment potential; dark blue = low recruitment potential). The light gray dots represent the status estimated for 2013 under the low-recruitment scenario, corresponding to bootstrap estimates of uncertainty. The dark blue lines give the historical point estimates for the low recruitment, and the light blue gives the historical trend for the high recruitment. SSB = spawning stock biomass.

The SCRS has long recognized the sensitivity of the conclusions regarding stock status but has been unable to resolve the issue, in part because of the lack of information for the early years of the fishery, when exploitation was low and stock status was presumably higher, and in part because of the large interannual variation in recruitment that typically plagues attempts to discern the nature of the spawner-recruit relationship in wild populations. Thus, the SCRS has typically included language to the effect of "The Committee has insufficient evidence to favor either scenario over the other and notes that both are plausible (but not extreme) lower and upper bounds on rebuilding potential." Nevertheless, the conclusion that overfishing has ended appears to be robust across recruitment scenarios, suggesting that the western stock will eventually rebuild to near the true level of MSY even if the high-recruitment scenario is more correct, albeit well after ICCAT's prescribed rebuilding period.

**Table 1.1.** Kobe II matrices (updated during the 2014 stock assessment) giving the probability (in %) that the spawning stock biomass will exceed the level that will produce MSY ($SSB > SSB_{MSY}$, not overfished) in any given year for various constant catch levels under the low-recruitment and high-recruitment scenarios. The current TAC of 1,750 t (Rec. 13-09) is indicated in bold.

| TAC (mt) | 2015 | 2016 | 2017 | 2018 | 2019 |
|---|---|---|---|---|---|
| Low recruitment | | | | | |
| 0 | 100.0 | 100.0 | 100.0 | 100.0 | 100.0 |
| 1500 | 100.0 | 100.0 | 100.0 | 100.0 | 100.0 |
| 1700 | 100.0 | 100.0 | 100.0 | 100.0 | 100.0 |
| **1750** | **100.0** | **100.0** | **100.0** | **100.0** | **100.0** |
| 1800 | 100.0 | 100.0 | 100.0 | 100.0 | 100.0 |
| 2000 | 100.0 | 100.0 | 100.0 | 100.0 | 100.0 |
| 2250 | 100.0 | 100.0 | 100.0 | 100.0 | 100.0 |
| 2500 | 100.0 | 100.0 | 100.0 | 100.0 | 100.0 |
| 2750 | 100.0 | 100.0 | 100.0 | 100.0 | 100.0 |
| 3000 | 100.0 | 100.0 | 100.0 | 100.0 | 100.0 |
| 3250 | 100.0 | 100.0 | 100.0 | 100.0 | 100.0 |
| 3500 | 100.0 | 100.0 | 100.0 | 100.0 | 99.8 |
| High recruitment | | | | | |
| 0 | 1.2 | 1.4 | 1.4 | 1.6 | 6.0 |
| 1500 | 1.2 | 1.2 | 1.2 | 1.2 | 1.6 |
| 1700 | 1.2 | 1.2 | 1.2 | 1.2 | 1.6 |
| **1750** | **1.2** | **1.2** | **1.0** | **1.2** | **1.6** |
| 1800 | 1.2 | 1.2 | 1.0 | 1.2 | 1.6 |
| 2000 | 1.2 | 1.2 | 1.0 | 1.2 | 1.4 |
| 2250 | 1.2 | 1.2 | 0.8 | 0.4 | 1.2 |
| 2500 | 1.2 | 1.2 | 0.6 | 0.4 | 1.2 |
| 2750 | 1.2 | 1.0 | 0.4 | 0.4 | 1.2 |
| 3000 | 1.2 | 0.8 | 0.4 | 0.4 | 0.8 |
| 3250 | 1.2 | 0.8 | 0.4 | 0.2 | 0.8 |
| 3500 | 1.2 | 0.8 | 0.4 | 0.2 | 0.6 |

*Note*: TAC = total allowable catch.

### *Eastern Atlantic and Mediterranean Stock*

The 2014 assessment results estimated that SSB declined from ~300,000 t prior to the 1970s to ~150,000 t in the mid-2000s. The estimates for more recent years, however, suggest that SSB has increased rapidly to ~500,000 t in 2013. Estimates of the fishing mortality rate (F) on younger fish (ages 2–5) increased until 2008 and then fell sharply, presumably owing to the minimum size regulations under Recommendation 06-05. The fishing mortality rate on older fish (ages 10+) appeared to decrease during the first two decades and then rapidly increased after 1980 as the fishery grew to supply farms fattening bluefin tuna for the growing Japanese sashimi market. Recent levels appear to have declined owing to stricter adherence to the various management measures in place (including much lower TACs).

Although the data and resulting estimates are regarded as highly uncertain, they are qualitatively consistent with the improvement in stock status expected under current regulations. Whereas the perception of the stock status relative to the MSY proxy ($F_{0.1}$) is sensitive to the assumptions regarding the future selectivity pattern and recruitment levels, overall there was little indication of overfishing and a reasonable chance that the stock may no longer be overfished. For example, under the medium-recruitment scenario, the estimates of $F_{2013}$ were below the reference target in the majority of runs (the median estimates of $F_{2013}/F_{0.1}$ were 0.4 and 0.36 for the reported and inflated catch scenarios, respectively), and the estimates for SSB were generally above the level expected at $F_{0.1}$ ($SSB_{2013}/SSB_{0.1} = 1.10$ and 1.11 for reported and inflated catch scenarios, respectively). The projections of stock status suggest that if future catch is limited to 30,000 t or less, there is a greater than 60% probability that F would remain below $F_{0.1}$ and that SSB would meet or exceed $SSB_{F0.1}$ by the end of 2022 (Figure 1.7 and Table 1.2).

It is important to note that the rapid rise in estimates of SSB is driven by the recent trends in the two fishery-dependent indices of abundance for older fish—Japanese longlines and Spanish/Moroccan traps. To fit these indices, the VPA estimated that the year classes in 2004–7 were substantially larger than the 2003 year class. However, no independent evidence, anecdotal or otherwise, has been offered to substantiate the existence of such extraordinarily large year classes. For example, the 2003 year class can be easily tracked in the size frequencies of catches from the Japanese longline and other fisheries (Suzuki et al. 2013). If the 2004–7 year classes were as large as estimated, they should have been evident in the most recent size-frequency information. Inasmuch as there are reasons to suspect that the sharp increase in the trap and longline indices may in part be a consequence of the reduced spatial scales in which they operate and perhaps other changes in fishing practices in response to the reduced quotas, the SCRS advised caution in interpreting the results until the high estimates of recruitment could be better verified. The SCRS also noted that two other assessment models and some sensitivity analyses of the VPA did not estimate such high recruitments or as rapid an increase in recent SSB. Perhaps more importantly, the

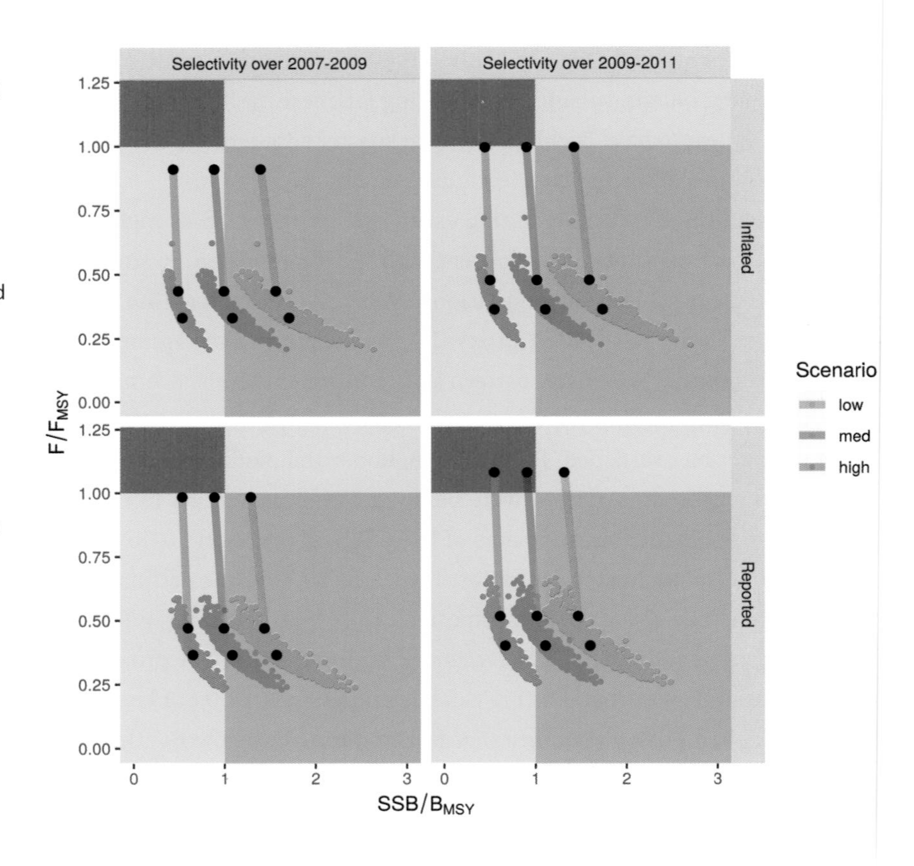


**Figure 1.7.** Stock status from 2011 to the terminal year (2013) (*black dots*) estimated from virtual population analysis continuity run from the 2014 stock assessment with reported and inflated catch (*upper and lower panels*) and considering low, medium, and high recruitment levels (*blue, green, and red lines*). Blue, green, and red dots represent the distribution of the terminal year obtained through bootstrapping for the corresponding three recruitment levels. The 2013 spawning stock biomass (SSB) and F relative to reference points were calculated with the selectivity pattern over 2007–9, which was the same period as the 2010 stock assessment (*left panel*). 2013 SSB and F relative to the reference points with the selectivity pattern over 2009–11, which was the same period as the 2012 stock assessment (*right panel*).

two key indicators of stock abundance (Moroccan trap and Japanese longline indices) have trended downward since the last assessment, suggesting that the projected increases may be overly optimistic.

## Current Management

The extraordinary amount of illegal, unreported, and unregulated (IUU) fishing for bluefin tuna in the Mediterranean Sea brought to light by the nongovernmental organizations and the SCRS eventually led to a number of important measures to better track and control the catches of ABTs and to a strengthening of the function of ICCAT's Management Measures Compliance Committee. Many of these measures are detailed in Recommendation 14-04, a 35-page document with 103 operative paragraphs and 11 annexes (in contrast, Recommendation 14-05 for western ABT, where substantial IUU has not been identified by the SCRS, is only 5 pages long with 28 operative paragraphs and no annexes). The new measures include a requirement for CPCs to reduce their fishing effort commensurate with their individual allocation, including the establishment of individual quotas for vessels

**Table 1.2.** Stock status projection from the 2014 eastern assessment of the probability (in %) that the fishing mortality rate would remain below the maximum sustainable yield proxy and biomass would remain above that which would support maximum sustainable yield

| TAC (mt) | 2017 | 2018 | 2019 | 2020 | 2021 | 2022 |
|---|---|---|---|---|---|---|
| 0 | 77 | 84 | 91 | 96 | 98 | 100 |
| 2,000 | 76 | 84 | 91 | 96 | 98 | 99 |
| 4,000 | 76 | 84 | 91 | 95 | 98 | 99 |
| 6,000 | 76 | 84 | 90 | 94 | 98 | 99 |
| 8,000 | 76 | 83 | 90 | 94 | 98 | 99 |
| 10,000 | 76 | 83 | 90 | 94 | 97 | 99 |
| 12,000 | 76 | 83 | 89 | 94 | 97 | 99 |
| 14,000 | 76 | 82 | 89 | 93 | 97 | 98 |
| 16,000 | 76 | 82 | 89 | 93 | 96 | 98 |
| 18,000 | 76 | 82 | 88 | 93 | 96 | 98 |
| **19,296** | **76** | **82** | **88** | **93** | **96** | **98** |
| 20,000 | 76 | 82 | 88 | 92 | 95 | 98 |
| 22,000 | 76 | 81 | 87 | 92 | 95 | 97 |
| 24,000 | 76 | 81 | 87 | 92 | 95 | 97 |
| 26,000 | 75 | 81 | 87 | 91 | 94 | 97 |
| 28,000 | 75 | 81 | 86 | 90 | 94 | 96 |
| 30,000 | 75 | 80 | 86 | 90 | 93 | 96 |

*Note*: TAC = total allowable catch.

over 24 meters long. The commission also required CPCs fishing for eastern Atlantic and Mediterranean bluefin tuna to develop annual fishing plans that identify the quota allocated to each fishing fleet, how the quotas were allocated, and the measures taken to ensure compliance with the quotas. Several indirect controls were also developed to control fishing effort and to reduce potential IUU, the most important of which were reductions in the duration of the fishing seasons allowed for each fleet. Of particular significance was the restriction of purse seine fleets to just one month, as purse seiners account for approximately 80% of the total bluefin tuna catch in the Mediterranean Sea.

Other measures were introduced to better track bluefin tuna catches. For example, ICCAT prohibits any domestic trade, landing, imports and exports, caging operations, and reexports that are not accompanied by an accurate and validated bluefin catch document (BCD), which was made electronic (eBCD) in 2016.

The total allowable catch for the East Atlantic and Mediterranean Sea from 2010 to 2014 was drastically reduced from the prior decade (ranging from 12,900 to 13,400 t [Recs. 09-06, 13-07]), which in combination with other regulations appears to have led to a substantial improvement in the status of the eastern stock. Nevertheless, the uncertainty associated with the 2014 stock assessment led the SCRS to be somewhat equivocal in its advice. The SCRS indicated that "maintaining the 2014 TAC (13,400 t.) or gradually increasing it should not undermine the success of the rebuilding plan for the eastern Atlantic and Mediterranean stock." The SCRS was unable to reach consensus on the upper limit for such an increase that would not jeopardize the recovery of this stock but did advise that "a gradual increase (in steps over, e.g., 2 or 3 years) of the catch to the level of the most precautionary MSY estimate would allow the population to increase even in the most conservative scenario (low-recruitment scenario)." The most precautionary MSY for the eastern Atlantic and Mediterranean stock was estimated by the SCRS to be 23,256 t (corresponding to the low-recruitment scenario with the reported catch data). On the basis of this advice, ICCAT set the TACs for 2015, 2016, and 2017 at 16,142 t, 19,296 t, and 23,155 t, respectively (Rec. 14-04), with an additional quota of 500 t for Algeria in 2017.

The results of the western Atlantic stock assessment also showed an increase in SSB over the last few years of the time series used. The SCRS indicated that the strong 2002–3 year classes and the reduction in fishing mortality contributed to this trend. Considering the uncertainties regarding the future productivity of the stock (i.e., future potential recruitment levels) and the possible effects of mixing with the eastern stock, the SCRS advised that a TAC of 2,250 t was not expected to result in stock growth by 2019, and as such, it should not be exceeded. In contrast, the 2013 TAC of 1,750 t would allow for the SSB to increase more quickly, which in turn might help to resolve the high- and low-recruitment hypothesis. Considering the SCRS recommendation for the western ABT stock, ICCAT increased the TAC to 2,000 t for 2015 and 2016 (Rec. 14-05).

## The Future: Resolving the Uncertainties for Better Assessments and Management

Scientists and stakeholders have identified a number of uncertainties affecting the assessment and management of ABT (e.g., Fromentin et al. 2014a, Leach et al. 2014). Among the earliest and most frequently cited concerns has been the issue of stock structure and the degree of intermixing. Since before the inception of ICCAT, it was recognized that bluefin tuna frequently made transatlantic migrations and that there were at least two major spawning grounds (one in or near the Gulf of Mexico and one in or near the Mediterranean Sea). Today, the weight of the evidence appears to confirm the hypothesis of two main stocks, but questions remain regarding alternative spawning grounds (e.g., Mather et al. 1995, Piccinetti et al. 2013, Richardson et al. 2016) and potential subpopulations or contingents within the Mediterranean (see Arrizabalaga et al. 2016). Moreover, several electronic tagging studies suggest that the rate of interchange may be higher than previously thought (Block et al. 2005, Walli et al. 2009, Galuardi et al. 2010, Carruthers et al. 2016). Otolith microchemistry studies show that fish originating from the Mediterranean account for a substantial proportion of catches in some western fisheries (e.g., Rooker et al. 2008, Siskey et al. 2016) and that, in some years, fish originating in the Gulf of Mexico may account for a significant proportion of the catch in some eastern Atlantic areas (Di Natale and Tensek 2016).

ICCAT and others have recommended moving toward the use-assessment models that account for stock mixing explicitly. However, the efficacy of these mixing models is compromised by the generally poor quality of the fishery data in the Mediterranean Sea since the advent of farming. If the assessments of the western and eastern populations are linked through a mixing model, then the uncertainties associated with the Mediterranean fishery data will be propagated through to the western stock assessment. Thus, it is unclear whether the assessment of the western stock would actually be improved by the use of a mixing model, even if the mechanism and magnitude of mixing were known perfectly. If, in addition, the mixing model were misspecified, the resulting assessment could be more biased than an assessment that does not account for mixing at all (Porch et al. 1998).

The future recruitment of young fish to the population has been another major source of uncertainty for ABT assessments. As for many other exploited species, the relationship between spawners and the number of young fish recruiting to the population has been difficult to discern. The combination of high natural variations in recruitment and lack of contrast in the available data (the time series of catch and relative abundance data begin well after both stocks were reduced by exploitation) poses a difficult statistical challenge in and of itself. Considering also the error-prone nature of the available data and the nonstationary nature of the spawner-recruit relationship, one must conclude that the true nature of the spawner-recruit relationship will not be resolved for some time. Past assessments of the eastern and western Atlantic stocks have tackled this issue by examining the implications of alternative recruitment hypotheses that are considered to bracket the range of possibilities (see section entitled "A Brief Review of Current Methods"). However, a more fruitful course may be to focus instead on adopting biological reference points and management procedures that are robust to this and other sources of uncertainty (Porch and Lauretta 2016).

A related issue, and a subject of considerable controversy, is how best to index the number of eggs produced by each stock. The assessments for both assume that the number of eggs produced is proportional to mature biomass. Recent studies seem to support that approach in that they suggest that batch fecundity per unit of body weight is fairly constant (e.g., Correiro et al. 2005, Aranda et al. 2013, Knapp et al. 2014). However, the assessments have differed markedly in their choice of maturity schedules, with the western assessment assuming that fish do not contribute substantially to the spawning potential of the stock until after age 9 (on the basis of the age composition of fish caught on the Gulf of Mexico spawning ground) and the eastern assessment assuming that spawning occurs as early as age 3 (on the basis of histological analyses of fish caught on Mediterranean spawning grounds). The SCRS has long recognized that actual maturity schedules are unlikely to be so disparate between the stocks, particularly given the similarity in their rates of growth. Indeed, there is some recent evidence that suggests that some western-origin fish may mature as early as age 3 (Heinisch et al. 2014, Richardson et al. 2016). However, in the case of batch spawners like bluefin tuna, the important quantity is not maturity but spawning frequency (and perhaps egg quality). The use of mature biomass as a proxy for spawning output presumes that all mature fish are equally successful, i.e., young fish spawn just as often with as high a quality of eggs as older fish. This contention has not been well established for large pelagic fish. To the contrary, histological and close-kin genetic analyses of southern bluefin tuna suggest that older fish contribute substantially more to the spawning output of the population than would otherwise be expected because of their weight (Farley et al. 2015, Bravington et al. 2016).

One of the most influential parameters in a stock assessment is the natural mortality rate *M*, which is notoriously difficult to estimate within a stock assessment unless the data are available from a well-designed mark-recapture study, or the available indices of abundance and age-composition data cover a period when fishing was negligible. Inasmuch as neither situation has been true for ABT, the SCRS has based $M$ values for the western stock on results of a tagging study (Farber 1980) and for the eastern stock on estimates for southern bluefin tuna (Fonteneau and Maguire 2014). Since the growth rates and habitats are similar for the two stocks, the SCRS has recently proposed replacing these assumptions with a single common vector based on the Lorenzen (2000) mortality function of weight ($M = 3.0.W^{-0.288}$, where $W$ is the weight-at-age of individual fish) rescaled so that the mortality rate on the oldest ages plateaus at 0.1 (the value used previously for the eastern stock).

Two large research programs have developed in recent years targeting international research on ABT to address these uncertainties, among others. The largest of these is the ICCAT GBYP, with an annual budget of ~2 million euros (www.iccat.int/GBYP/en/). The program was officially adopted by ICCAT in 2008 and implemented in 2010 with the following priorities:

1. Improve basic data collection.
2. Improve understanding of key biological and ecological processes.
3. Improve assessment models and provisioning of scientific advice on stock status.

The GBYP has generally emphasized research in the eastern Atlantic and the Mediterranean Sea, in tacit recognition of the Bluefin Tuna Research Program (BTRP) administered by the United States, which has similar goals but focuses on the western Atlantic (albeit with a smaller budget of ~$650,000 per annum).

These GBYP, BTRP, and other research programs have produced a great deal of new information that promises to fundamentally change the assessment and management of ABT. Among the greatest successes have been the collection of tissues; the development of methods to determine the origin of fish from genetics, otolith microchemistry, otolith shapes (e.g., Rooker et al. 2014, Brophy et al. 2015, Fraile et al. 2015, Puncher et al. 2015, Rodríguez-Ezpeleta et al. 2016; see also this volume, chapter 2); and the calibration of techniques to determine ages from both otoliths and spines (Rodríguez-Marín et al. 2012, 2016). For the first time, it has become feasible to prepare age-length keys by stock of origin, which can be applied to length samples to discern the stock of origin and age composition of the catch (Anonymous 2016b).

Another important success has been the recovery of data from the eastern ABT fisheries using market data, stereoscopic cameras, and other sources of information to recreate the magnitude and size frequency of the catch since the farming era (Anonymous 2016b). Several new indices of abundance were also developed for possible use in the 2017 assessment, including an acoustic survey of adult bluefin in the Gulf of St. Lawrence (Melvin 2015), an aerial survey for juvenile bluefin tuna in the northwestern Mediterranean Sea (Bauer et al. 2015), a larval survey in the Balearic Sea (Ingram et al. 2015), and an aerial survey of adult bluefin covering much of the Mediterranean Sea (Di Natale et al. 2017). In addition, scientists from Canada, Japan, Mexico, and the United States have jointly developed the most comprehensive collection of set-by-set longline data for western ABT yet compiled, and plans are under way to develop a combined index that will cover a large fraction of the fishable habitat.

Arguably the greatest investment has been made in various mark-recapture experiments, because they potentially provide needed information on most of the variables of interest for stock assessments (population size, mortality rates, movements, growth, etc.). More than 24,000 conventional tags and nearly 1,500 electronic tags have been released across the different programs since 2010, and nearly 1,000 electronic tag tracks have been compiled, which the SCRS will use to inform movement models (Hanke et al. 2016).

Efforts are under way to develop more realistic models that can better utilize all the new information described above for stock assessments and management strategy evaluations. One approach (Cadrin et al. unpublished) incorporates stock mixing implicitly by applying the VPA to the catch at age of western-origin fish only (parsing the total Atlantic catch using otolith-derived stock composition information). This approach effectively focuses on the western stock and avoids the need to estimate the level of intermixing between stocks, but the data are rather sparse prior to 2010 and the fraction of the catch that is of western origin is not well determined. Other recent modeling efforts are focusing on developing statistical catch-at-length models to make better use of the available size and age data and to move away from cohort slicing (Irie and Takeuchi 2015, Butterworth and Rademayer 2015, Walters and Calay unpublished).

Perhaps the most valuable contribution of the GBYP and the BTRP is that they have demonstrated the value of well-coordinated, centralized research and data collection programs. A wide array of research initiatives have been implemented with unprecedented levels of collaboration among CPCs, and important insights have been gained into meeting the logistical challenges posed by what is arguably the world's most geopolitically complex regional fisheries management organization. Nevertheless, as of this writing, few CPCs collect adequate size, age, and other biological information for ABT, and funding for research can be unpredictable. Priority should be placed on developing mechanisms to consistently fund routine data collection and research activities on a long-term basis (Polacheck et al. in press).

### Charting the Course for Better Management

The key to the success of the recovery plan for bluefin tuna has been an unprecedented degree of international cooperation, in terms of both management and research. Whereas in former days ICCAT developed a reputation for finding ways to work around the scientific advice (Anonymous 2009c), it appears to have successfully arrested overfishing and may become the first tuna regional fisheries management organization (tRFMO) to successfully rebuild a bluefin tuna stock. Even so, the true status of both the eastern and western stocks is uncertain for the many reasons already discussed, and among ICCAT members there remains considerable pressure to substantially increase the quota and renegotiate the allocations.

Dissatisfaction with the process for developing the allocation arrangements has led several countries over the years to lodge objections to eastern Atlantic and Mediterranean management recommendations. Under Article VII of the ICCAT convention, objecting parties are not legally bound by the terms of the recommendation. Libya and Morocco objected to the recommendation adopted in 1998 and set their own catch limits well above the levels allocated by ICCAT, until their concerns were eventually addressed in the 2002 recommendations. Algeria, Turkey, and Norway objected to the 2010 recommendation, the latter owing to concerns about the transparency of decision-making on allocations. Turkey has been lodging objections to eastern bluefin tuna recommendations since the mid-2000s over concerns that its allocation is unfair and not in line with its historical share. Despite its objections, Turkey voluntarily complied with its assigned allocation until the adoption of Recommendation 14-04, which substantially raised the total quota for eastern bluefin but did not change the allocation arrangement. Turkey objected to Recommendation 14-04 and established an autonomous quota consistent with what it viewed as its historical share—a level substantially higher than its ICCAT-assigned quota. If other CPCs take a similar path, improvements to the scientific advice will be of little help; the management measures adopted by ICCAT will be rendered impotent, and the stock may once again be in danger of collapse. Indeed, ICCAT is at a crossroads, in terms of how it governs itself and in terms of how it uses scientific advice. The question of governance, of course, involves matters of state and is beyond the scope intended for this chapter. The rest of this chapter therefore focuses on the use of scientific advice, which boils down to a decision on how best to manage under uncertainty.

There is a long history of literature devoted to managing fisheries under uncertainty, with most approaches falling roughly onto two main paths: (1) providing an accurate accounting of the main uncertainties through the stock-assessment process, and (2) developing management procedures that are robust to likely but unquantified sources of uncertainty. The first SCRS working groups to address the matter met in 1998 and 1999 (Anonymous 1999b, Gavaris et al. 2009). They suggested a simulation framework for testing the performance of several proposed harvest-control rules (HCRs) and stressed the need for joint manager-scientist meetings to make further progress, but the meetings envisioned were never held (until very recently). In 2007 a meeting was held between representatives from all five tRFMOs (in Kobe, Japan), where recommendations were made to "standardize the presentation of stock assessments and to base management decisions upon the scientific advice, including the application of the precautionary approach." Subsequent meetings of the Joint Tuna RFMOs refined this advice and adopted the so-called Kobe strategy matrix, which presents the probability that a candidate management measure (e.g., a TAC) will meet the intended management objective in a specified time frame. This and other developments effectively set ICCAT and the SCRS along the first path, attempting to quantify the uncertainty, with the implicit promise that more realistic assessment models and more informative data will reduce that uncertainty and eventually allow higher catch limits (closer to the MSY).

The limited financial resources available for stock assessments and supporting activities make it challenging to fully account for all of the perceived sources of uncertainty, and it becomes necessary to conduct cost-benefit analyses to determine the type of research and data collection activities that are most likely to improve the stock assessment. There are also practical concerns regarding the assessment process itself and how to balance the competing needs for thoroughness, transparency, and timeliness. The SCRS has begun to move away from relying on one or two relatively simple assessment tools like surplus production models and VPAs, toward a more thorough framework built on multimodel inferences and more flexible statistical models.

In principle, this more comprehensive framework can take advantage of more types of data and do a better job incorporating important uncertainties. However, it is more complex and difficult to use in a working group environment where analyses and reports are expected to be completed in just a few days. Moreover, in many cases, there are only a handful of analysts available with the necessary skills. Therefore, much of the work must be accomplished outside the working group, which can affect both timeliness and transparency. The SCRS has tried to ensure the transparency of its work by requiring software, including code and documentation, to be made available through the ICCAT software catalog. It also requires that all data used in the analyses be made available to all registered participants of the meeting. Nevertheless, the details of the assessment process can be difficult to oversee from afar and may be hard to reproduce when key analysts no longer participate. Thus, in addition to consistent mechanisms for providing the needed data, ICCAT needs to ensure that the relevant expertise is consistently available if it wants to continue along the path of using more complex models and better data.

The biggest challenge for the Kobe matrix style of providing and using scientific advice, however, lies in the ability to accurately quantify the true uncertainty—or at least to do so to the satisfaction of those who must make decisions based on it. The array of possible sources of uncertainty is formidable, necessitating complex models that require more data on a continuing basis than current resources can support. The added complexity also makes stock assessments more time-consuming to conduct, which mitigates against providing timely advice. The SCRS has for many years advised ICCAT that the current demand for bluefin tuna stock assessments every two or three years compromises the ability of bluefin tuna scientists to conduct the research necessary to improve the assessment, and moving to more complex assessment models exacerbates the situation because they take much longer to implement and review.

The second approach to managing under uncertainty is to develop management procedures that are robust to the main perceived sources of uncertainty. It is of course impractical to test the performance of candidate management procedures on real fisheries, but it is relatively easy to test them through simulations of closed-loop feedback systems, otherwise known as management strategy evaluation (MSE). Ideally, an MSE can be used to identify a management procedure (MP) that can run for several years on autopilot, without the need for managers to agree on measures based on frequent updates of the stock assessment. MSEs can also be used for strategic purposes, for example to explore the robustness of existing or proposed elements of a management regime or to identify the benefits of a proposed scientific data collection program. The Commission for the Conservation of Southern Bluefin Tuna (CCSBT) used an MSE to develop an MP for southern bluefin tuna, and the Indian Ocean Tuna Commission (IOTC) has done the same for skipjack tuna (*Katsuwonus pelamis*). The other tRFMOs are also starting to conduct MSEs for various stocks. Kell et al. (2003), for example, used MSE to evaluate the current advice framework at ICCAT as an "implicit" MP (i.e., a set of rules for management of a resource that contains all the elements of an MP but is not run on autopilot). Several HCRs have been simulation tested for North Atlantic albacore but not yet adopted by ICCAT because more simulation trials need to be run to ensure they are sufficiently robust (Anonymous 2016a).

Conducting an MSE is a long process and requires six steps: (1) identifying management objectives; (2) selecting hypotheses for the operating model (OM); (3) conditioning the OM based on data and knowledge, and possibly weighting and rejecting hypotheses; (4) identifying candidate management strategies; (5) running the MP as a feedback control to simulate the long-term impact of management; and (6) identifying the MPs that robustly meet management objectives. The first and last steps require a continuing dialogue between managers and stakeholders. The Kobe framework has helped in this respect, as have the Kobe phase plot and matrix. However, fully addressing the effect of uncertainty on achieving management objectives requires a move toward risk management. An MSE can help make that step, especially if it helps the tRFMOs to consider social and economic as well as biological objectives. This may require expanding the objectives of ICCAT to consider a range of performance statistics, such as the variability in catch as well as its magnitude, and balancing short-term versus long-term objectives (see the performance indicators of Appendix 2 in Rec. 16-21).

Considerable investments have been made in developing OMs that represent credible hypotheses for ABT population and fishery dynamics and conditioning those OMs on data from the GBYP, the BTRP, and ICCAT (Kerr et al. 2013, Carruthers et al. 2016, Anonymous 2016b, Kerr et al. 2017). A main source of uncertainty important for management is the stock structure, and so the OM is being conditioned so that the robustness of current advice based on an eastern and western stock can be evaluated and then robust alternatives proposed and evaluated. It is hoped that the results of this work will lead the way in developing an effective management procedure that bridges the gap between timely and thorough advice regarding ABT.

## Acknowledgments

The authors thank the ICCAT Secretariat, the SCRS, and its associated working groups for their many contributions toward gaining a better understanding of Atlantic bluefin tuna and improving the scientific advice offered to ICCAT. This chapter draws heavily from the reports produced by these groups and the helpful comments of some of its members, including A. Di Natale, J. Cort, A. Gordoa, L. Kell, A. Kimoto, P. Lino, D. Secor, and K. Blankenbecker. The opinions expressed in this chapter should be attributed to the authors alone; they do not necessarily reflect the views of the SCRS, ICCAT, or its members, and in no way do they anticipate the future policies of ICCAT.

### REFERENCES

Al-Idrisi al-Qurtubi al-Hasani al-Sabti A.A.A.M. 1154a. *Tabula Rogeriana.*

Al-Idrisi al-Qurtubi al-Hasani al-Sabti A.A.A.M. 1154b. *De geographia universali or Kitāb Nuzhat al-mushtāq fī dhikr al-amṣār wa-al-aqṭār wa-al-buldān wa-al-juzur wa-al-madā' in wa-al-āfāq.* Rome (Latin translation of the original work written in Arabic ~1154).

Anonymous. 1971. *Report of the Standing Committee on Research and Statistics (SCRS).* ICCAT Report for Biennial Period 1970–71, Part II.

Anonymous. 1980. *Report of the Standing Committee on Research and Statistics (SCRS).* ICCAT Report for Biennial Period 1978–79, Part II (1979).

Anonymous. 1982. *Report of the Standing Committee on Research and Statistics (SCRS).* ICCAT Report for the Biennial Period 1980–81, Part II (1981).

Anonymous. 1994. *Report of the Standing Committee on Research and Statistics (SCRS).* ICCAT Report for the Biennial Period 1992–1993, Part II (1993).

Anonymous. 1995. *Report of the Standing Committee on Research and Statistics (SCRS).* ICCAT Report for the Biennial Period 1994–1995, Part I (1994).

Anonymous. 1999a. Report of the ICCAT SCRS Bluefin Tuna Stock Assessment Session. *ICCAT Collective Volume of Scientific Papers* 49:1–191.

Anonymous. 1999b. Report of the ICCAT Ad Hoc Working Group Meeting on the Precautionary Approach. *ICCAT Collective Volume of Scientific Papers* 49 (4): 243–60.

Anonymous. 2005. Report of the 2004 data exploratory meeting for the East Atlantic and Mediterranean bluefin tuna. *ICCAT Collective Volume of Scientific Papers* 58:662–99.

Anonymous. 2007. *Report of the Standing Committee on Research and Statistics (SCRS)*. ICCAT Report for the Biennial Period 2006–2007, Part I (2006), vol. 2. http://iccat.int/Documents/BienRep/REP_EN_06-07_I_2.pdf.

Anonymous. 2008a. *Report of the Standing Committee on Research and Statistics (SCRS)*. September 29–October 3, 2008, Madrid, Spain. https://www.iccat.int/Documents/Meetings/Docs/2008_SCRS_ENG.pdf.

Anonymous. 2008b. *Report of the Standing Committee on Research and Statistics (SCRS)*. Report for the Biennial Period 2006–2007, Part II (2007), vol. 2. http://iccat.int/Documents/BienRep/REP_EN_06-07_II_2.pdf.

Anonymous. 2009a. Report of the world symposium for the study into the stock fluctuation of northern bluefin tunas (*Thunnus thynnus* and *Thunnus orientalis*), including the historical periods. *ICCAT Collective Volume of Scientific Papers* 63:1–49.

Anonymous. 2009b. *Report of the Standing Committee on Research and Statistics (SCRS)*. Report for the Biennial Period 2008–2009, Part I (2008), vol. 2. http://iccat.int/Documents/BienRep/REP_EN_08-09_I_2.pdf.

Anonymous. 2009c. *Report of the independent performance review of ICCAT*. www.iccat.int/Documents/Other/PERFORM_%20REV_TRI_LINGUAL.pdf.

Anonymous. 2009d. Report of the 2008 ICCAT Atlantic bluefin tuna stock assessment session. June 23 to July 4, 2008, Madrid, Spain. *ICCAT Collective Volume of Scientific Papers* 64 (1): 1–352.

Anonymous. 2010. *Report of the Standing Committee on Research and Statistics (SCRS)*. Report for the Biennial Period 2008–2009, Part II (2009), vol. 2. http://iccat.int/Documents/BienRep/REP_EN_08-09_II_2.pdf.

Anonymous. 2015a. *Report of the Standing Committee on Research and Statistics (SCRS)*. https://www.iccat.int/Documents/Meetings/SCRS2015/SCRS_PROV_ENG.pdf.

Anonymous. 2015b. Report of the 2014 ICCAT Atlantic bluefin tuna stock assessment session, September 22–27, 2014, Madrid, Spain. *ICCAT Collective Volume of Scientific Papers* 71 (2): 692–945.

Anonymous. 2016a. *Report of the 2016 ICCAT North and South Atlantic albacore stock assessment meeting*. April 28 to May 6, 2016, Madeira, Portugal. http://iccat.int/Documents/Meetings/Docs/2016_ALB_REPORT_ENG.pdf.

Anonymous. 2016b. Report of the 2016 ICCAT bluefin data preparatory meeting, July 25–29, 2016. *ICCAT Collective Volume of Scientific Papers*. https://www.iccat.int/Documents/Meetings/Docs/2016_BFT_DATA_PREP_ENG.pdf.

Aranda, G., A. Medina, A. Santos, F.J. Abascal, and T. Galaz. 2013. Evaluation of Atlantic bluefin tuna reproductive potential in the western Mediterranean Sea. *Journal of Sea Research* 76:154–60.

Aristotelis. 1635. De animalibus. In *Stagiritae Peripatetico Rum: Principis de Historia Animalium*, edited by T. Goza. Venezia, Italy.

Arrizabalaga, H., I. Arregi, A. Medina, N. Rodríguez-Ezpeleta, J.M. Fromentin, and I. Fraile. 2016. Life history and migrations of Mediterranean bluefin populations. Bluefin Futures Symposium, January 18–20. Monterey, California.

Bauer, R.K., S. Bonhommeau, B. Brisset, and J.-M. Fromentin. 2015. Aerial surveys to monitor bluefin tuna abundance and track efficiency of management measures.

Porch, C.E., S.C. Turner, and J.E. Powers. 2001. Virtual population analyses of Atlantic bluefin tuna with alternative models of transatlantic migration: 1970–1997. *ICCAT Collective Volume of Scientific Papers* 52 (3): 1022–45.

Powers, J., and V. Restrepo. 1992. Additional options for age-sequenced analysis. *ICCAT Collective Volume of Scientific Papers* 39 (2): 540–53.

Puncher, G.N., V. Onar, N.Y. Toker, and F. Tinti. 2014. A multitude of Byzantine era bluefin tuna and swordfish bones uncovered in Istanbul, Turkey. *ICCAT Collective Volume of Scientific Papers* 71 (4): 1626–31.

Puncher, G.N., H. Arrizabalaga, F. Alemany, A. Cariani, I.K. Oray, F.S. Karakulak, and F. Tinti. 2015. Molecular identification of Atlantic bluefin tuna (*Thunnus thynnus*, Scombridae) larvae and development of a DNA character-based identification key for Mediterranean scombrids. *PLOS ONE* 10 (7): e0130407.

Puncher, G.N., A. Cariani, E. Cilli, F. Massari, P.L. Martelli, A. Morales, V. Onar, N.Y. Toker, T. Moens, and F. Tinti. 2016. Unlocking the evolutionary history of the mighty bluefin tuna using novel paleogenetic techniques and ancient tuna remains. *ICCAT Collective Volume of Scientific Papers* 72 (6): 1429–39.

Punt, A.E., and D.S. Butterworth. 1995. Use of tagging data within a VPA formalism to estimate migration rates of bluefin tuna across the north Atlantic. *ICCAT Collective Volume of Scientific Papers* 44 (1): 166–82.

Ravier, C., and J.-M. Fromentin. 2001. Long-term fluctuations in the Eastern Atlantic and Mediterranean bluefin tuna population. *ICES Journal of Marine Science* 58:1299–1317.

Richardson, D.E., K.E. Marancik, J.R. Guyon, M.E. Lutcavage, B. Galuardi, C.H. Lam, H.J. Walsh, S. Wildes, D.A. Yates, and J.A. Hare. 2016. Discovery of a spawning ground reveals diverse migration strategies in Atlantic bluefin tuna (*Thunnus thynnus*). *Proceedings of the National Academy of Sciences USA* 113:3299–3304.

Rodríguez-Ezpeleta, N., N. Díaz-Arce, F. Alemany, S. Deguara, J. Franks, J.R. Rooker, M. Lutcavage, et al. 2016. A genetic traceability tool for differentiation of Atlantic Bluefin tuna spawning grounds. *ICCAT Collective Volume of Scientific Papers* SCRS/2016 /P/032.

Rodríguez-Marín, E., J. Neilson, P.L. Luque, S. Campana, M. Ruiz, D. Busawon, and J.M.O. de Urbina. 2012. BLUEAGE, a Canadian-Spanish joint research project: Validated age and growth analysis of Atlantic bluefin tuna (*Thunnus thynnus*). *ICCAT Collective Volume of Scientific Papers* 68 (1): 254–60.

Rodríguez-Marín, E., P. Quelle, M. Ruiz, and P.L. Luque. 2016. Standardized age-length key for east Atlantic and Mediterranean bluefin tuna based on otoliths readings. *ICCAT Collective Volume of Scientific Papers* 72 (6): 1365–75.

Rooker, J.R., and D.H. Secor. 2016. Migration ecology and stock mixing of Atlantic bluefin tuna recorded within the microchemistry of otoliths. Bluefin Futures Symposium, January 18–20. Monterey, CA.

Rooker, J.R., D.H. Secor, G. DeMetrio, R. Schloesser, B.A. Block, and J.D. Neilson. 2008. Natal homing and connectivity in Atlantic bluefin tuna populations. *Science* 322: 742–44.

Rooker, J.R., H. Arrizabalaga, I. Fraile, D.H. Secor, D.L. Dettman, N. Abid, P. Addis, et al. 2014. Crossing the line: Migratory and homing behaviors of Atlantic bluefin tuna. *Marine Ecology Progress Series* 504:265–76.

Safina, C. 1998. *Song for the Blue Ocean: Encounters along the World's Coasts and beneath the Seas.* New York: Henry Holt.

Siskey, M.R., W.J. Wilberg, R.J. Allman, B.K. Barnett, and D.H. Secor. 2016. Forty years of fishing: Changes in age structure and stock mixing in Northwestern Atlantic bluefin tuna (*Thunnus thynnus*) associated with size-selective and long-term exploitation. *ICES Journal of Marine Science* 73 (10): 2518–28. doi:10.1093/icesjms/fsw115.

Spoto, S. 2002. *Sicilia Antica: Usi, costumi e personaggi dalla preistoria alla società greca, nell'isola culla della civiltà europea.* Edited by N.E. Compton. Rome: Newton & Compton.

Suzuki, Z., A. Kimoto, and O. Sakai. 2013. Note on the strong 2003 year-class that appeared in the Atlantic bluefin fisheries. *ICCAT Collective Volume of Scientific Papers* 69 (1): 229–34.

Takeuchi, Y., A. Sudal, and Z. Suzuki. 1999. Review of information on large bluefin tuna caught by Japanese longline fishery off Brasil from the late 1950s to the early 1960s. *ICCAT Collective Volume of Scientific Papers* 49:416–28.

Takeuchi, Y., K. Oshima, and Z. Suzuki. 2009. Inference on the nature of Atlantic bluefin tuna off Brazil caught by Japanese longline fishery around the early 1960s. *ICCAT Collective Volume of Scientific Papers* 63:186–94.

Tangen, M. 2009. The Norwegian fishery for Atlantic Bluefin Tuna. *ICCAT Collective Volume of Scientific Papers* 63:79–93.

Taylor, N., M.K. McAllister, G.L. Lawson, T. Carruthers, and B.A. Block. 2011. Atlantic bluefin tuna: A novel multistock spatial model for assessing population biomass. *PLOS ONE* 6 (12): e27693. doi:10.1371/journal.pone.0027693.

Tiews, K. 1978. On the disappearance of bluefin tuna in the North Sea and its ecological implications for herring and mackerel. Rapport des Procès verbaux des Réunions du Conseil Permanent International pour l'Exploration de la Mer 172:301–9.

Tusa, S. 1999. *La Sicilia nella Preistoria* [1983]. Palermo, Italy: Sellerio.

Walli. A., S.L.H. Teo, A. Boustany, C.J. Farwell, T. Williams, H. Dewar, E. Prince, and B.A. Block. 2009. Seasonal movements, aggregations and diving behavior of Atlantic bluefin tuna (*Thunnus thynnus*) revealed with archival tags. *PLOS ONE* 4 (7): e6151. doi:10.1371/journal.pone.0006151.

Young, E.B. 1975. Canadian statement for Panel 2. Regular meeting of ICCAT Council, Madrid, November 22, 1974. ICCAT Report for the biennial period 1974–1975, Part I (1974), pp. 50–53.

CHAPTER 2

# Otolith Microchemistry

## Migration and Ecology of Atlantic Bluefin Tuna

Jay R. Rooker and David H. Secor

## Introduction

Atlantic bluefin tuna (*Thunnus thynnus*) exhibit strong population structure, with two dominant spawning areas in the Mediterranean Sea and the Gulf of Mexico, producing recruits that occupy areas within these marginal seas or adjacent waters in the North Atlantic Ocean. This view that there are two dominant stocks in the Atlantic Ocean was initially supported by the distribution of fisheries, conventional tagging data, and ichthyoplankton surveys (e.g., Mather et al. 1995). Although early conventional tagging data indicated a small degree of mixing between the two populations, particularly at young ages (Rooker et al. 2007), the International Commission for the Conservation of Atlantic Tunas (ICCAT) codified population structure with separate assessments for eastern and western stocks (NRC 1994), which have continued for more than 30 years. The premise of two stocks with limited connectivity or mixing has been challenged (e.g., Block et al. 2005, Fromentin and Powers 2005), and it is now widely accepted that significant mixing between eastern and western stocks is important, with individuals commonly migrating across the 45°W management boundary (Galuardi et al. 2010, Rooker et al. 2014, Graves et al. 2015). Given that the population dynamics of Atlantic bluefin tuna appear highly sensitive to the migration and mixing of individuals (Taylor et al. 2011, Kerr et al. 2012), findings of increased connectivity underscore the need to better characterize the annual migratory patterns and stock mixing of Atlantic bluefin tuna throughout their range.

A variety of approaches have been used to investigate the movements, migrations, population structure, and mixing of Atlantic bluefin tuna and other tuna populations, including conventional tags, electronic tags, genetics, pollutant tracers, and chemical markers in calcified structures or hard parts (Gunn and Block 2001, Rooker et al. 2007, Teo and Boustany 2015). Of these, natural chemical markers in otoliths (ear stones) appear to show the greatest potential for determining natal origin, stock integrity, and the collective movements of Atlantic bluefin tuna (Magnuson et al. 2001, Rooker et al. 2008a,b). This is because otoliths precipitate material (primarily $CaCO_3$) as the tuna grows, and the chemical composition of newly accreted otolith material is reflective of the physicochemical conditions of the seawater inhabited by the individual (Rooker et al. 2001a, Elsdon and Gillanders 2003). As a result, the chemical composition of otolith material deposited within the first annulus (i.e., during the first year of life) is often linked to the seawater chemistry at the individual's place of origin. Additionally, the movement ecology of individual tunas can be inferred by changes in chemical composition of otoliths during the early life period (Baumann et al. 2015), although this approach has not yet been applied to Atlantic bluefin tuna. To date, two classes of chemical markers—trace elements and stable isotopes—have been widely used to investigate the origin and stock mixing of both tropical (Wells et al. 2012, Rooker et al. 2016) and temperate tunas, including recent research on Atlantic bluefin tuna (Rooker et al. 2014, Secor et al. 2015, Siskey et al. 2016).

This chapter is a synthesis on stock composition and migration studies on Atlantic bluefin tuna using otolith chemistry. These studies have spanned nearly 20 years and have culminated with the operational use of otolith-based population assignments in stock-assessment models (Taylor et al. 2011, Kerr et al. 2017). Both the development of the approach and the application of these novel tracers are reviewed, and a synopsis of the stock composition (i.e., mixing levels) of Atlantic bluefin tuna is presented for different regions of the North Atlantic Ocean. Three primary questions related to the migration ecology of Atlantic bluefin tuna are addressed: (1) What is the degree of transatlantic movement, and which geographic regions and fisheries represent hot spots for stock mixing? (2) Following transatlantic movement, is homing to natal sites in the east (the Mediterranean Sea) and west (the Gulf of Mexico) prevalent? (3) Can natural chemical markers be used to document age-specific egress/ingress from nursery areas? We end the chapter by discussing the potential impacts of different migration and mixing scenarios on the current assessment frameworks used to manage Atlantic bluefin tuna. In this retrospective, we recognize that mixing represents an emergent property influenced by transatlantic movements and the relative productivity of populations. Further, additional spawning areas and population structure (subpopulations and contingents) likely exist (Fromentin et al. 2014, Richardson et al. 2016) and have yet to be evaluated with otolith chemistry.

## History of the Approach

The earliest attempt to use chemical markers in hard parts to discriminate Atlantic bluefin tuna stocks was performed by Calaprice (1985). This study demonstrated that trace elements and stable isotopes in the vertebrae of Atlantic bluefin tuna showed potential for identifying individuals of eastern and western origin. Still, uncertainties in predictions of geographic origin and stock mixing from this research were relatively high, and the National Research Council (1994) recommended that new analytical approaches were necessary to further advance this line of research. In 1997 a workshop of international experts identified rigorous analytical methods and standardization as chief technological constraints for applying this approach to Atlantic bluefin tuna (Secor and Chesney 1998). Initial research followed related work on its congener in the Pacific Ocean (*T. orientalis;* Rooker et al. 2001b), and it focused on trace element comparisons between otoliths from eastern and western nurseries and evaluation of decontamination, accuracy, and standardization protocols (Rooker et al. 2001a, 2002; Secor et al. 2002). Using a suite of influential elements (Li, Mg, Mg, Sr, Ba) in the whole otoliths of juvenile (age 0, age 1) Atlantic bluefin tuna, Rooker et al. (2003) demonstrated that classification of individuals to eastern and western nurseries was possible. However, classification success was relatively modest (60%–85%), indicating that the discriminatory power of the approach needed to improve prior to full-scale investigations of stock mixing.

Concurrent with these investigations, a second class of chemical markers in otoliths, stable oxygen and carbon isotope ratios ($\delta^{18}O$, $\delta^{13}C$), was being used in conjunction with trace elements to improve the classification of marine fish populations to natal sites (e.g., Thorrold et al. 2001). In response, Rooker and Secor (2004) assessed the utility of $\delta^{18}O$ and $\delta^{13}C$ for discriminating yearling Atlantic bluefin tuna of eastern and western origin and showed that classification success was nearly 100% with these chemical markers alone. This research also demonstrated that the successful discrimination of eastern and western yearlings was almost entirely due to differences in one marker: otolith $\delta^{18}O$. The observed differences in otolith

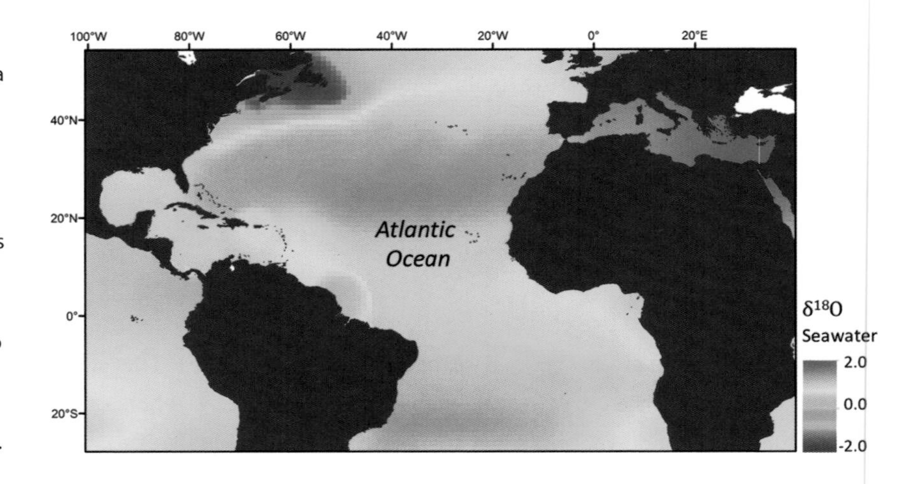


**Figure 2.1.** Gridded global seawater $\delta^{18}O$ data (1° resolution) downloaded from the National Aeronautics and Space Administration Goddard Institute for Space Studies (http://data.giss.nasa.gov/o18data/) and displayed in ArcGIS 10.2 (Esri Inc.) to visualize spatial differences in seawater $\delta^{18}O$ across the Atlantic Ocean. Originally described by LeGrande and Schmidt 2006, this data set uses relationships between regional seawater $\delta^{18}O$ and salinity, as well as $PO_4$, to delineate water mass margins and interpolate seawater $\delta^{18}O$ values.

$\delta^{18}O$ were not entirely unexpected, given the observed geographic variation in seawater $\delta^{18}O$ across the Atlantic Ocean, with significantly higher values observed in the Mediterranean Sea relative to the Gulf of Mexico spawning areas (LeGrande and Schmidt 2006; Figure 2.1). The promise of otolith $\delta^{18}O$ as a birth certificate led some scientists to speculate that this chemical marker represented the "Rosetta Stone" for explaining longstanding questions about stock mixing and migration. However, the impact of interannual variability in otolith $\delta^{18}O$ and $\delta^{13}C$, which is known to influence the utility of these natural markers, was not fully assessed in Rooker and Secor (2004) because only two year classes of yearling Atlantic bluefin tuna were included.

Since the initial discovery of otolith $\delta^{18}O$ as a discriminator of nursery origin for Atlantic bluefin tuna, subsequent papers have confirmed the promise of this marker for retrospective determination of stock origin. In fact, yearlings from many year classes have been pooled to support more robust classification procedures that could apply to all year classes, but this has resulted in greater overlap in stable isotopes between principal nurseries. Still, classification success remained sufficiently high (87%) to support stock composition analysis (Rooker et al. 2008a,b). This juvenile baseline was later updated with improved and consistent methods for milling otolith material and was comprised of yearlings sampled in the Mediterranean Sea and US natal regions over a longer period (1998–2011) (Rooker et al. 2014; Figure 2.2). For the years represented by this baseline, western yearlings showed much greater variability in otolith $\delta^{18}O$ than eastern yearlings (Figure 2.2). We speculate that overlap in the distribution of otolith

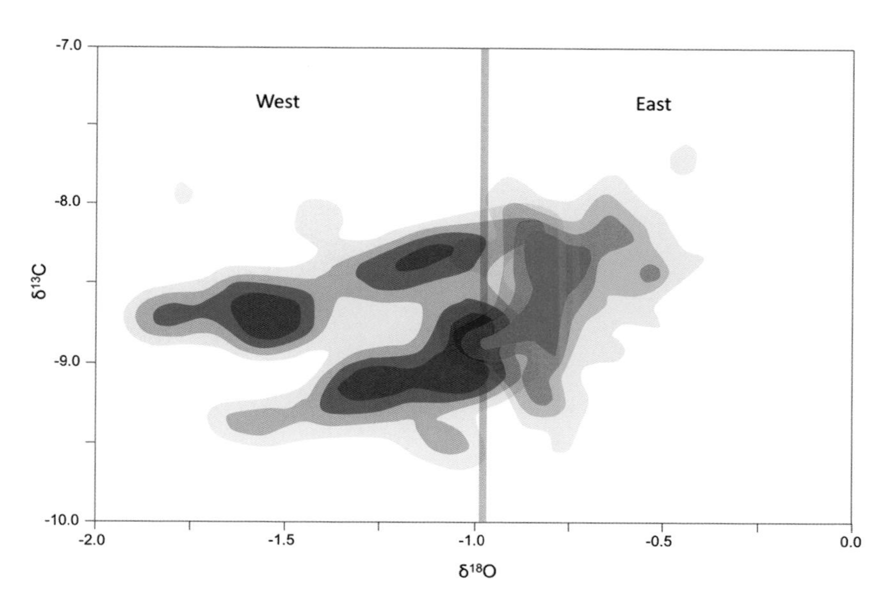


**Figure 2.2.** Otolith $\delta^{18}O$ and $\delta^{13}C$ values expressed as contour plots for yearling (12+ months) Atlantic bluefin tuna from eastern (*red*) and western (*blue*) nurseries. Contour plots shown at four levels (from dark to light fill colors, 20%, 40%, 60%, and 80%). Data for plot derived from Rooker et al. 2014. Otolith $\delta^{18}O$ threshold value of approximately −0.9 (*denoted by line*) represents the tentative boundary between eastern and western occurrence.

$\delta^{18}O$ values between eastern and western yearlings may be due in part to transatlantic migrations during the yearling period, particularly from the Mediterranean Sea population, which has been documented through past conventional tagging studies (Rooker et al. 2007). Further, modalities in otolith $\delta^{18}O$ for the western sample are evident and possibly indicative of additional spawning areas in the western Atlantic Ocean (Richardson et al. 2016) or of the presence of contingents with distinct migration pathways or egress patterns (e.g., different residency periods in the Gulf of Mexico nursery).

## Transatlantic Movement and Stock Mixing

Otolith $\delta^{13}C$ and $\delta^{18}O$ values from milled cores (corresponds to yearling period) of Atlantic bluefin tuna collected from several regions of the North Atlantic Ocean (western, central, and eastern) were used to assess stock mixing and transatlantic migrations. An overview of observed stock composition and mixing within each of the three major regions follows.

### Northwestern Atlantic Ocean

Atlantic bluefin tuna collected in the 1990s and the early 2000s from US fisheries exhibited high levels of mixing, particularly for school (~< 60 kg, < 148 cm straight fork length [SFL], < 152 cm curved fork length [CFL]) and medium (~60–140 kg, 148–198 SFL cm, 152–204 cm CFL) category

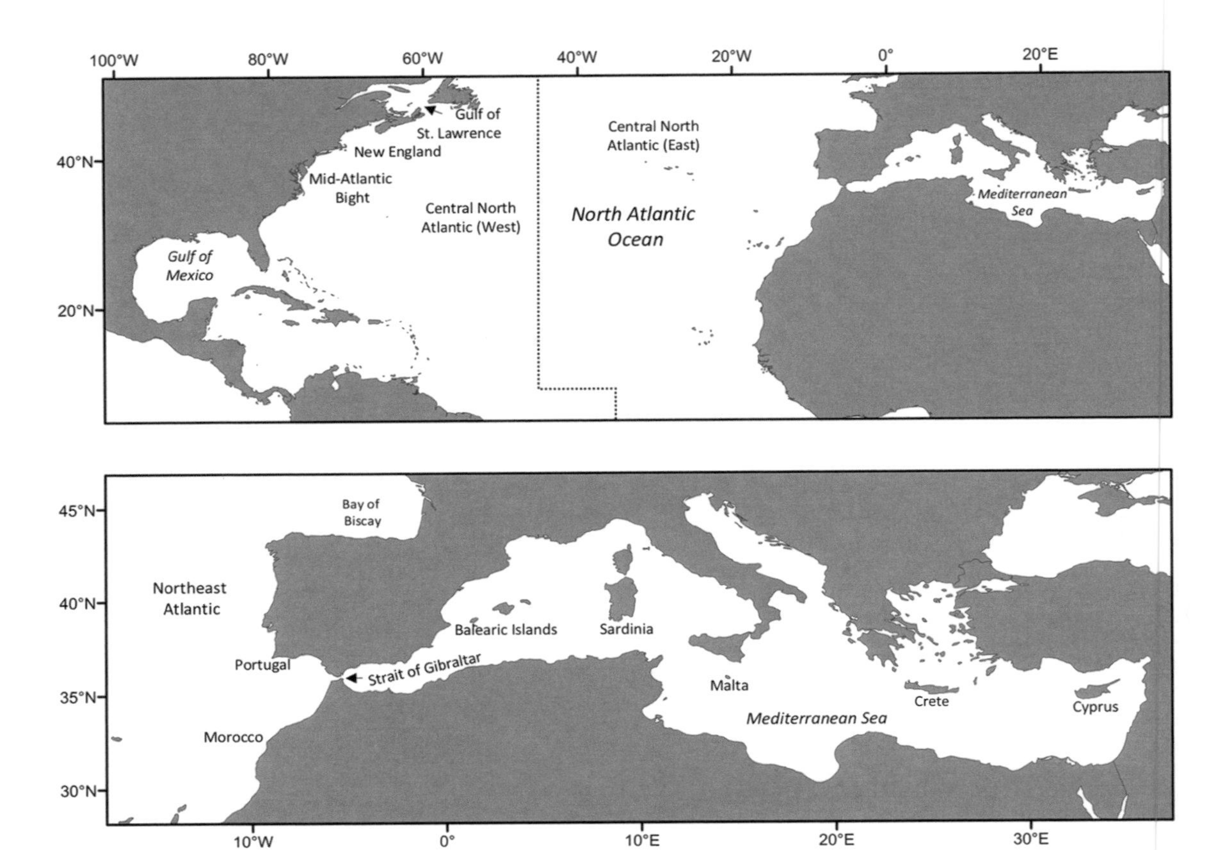


**Figure 2.3.** Map of primary collection areas for Atlantic bluefin tuna. Management boundary (45°W longitude) denoted with dashed line.

fish. Initial estimates of stock mixing in the US Atlantic Ocean (Figure 2.3) were published by Rooker et al. (2008a) using a six-year baseline sample of otolith $\delta^{13}C$ and $\delta^{18}O$ values for yearling Atlantic bluefin tuna. Findings from this study indicated that a large fraction of the Atlantic bluefin tuna from US fisheries in the mid-Atlantic were of eastern (Mediterranean) origin, with eastern migrants accounting for 57% and 44% of the school and medium category fish sampled, respectively (Table 2.1). The presence of eastern migrants in samples of giant category Atlantic bluefin tuna (~> 140 kg; > 198 cm SFL; > 204 cm CFL) from both the mid-Atlantic (35%) and the Gulf of Maine (5%) was lower, suggesting that transatlantic movement and stock mixing may be more prevalent for younger or smaller individuals. Using the new baseline produced by Rooker et al. (2014), Siskey et al. (2016) evaluated a new sample of Atlantic bluefin tuna from the mid-Atlantic and New England during this same period (1996–2002)

**Table 2.1.** Summary of stock composition for Atlantic bluefin tuna collected from several regions within both the eastern and western management areas

| Region | Year(s) sampled | N | Size class | % West | % East | 1SD (%) | Reference |
|---|---|---|---|---|---|---|---|
| West | | | | | | | |
| Gulf of Mexico | 1976–1978 | 102 | G | 100.0 | 0.0 | 0.0 | Siskey et al. 2016 |
| Gulf of Mexico | 2004–2007 | 42 | G | 99.3 | 0.7 | 1.7 | Rooker et al. 2008a |
| Gulf of Mexico | 1999–2011 | 183 | G | 100.0 | 0.0 | 0.5 | Secor et al. 2014 |
| Gulf of Mexico | 2009–2014 | 203 | G | 100.0 | 0.0 | 0.0 | Siskey et al. 2016 |
| US Atlantic (mid-Atlantic) | 1974–1977 | 102 | S,M,G | 100.0 | 0.0 | 0.0 | Siskey et al. 2016 |
| US Atlantic (mid-Atlantic) | 1996–2002 | 154 | S, M,G | 54.3 | 45.7 | 7.2 | Rooker et al. 2008b |
| US Atlantic (mid-Atlantic) | 1996–2000 | 76 | S,M,G | 37.0 | 63.0 | 7.7 | Siskey et al. 2016 |
| US Atlantic (mid-Atlantic) | 2010–2014 | 854 | S,M,G | 90.0 | 10.0 | 1.7 | Siskey et al. 2016 |
| US Atlantic (New England) | 1996–2002 | 153 | S,M,G | 59.0 | 41.0 | 5.6 | Siskey et al. 2016 |
| US Atlantic (New England) | 1996–1998 | 72 | S,M,G | 94.8 | 5.2 | 5.3 | Rooker et al. 2008b |
| US Atlantic (New England) | 2010–2014 | 318 | M,G | 99.0 | 1.0 | 1.2 | Siskey et al. 2016 |
| US Atlantic (mid-Atlantic, New England) | 2015 | 175 | S,M,G | 28.4 | 71.6 | 5.0 | Barnett et al. 2016 |
| CA Atlantic (Gulf of St. Lawrence) | 1975–2007 | 224 | G | 100.0 | 0.0 | 0.1 | Schloesser et al. 2010 |
| CA Atlantic (Gulf of St. Lawrence) | 2011–2012 | 191 | G | 100.0 | 0.0 | 0.2 | Busawon et al. 2014 |
| CA Atlantic (Maritime) | 2011–2012 | 151 | M,G | 85.0 | 15.0 | 5.4 | Busawon et al. 2014 |
| Central North Atlantic | | | | | | | |
| West of 45° management boundary | 2010–2011 | 25 | M,G | 56.0 | 44.0 | 16.8 | Rooker et al. 2014 |
| East of 45° management boundary | 2010–2011 | 177 | M,G | 15.1 | 84.9 | 4.9 | Rooker et al. 2014 |
| East | | | | | | | |
| Mediterranean Sea (Malta, Spain) | 2003–2007 | 132 | M,G | 4.2 | 95.8 | 3.1 | Rooker et al. 2008a,b |
| Mediterranean Sea (Spain) | 2011 | 13 | M,G | 0.0 | 100.0 | 0.0 | Rooker et al. 2014 |
| Mediterranean Sea (Sardinia) | 2011 | 20 | M,G | 0.0 | 100.0 | 0.0 | Rooker et al. 2014 |

*(continued)*

**Table 2.1.** *continued*

| Region | Year(s) sampled | N | Size class | % West | % East | 1SD (%) | Reference |
|---|---|---|---|---|---|---|---|
| Mediterranean Sea (Malta) | 2011 | 82 | M,G | 0.0 | 100.0 | 0.0 | Rooker et al. 2014 |
| Mediterranean Sea (Cyprus) | 2011 | 48 | M,G | 0.9 | 99.1 | 2.9 | Rooker et al. 2014 |
| Northeast Atlantic (Bay of Biscay) | 2009–2011 | 217 | S,M | 0.7 | 99.3 | 0.6 | Fraile et al. 2015 |
| Northeast Atlantic (Portugal, Strait of Gibraltar) | 2010–2011 | 109 | M,G | 0.0 | 100.0 | 0.0 | Rooker et al. 2014 |
| Northeast Atlantic (Morocco) | 2011–2012 | 81 | M,G | 6.1 | 93.9 | 4.7 | Rooker et al. 2014 |

*Notes*: Percent composition is based on maximum likelihood estimates (MLE). Standard deviation (SD) of MLE-based proportions is also given for each study. Sampling years, sample sizes (N), and size/weight categories of fish are shown for each study. Size classes (approximate weights and curved fork lengths [CFL]) of individuals included in each study: School (S, < 60 kg; xx CFL cm), Medium/Large (M, 60–140 kg; xx CFL cm), Giant (G, > 140 kg; xx CFL cm).

and also observed high levels of eastern stock contributions for smaller size-class fish: 73%, 51%, 32%, and 0% eastern fish for sizes 50–100 cm, 101–200 cm, 201–250 cm, and 251–300 cm CFL, respectively.

More recently, Siskey et al. (2016) investigated stock mixing of Atlantic bluefin tuna under varying periods of fishing and recruitment levels for the western stock over a 40-year period. In contrast to the higher levels of eastern stock contributions for collections in the 1990s and early 2000s, mixing was undetected for samples collected in the 1970s, regardless of size. This period preceded a major increase in fisheries targeting giants for the sushi market and was characterized by several strong year classes of western fish (Mather et al. 1995) Thus, any eastern-origin fish migrating into the western management area would likely have been rare relative to the more abundant western fish from these strong year classes. For more recent sampling (2009–2014), Siskey et al. (2016) observed a return to minimal stock mixing (<5% across all size classes), potentially an indication of recovery for the western stock (Table 2.1; Figure 2.4). Nevertheless, other recent samples of US fisheries reported higher levels of stock mixing. Secor et al. (2015) reported 24% mixing by Mediterranean individuals in the North Carolina–Virginia winter fishery sampled in 2011–13, and Barnett et al. (2016) observed large numbers of eastern migrants (>70%) in a 2015 sample of US summer-caught Atlantic bluefin tuna, which was dominated by smaller school and medium category fish (Table 2.1).

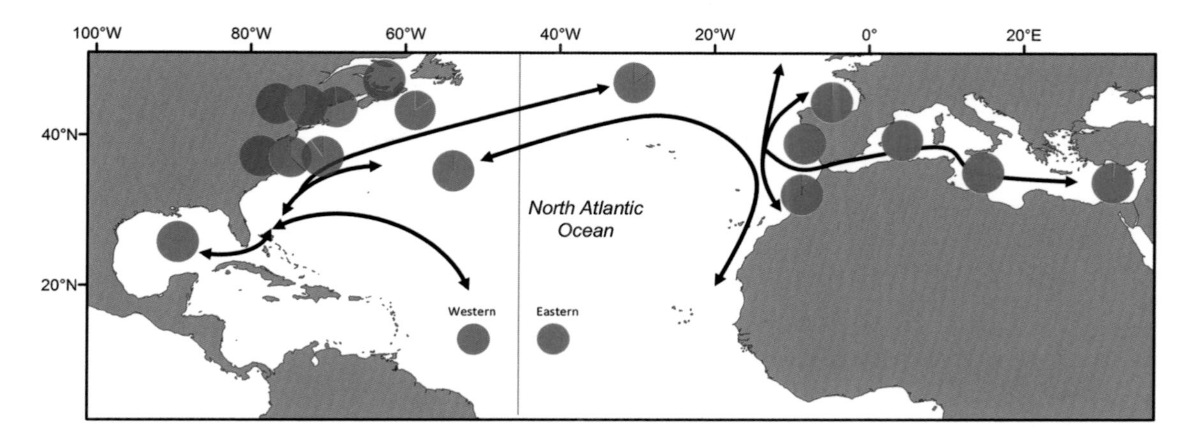


**Figure 2.4.** Most recent estimates of stock composition available for Atlantic bluefin tuna in key spawning, foraging, and fishing areas based on otolith $\delta^{18}O$ and $\delta^{13}C$. Estimates derived from several studies: Gulf of Mexico, mid-Atlantic, and New England (Siskey et al. 2016), Gulf of St. Lawrence and Canadian Maritime (Busawon et al. 2014), Central North Atlantic, Northeastern Atlantic Ocean (Morocco, Portugal), Mediterranean Sea (Rooker et al. 2014), and Bay of Biscay (Fraile et al. 2015). Generalized migration routes (*arrows*) are from Fromentin et al. 2014. See Table 2.1 for details about samples and mixing estimates derived for each study. Overlapping bubbles in the mid-Atlantic and New England denote retrospective estimates from three time periods (~1970s, 1990s, 2010s, *left to right*).

Stock mixing of Atlantic bluefin tuna in Canadian waters has been investigated in the Gulf of St. Lawrence, because this region represents an important foraging area for giant (adult) category fish (Mather et al. 1995). The origin of Atlantic bluefin tuna in the Gulf of St. Lawrence was first investigated by Rooker et al. (2008a), and this study indicated that ~100% of the giant category fish collected in this region were of western origin (Table 2.1). The importance of the Gulf of St. Lawrence for western fish was confirmed by Schloesser et al. (2010), with their study reporting that giants collected over multiple decades (1970s, 1980s, and 2000s) were almost exclusively (99%–100%) of western origin. However, there is recent evidence of eastern migrants in Canadian waters, with moderate mixing of eastern-origin fish (15%) observed in areas off Nova Scotia outside the Gulf of St. Lawrence (Busawon et al. 2014). The presence of eastern fish in the Gulf of St. Lawrence was also recently reported from electronic tag deployments, and fishery data from the same time period show an increased presence of small fish in this region (Busawon et al. 2014, Wilson et al. 2015).

Overall, the Northwestern Atlantic Ocean (NWAO) is a mixing hot spot for Atlantic bluefin tuna, with otolith chemistry research clearly showing that the US fisheries are dependent to some degree on production from the Mediterranean Sea (Figure 2.4). The extent of transatlantic movement and stock mixing in these waters appears to be highly variable across size/age classes, years, and decades. These results are supported in part by other approaches. Conventional tagging studies first indicated very low rates of mixing in the 1970s, whereas more recent electronic tagging studies showed transatlantic migration rates—mostly eastern-origin fish returning to the Mediterranean Sea to spawn—exceeding 15% (Kurota et al. 2009). Electronic tagging also suggested that a significant number of individuals in the North Carolina winter fishery migrated from the Mediterranean Sea as juveniles and returned as adults (Block et al. 2005). Higher mixing of juvenile (school category) Atlantic bluefin tuna (~ages 1–4) in the NWAO was also reported by scientists using organochlorine tracers (e.g., chlorodane:PCB ratios), with eastern-origin fish accounting for a significant fraction of the US recreational fishery in the mid-Atlantic from 2000 to 2008 (33%–88%; Dickhut et al. 2009) and from 2011 to 2012 (18%–33%; Graves et al. 2015).

### Northeastern Atlantic Ocean

The origin of Atlantic bluefin tuna in several regions of the Northeastern Atlantic Ocean (NEAO) has generally shown low rates of mixing from the Gulf of Mexico population, with the exception of a few areas outside the Mediterranean Sea. The stock composition of medium and giant category fish from trap fisheries in two regions of the NEAO (Morocco and Portugal) was determined recently by Rooker et al. (2014), and a difference was observed regarding the presence of western migrants in the two fisheries. No western migrants were detected in the sample from the Portuguese traps located west of the Strait of Gibraltar (Table 2.1; Figure 2.4). Conversely, a small fraction of western migrants (6%) were detected in the trap fishery from the northwest coast of Africa off Morocco, suggesting that western-origin fish also display transatlantic movement and inhabit waters adjacent to the Mediterranean Sea spawning area. Fraile et al. (2015) examined Atlantic bluefin tuna present in the regional bait boat fleet operating north of these trap fisheries in the Bay of Biscay and also detected western migrants in this fishery, although they accounted for only 1% and 3% of the entire sample of juveniles and adults examined. Similar to stock-mixing estimates reported above for the NWAO and NEAO, contribution rates varied among years examined, with the presence of western migrants detected in the 2009 sample (2%–5%) only from the Bay of Biscay, whereas 2010 and 2011 samples were comprised entirely of fish produced from the Mediterranean Sea.

### Central North Atlantic Ocean: East and West of 45°W

The first attempt to assess the origin of Atlantic bluefin tuna targeted by high seas fishing fleets (e.g., Japanese longline vessels) operating on each side of the 45°W management boundary was performed by Rooker et al. (2014). Significant stock mixing of Atlantic bluefin tuna was reported in this study, with both stocks commonly crossing the management boundary. The estimated contribution of western-origin fish in the Central North Atlantic Ocean (CNAO) fishery was 21% for specimens collected from 2010 to 2011, with the majority of Atlantic bluefin tuna (79%) in this region classified as eastern-origin fish (Table 2.1). Migrants from both spawning areas were detected on each side of the management boundary; 44% of the Atlantic bluefin tuna collected west of 45°W were classified as eastern-origin fish, whereas 15% of the Atlantic bluefin tuna collected east of 45°W were classified as western-origin fish. This finding confirms that the CNAO is an important mixing zone for Atlantic bluefin tuna, with both eastern and western stocks readily "crossing the line" and entering the other management zone (Figure 2.4).

## Natal Homing to Gulf of Mexico and Mediterranean Sea

Natal homing—defined here as the return of adults to their place of origin—has been addressed by examining the otolith core $\delta^{18}O$ and $\delta^{13}C$ values of adult Atlantic bluefin tuna collected in each spawning area. Because chemical markers in otolith cores are nursery-specific tags that remain unchanged during the life of an individual, they are particularly suitable for investigations of natal homing. Central questions of philopatry (multigenerational return to the same spawning region) or spawning fidelity (multiple returns by an adult to the same spawning region, regardless of natal origin) are better addressed by genetic markers and electronic tagging approaches (Block et al. 2005, Secor 2015), although otolith chemistry is applicable in these types of studies.

Findings from several studies using stable isotopes indicate that despite the presence of transatlantic movement and stock mixing, homing to the Gulf of Mexico and Mediterranean Sea spawning areas exceeds 95%. Initial estimates of nursery origin of adults (giants) from both spawning areas were reported by Rooker et al. (2008a), and nearly all of the adults captured in the Gulf of Mexico (99%) and the Mediterranean Sea (96%) were from the same region in which they were collected, supporting the premise of natal homing. Investigations of giant Atlantic bluefin tuna collected in the Gulf of Mexico over longer periods (1974–2014, year classes 1944–2006; Siskey et al. 2016) found similar results, with all adults (100%) classified to the western stock. In the east, the origin of medium and giant category Atlantic bluefin tuna collected at both the entrance to (Strait of Gibraltar) and multiple locations within the Mediterranean Sea spawning area (Balearic Islands, Malta, Sardinia, and Cyprus) showed that fish in these locations were almost exclusively of eastern origin (99%–100%; Rooker et al. 2014). The presence of nearly homogenous populations of western-origin adults in the Gulf of Mexico and eastern-origin adults in the Mediterranean Sea clearly demonstrates that Atlantic bluefin tuna exhibit strong natal homing and that straying by either stock into the opposite spawning area appears to be uncommon (Figure 2.4). These results are consistent with genetic studies that show strong separation between larvae and young-of-the-year Atlantic bluefin tuna captured in the two spawning areas (Carlsson et al. 2007, Boustany et al. 2008).

## Age-Specific Migration

Life history transects of otolith $\delta^{18}O$ from the core (nursery habitat) to the margin (recent habitat) serve as an emerging and promising tool for determining age-specific movements and natal homing patterns of adult Atlantic bluefin tuna. Using an otolith $\delta^{18}O$ threshold value of approximately −0.9 as the boundary between eastern (>−1.0) and western (<−0.9) habitation (Figure 2.2), Schloesser et al. (2010) provided preliminary evidence that the approach could be used to document resident and migratory patterns of Atlantic bluefin tuna. Here, we show the application of otolith $\delta^{18}O$ life history transects to identify different migration patterns displayed by Atlantic bluefin tuna using data from a pilot study. Three unique migratory contingents have been identified: NEAO/Mediterranean residents, NWAO/Gulf of Mexico residents, and transatlantic migrants.

Otolith $\delta^{18}O$ values of Atlantic bluefin tuna that fall within the expected range of only one region (east or west) during the life of an individual were classified as residents of this region. Examples of NEAO/Mediterranean res-

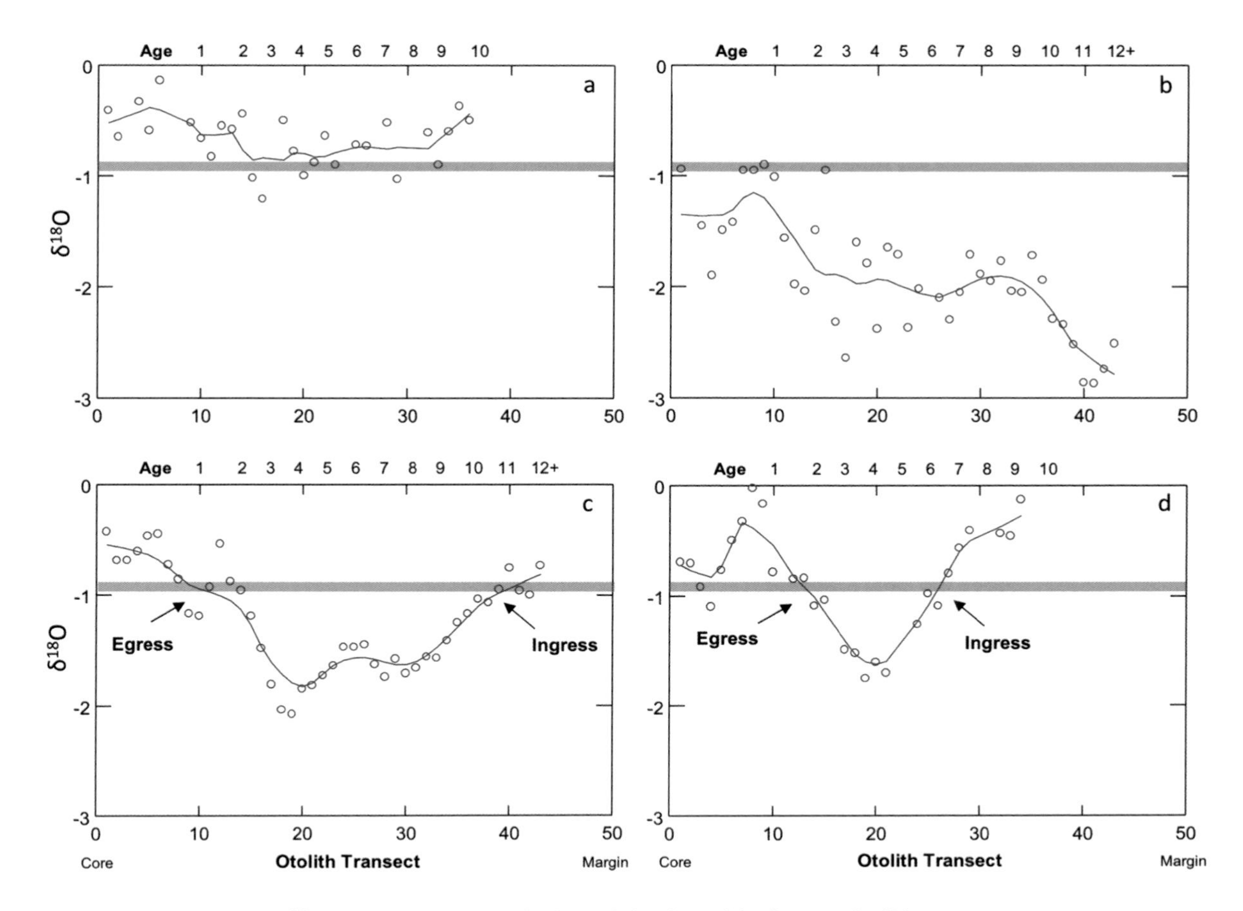


**Figure 2.5.** Otolith $\delta^{18}O$ life history transects for four adult Atlantic bluefin tuna. Otolith $\delta^{18}O$ threshold value –0.9 denotes the general boundary between eastern and western habitation. Three unique migration behaviors are shown: (*a*) Northeastern Atlantic Ocean / Mediterranean resident collected in Mediterranean Sea, (*b*) Northwestern Atlantic Ocean / Gulf of Mexico resident collected in the Gulf of Mexico, and (*c, d*) transatlantic migrants captured in the Mediterranean Sea. Putative timing of Mediterranean Sea egress and ingress denoted for the two specimens displaying transocean migrations. Otolith microstructure analysis was used to determine the age and distance of each annulus from the otolith core. Estimated age is provided at the top of each plot.

ident and NWAO/Gulf of Mexico resident (Figure 2.5a and b, respectively) do not appear to undergo transatlantic movement, because otolith $\delta^{18}O$ values remain within the expected range of eastern and western yearling values during the entire life of the individual. Intriguingly, the individual classified as a western resident (Figure 2.5b) shows a conspicuous decline in otolith $\delta^{18}O$ over time after age 1, which likely corresponds to an ontogenetic shift by this individual, which is possibly spending more time in the NWAO (i.e., Canadian waters) with increasing age because the lowest seawater $\delta^{18}O$ values in their presumed range occur here (see Figure 2.1). In contrast, the two other adult Atlantic bluefin tuna show otolith $\delta^{18}O$ profiles that are indicative of movement away from the Mediterranean Sea, followed by a return to this region later in life (Figure 2.5, c and d). Otolith $\delta^{18}O$ life history transects of both show potential egress from the Mediterranean Sea/NEAO around age 1 or age 2 into more depleted $\delta^{18}O$ waters in the NWAO, as indicated by reduced otolith $\delta^{18}O$ values $<-1.0$ after age 2. At approximately age 10 (Figure 2.5c) and age 7 (Figure 2.5d), otolith $\delta^{18}O$ values are again $>-0.9$, suggestive of movement back into the NEAO followed by homing to the Mediterranean Sea, which was the point of capture (otolith margin) for the specimens shown. Observed patterns of age-specific egress from the Mediterranean Sea followed by a return to this region several years later are in accord with otolith chemistry observations on both transatlantic movement and natal homing.

## Application of Otolith Chemistry to Assessment and Management

A broad international community of fisheries scientists is now working to integrate otolith chemistry results into assessments and modeling frameworks that explicitly include stock mixing (SCRS 2016). This database, compiled by ICCAT's Atlantic-wide research program for bluefin tuna (GBYP) now contains more than 5,500 records of individual population assignments supplied by European Union, Japanese, Canadian, and US scientists. Age-based assessments, including virtual population analysis (VPA) and statistical catch-at-age models are now being applied by assessment scientists in a manner that explicitly includes stock-of-origin information emanating from thousands of analyzed otoliths collected from fisheries throughout the North Atlantic Ocean (Table 2.1). The obvious benefit here is that the same individual (otolith) analyzed for origin can be assigned an age (Siskey et al. 2016), so that fleets, fishing selectivity, and catch-per-unit-effort (CPUE) series are explicitly attributed to age class, year class, maturity, and stock of origin. Clearly, this application depends on representative sampling of principal fisheries and population components (e.g., spawners in the Gulf of Mexico and the Mediterranean Sea). Still, the approach is feasible, and ICCAT has invested heavily in developing the capacity to analyze thousands of otoliths and develop assessment frameworks that can incorporate such information (Pallarés et al. 2013, Carruthers 2015). This does not discount the contribution of other central sources of information such as electronic tags and other types of natural tags (e.g., genetic, contaminant), which can resolve patterns in seasonal and regional selectivity through fitting movement matrices (Taylor et al. 2011). Indeed, assessment scientists have efficiently utilized information from diverse applications in developing assessment parameters through the application of likelihood frameworks (e.g., Kurota et al. 2009)

## Addressing Key Assessment Uncertainties

Assessment and management of Atlantic bluefin tuna is hampered by three uncertainties:

1. What stock boundaries best serve assessment and management aims, and can we scientifically support fixed boundaries for highly migratory Atlantic bluefin tuna?
2. Is stock productivity, particularly for the western stock, underlain by a single stock-recruitment function or multiple stock-recruitment functions?
3. How does stock mixing bias the western and eastern stock assessments?

Although these questions remain unresolved, otolith chemistry has provided important insights into each.

Analysis of chemical markers in otoliths has been particularly important in confirming that the 45°W management boundary in the Central Atlantic Ocean partially divides the two populations of Atlantic bluefin tuna: adults seasonally home to spawning areas on either side of that boundary (Rooker et al. 2008a, 2014; Siskey et al. 2016). However, the high level of mixing in US Atlantic fisheries supports the view that many Atlantic bluefin tuna cross the line, particularly juveniles emanating from natal sites in the Mediterranean Sea. Further, Atlantic bluefin tuna aggregations in the CNAO, targeted in high sea longline fleets, historically occurred as discrete fisheries on either side of the management boundary; however, over the past 20 years, this aggregation has been centered around the 45°W management boundary (SCRS 2014). To counter the artificial construct of a single boundary, assessment scientists have proposed more resolved stock boundaries that relate to specific population segments and seasonal migration behaviors for which otolith chemistry can provide support (Taylor et al. 2011, Kerr et al. 2017).

Assessments on each stock have shown independent abundance dynamics during the past three decades, but a dominant trend is the sustained depressed abundance and recruitment of the western stock at about one-third of its abundance in the 1970s (SCRS 2014). This has led to controversy on whether to reference the historical or the more recent abundance level in implementing harvest policy. In fact, two alternative stock-recruitment curves and their management implications are presented to ICCAT in each assessment cycle (e.g., SCRS 2014). Arguments center on (1) whether sustained high exploitation limits recovery to historical levels, (2) whether a regime change had occurred limiting juvenile production, and (3) whether the population has been altered in a manner (e.g., age truncation) that curtails recovery (Secor et al. 2015). In the late 2000s, a pulse of juvenile fish appeared in US fisheries, which was ascribed to a strong 2003 year class. If this year class was of western origin, then it goes against the latter two hypotheses related to a fundamental shift in western stock productivity. Through focused sampling of the size mode associated with this year class, Secor et al. (2015) used otolith chemistry to confirm that this year class was predominately comprised of western-origin juveniles. This result was confirmed further by Siskey et al. (2016), who showed overall lower mixing levels in juveniles and adults collected in US fisheries from 2009 to 2014, which is suggestive of higher recruitment emanating from the western stock. Still, a recent analysis indicates that a dominant year class produced in 2011 may now be moving through US fisheries, and initial evidence indicates that it is dominated by eastern-origin juveniles (Barnett et al. 2016).

With sufficient sampling of otoliths for population assignment across seasons, regions, fleets, and size classes, traditional assessment approaches such as VPA can be modified to evaluate the influence of regional fishing and stock mixing on spawner biomass across stocks (Butterworth and Punt 1994, Porch et al. 2000, Goethel et al. 2011). Indeed, it is probable that otolith chemistry data available in the GBYP database will enable scientists to parameterize VPAs according to stock of origin in future assessments (SCRS 2016). Beyond VPA approaches, more flexible statistical catch-at-age models can integrate information from electronic and conventional tagging, landings, CPUE series, and otolith chemistry into a single assessment that estimates yield and fishing mortality across fleets, regions, and seasons (Taylor et al. 2011). In an initial application of this modeling framework, otolith chemistry informed estimates of movement and yield pertinent to US fisheries, which in turn were used to evaluate the effect of fishery closures on Atlantic bluefin tuna recovery (Taylor et al. 2011).

Although it is feasible to refine statistical models such as those described above to include stock assignments in fitted catch, size, age, and CPUE data, these models are data intensive, often requiring parameters for which stock composition estimates are not available. In such instances, simulation approaches, known as operating models, can provide key insights by exploring the impact of alternate premises on population structure and of mixing on assessment outputs and management goals (Kerr and Goethel 2014). Scenarios in operating models can be guided by scientific consensus, new discoveries, or major controversies. Operating models should focus on important uncertainties, but in comparison to stock-assessment models, they do not require intensive data inputs and the narrow parameter constraints required for statistical convergence. Kerr et al. (2015) developed an operating model designed to evaluate how uncertainty in seasonal movements and stock productivity influenced stock mixing and fishery yields in NWAO fisheries. Tagging data informed movement matrices, which were used to model seasonal yields across fishing areas, and otolith chemistry was used to compare scenario results of eastern stock versus western stock yields across areas against observed mixing levels (Kerr et al. 2015). An ongoing emphasis within ICCAT is the development of management strategy evaluations, which develop operating models but then subject scenarios of mixing and other sources of uncertainty to monitoring, assessment, and harvest-control rules to evaluate the effectiveness of current management practices to principal sources of uncertainty (Carruthers 2015).

Future refinements and developments in otolith chemistry, including the numerical approaches for estimating stock composition (e.g., Hanke et al. 2016), will continue to aid in understanding the migration ecology of Atlantic bluefin tuna and provide information critical to their assessment and management. The use of this type of natural tracer is limited by the type of sampling required (lethal), cost constraints, and the type of information that can be generated—here, information related to seasonal exposures to ocean subbasins that differ in chemistry and temperature. As a result, other approaches such as population genomics will continue to be developed for Atlantic bluefin tuna, but they remain not yet operational in terms of the level of discrimination they can currently provide. This is likely to change in the near future, and molecular marker applications are highly relevant to assessing mixing, uncovering new population structures (subpopulations and contingents), and understanding migration behaviors. Further, molecular markers can directly assess stock abundance through close-kin genetics, and population genomics will likely be an important operational tool used in the near future to address Atlantic bluefin tuna migration ecology (Carruthers et al. 2016). The combination of this new, emerging technology with otolith chemistry and, potentially, information on other natural tracers will lead to more spatially explicit information on the origin, movement, and stock mixing of Atlantic bluefin tuna.

## Acknowledgments

This research relied on support from several agencies and programs, including ICCAT's Atlantic-wide research program for bluefin tuna (GBYP) from the European Community and the Bluefin Tuna Research Program from the National Oceanic and Atmospheric Administration's National Marine Fisheries Service. The authors thank all of the colleagues who provided valuable assistance over the years, with a special thanks to N. Abid, P. Addis, R. Allman, H. Arrizabalaga, B. Block, C. Brown, S. Cadrin, J. Cort, M. Dance, M. Deflorio, S. Deguara, G. De Metrio, D. Dettman, A. Di Natale, I. Fraile, B. Gahagan, J. Graves, A. Hanke, S. Karakulak, L. Kerr, A. Kimoto, J. Lee, P. Mace, D. Macías, J. Neilson, C. Porch, E. Prince, E. Rodríguez-Marín, O. Sakai, R. Schloesser, M. Siskey, V. Ticina, A. Traina, and S. Turner.

### REFERENCES

Barnett, B.K., D.H. Secor, and R. Allaman. 2016. Contribution of the Gulf of Mexico population to the U.S. Atlantic bluefin tuna fisheries in 2015. *ICCAT Collective Volume of Scientific Papers* 73 (6): 2007–12.

Baumann, H., R.J.D. Wells, J.R. Rooker, Z.A. Baumann, D.J. Madigan, H. Dewar, O.E. Snodgrass, and N.S. Fisher. 2015. Combining otolith microstructure and trace elemental analyses to infer the arrival of Pacific bluefin tuna juveniles in the California current ecosystem. *ICES Journal of Marine Science* 72:2128–38.

Block, B.A., S.L.H. Teo, A. Walli, A. Boustany, M.J.W. Stokesbury, C.J. Farwell, K.C. Weng, H. Dewar, and T.D. Williams. 2005. Electronic tagging and population structure of Atlantic bluefin tuna. *Nature* 434:1121–27.

Boustany, A.M., C.A. Reeb, and B.A. Block. 2008. Mitochondrial DNA and electronic tracking reveal population structure of Atlantic bluefin tuna (*Thunnus thynnus*). *Marine Biology* 156:13–24.

Busawon, D.S., J.D. Neilson, I. Andrushchenko, A. Hanke, D.H. Secor, and G. Melvin. 2014. Evaluation of Canadian sampling program for bluefin tuna, assessment of length-weight conversions, and results of natal origin studies 2011–2012. *ICCAT Collective Volume of Scientific Papers* 70 (1): 202–19.

Butterworth, D.S., and A.E. Punt. 1994. The robustness of estimates of stock status for the western North Atlantic bluefin tuna population to violations of the assumptions underlying the associated assessment models. *ICCAT Collective Volume of Scientific Papers* 42:192–210.

Calaprice, J.R. 1985. Chemical variability and stock variation in northern Atlantic bluefin tuna. ICCAT SCRS/85/36, 222–52.

Carlsson, J., J.R. McDowell, J.E.L. Carlsson, and J.E. Graves. 2007. Genetic identity of YOY bluefin tuna from the eastern and Western Atlantic spawning areas. *Journal of Heredity* 98 (1): 23–28.

Carruthers, T. 2015. Evaluating management strategies for Atlantic bluefin tuna. Report 2: Operating model development and data requirements. Report to ICCAT under Contract GBYP 02/2015. University of British Columbia, Vancouver, British Columbia.

Carruthers, T., J.E. Powers, M.V. Lauretta, A. Di Natale, and L. Kell. 2016. A summary of data to inform operating models in management strategy evaluation of Atlantic bluefin tuna. *ICCAT Collective Volume of Scientific Papers* 72 (7): 1796–1807.

Dickhut, R.M., A.D. Deshpande, A. Cincinelli, M.A. Cochran, S. Corsolini, R.W. Brill, D.H. Secor, and J.E. Graves. 2009. Atlantic bluefin tuna (*Thunnus thynnus*) population dynamics delineated by organochlorine tracers. *Environmental Science and Technology* 43:8522–27. doi:10.1021/Es901810e.

Elsdon, T.S., and B.M. Gillanders. 2003. Reconstructing migratory patterns of fish based on environmental influences on otolith chemistry. *Reviews in Fish Biology and Fisheries* 13:217–35.

Fraile, I., H. Arrizabalaga, and J.R. Rooker. 2015. Origin of bluefin tuna (*Thunnus thynnus*) in the Bay of Biscay. *ICES Journal of Marine Science* 72 (2): 625–34.

Fromentin, J.M., and J.E. Powers. 2005. Atlantic bluefin tuna: Population dynamics, ecology, fisheries and management. *Fish and Fisheries* 6:281–306.

Fromentin, J.M., G. Reygondeau, S. Bonhommeau, and G. Beaugrand. 2014. Oceanographic changes and exploitation drive the spatiotemporal dynamics of Atlantic bluefin tuna. *Fisheries Oceanography* 23 (2): 147–56.

Galuardi, B., F. Royer, W. Golet, J. Logan, J. Neilson, and M. Lutcavage. 2010. Complex migration routes of Atlantic bluefin tuna (*Thunnus thynnus*) question current population structure paradigm. *Canadian Journal of Fisheries and Aquatic Sciences* 67:966–76.

Goethel, G.R., T.J. Quinn, and S.X. Cadrin. 2011. Incorporating spatial structure in stock assessment: Movement modeling in marine fish population dynamics, *Reviews in Fisheries Science* 19 (2): 119–136.

Graves, J.E., A.S. Wozniak, R.M. Dickhut, M.A. Cochran, E.H. MacDonald, E. Bush, H. Arrizabalaga, and N. Goñi. 2015. Transatlantic movements of juvenile Atlantic bluefin tuna inferred from analyses of organochlorine tracers. *Canadian Journal of Fisheries and Aquatic Sciences* 72:625–33.

Gunn, J.S., and B.A. Block. 2001. Advances in acoustic, archival, and satellite tagging of tunas. In *Tuna: Physiology, Ecology, and Evolution*, edited by B.A. Block and E.D. Stevens, 167–224. San Diego, CA: Academic Press.

Hanke, A., D. Busawon, J.R. Rooker, and D.H. Secor. 2016. Estimates of stock origin for bluefin tuna caught in the western Atlantic fisheries from 1975 to 2013. *ICCAT Collective Volume of Scientific Papers* 72 (6): 1376–93.

Kerr, L.A., and D.R. Goethel. 2014. Simulation modeling as a tool for synthesis of stock identification information. In *Stock Identification Methods*, edited by S. Cadrin, L. Kerr, and S. Mariani, 501–33. 2nd ed. Cambridge, MA: Academic Press.

Kerr, L.A., S.X. Cadrin, and D.H. Secor. 2012. Evaluating population effects and management implications of mixing between eastern and western Atlantic bluefin tuna stocks. ICES CM 2012/N:13.

Kerr, L.A., S.X. Cadrin, D.H. Secor, and N. Taylor. 2015. Evaluating the effect of Atlantic bluefin tuna movement on the perception of stock units. SCRS/2014/170.

Kerr, L.A., S.X. Cadrin, D.H. Secor, and N.G. Taylor. 2017. Modeling the implications of stock mixing and life history uncertainty of Atlantic bluefin tuna. *Canadian Journal of Fisheries and Aquatic Sciences.* 74 (11): 1990–2004.

Kurota, H., M.K. McAllister, G.L. Lawson, J.I. Nogueira, S.L.H. Teo, and B.A. Block. 2009. A sequential Bayesian methodology to estimate movement and exploitation rates using electronic and conventional tag data: Application to Atlantic bluefin tuna (*Thunnus thynnus*). *Canadian Journal of Fisheries and Aquatic Sciences* 66 (2): 321–42.

LeGrande, A.N., and G.A. Schmidt. 2006. Global gridded data set of the oxygen isotopic composition in seawater. *Geophysical Research Letters* 33:L12604.

Magnuson, J.J., C. Safina, and M.P. Sissenwine. 2001. Whose fish are they anyway? *Science* 17:1267–68.

Mather, F.J., J.M. Mason, and C.A. Jones. 1995. Historical document: Life history and fisheries of Atlantic bluefin tuna. National Oceanic and Atmospheric Agency Technical Memorandum NMFS-SEFSC 370.

NRC (National Research Council). 1994. *An Assessment of Atlantic Bluefin Tuna*. Washington DC: National Academy Press.

Pallarés, P., J.M. Ortiz de Urbina, A. Fonteneau, J. Walters, A. Kimoto, E. Rodriguez-Marín, J. Neilson, et al. 2013. Report of the 2013 Bluefin Meeting on Biological Parameters Review. ICCAT. Tenerife, Spain. May 7–13. www.iccat.es/Documents/Meetings/Docs/2013-BFT_BIO_ENG.pdf.

Porch, C.E., S.C. Turner, and J.E. Powers. 2000. Virtual population analyses of Atlantic bluefin tuna with alternative models of transatlantic migration: 1970–1997. *ICCAT Collective Volume of Scientific Papers* 52 (3): 1022–45.

Richardson, D.E., K.E. Marancik, J.R. Guyon, M.E. Lutcavage, B. Galuardi, C.H. Lam, H.J. Walsh, S. Wildes, D.A. Yates, and J.A. Hare. 2016. Discovery of a spawning ground reveals diverse migration strategies of Atlantic bluefin tuna. *Proceedings of the National Academy of Sciences USA* 113:3299–3304.

Rooker, J.R., and D.H. Secor. 2004. Stock structure and mixing of Atlantic bluefin tuna: Evidence from stable $\delta^{13}C$ and $\delta^{18}O$ isotopes in otoliths. *ICCAT Collective Volume of Scientific Papers* 56 (3): 1115–20.

Rooker, J.R., V.S. Zdanowicz, and D.H. Secor. 2001a. Otolith chemistry of tunas: Assessment of base composition and post-mortem handling effects. *Marine Biology* 139:35–43.

Rooker, J.R., D.H. Secor, V.S. Zdanowicz, and T. Itoh. 2001b. Discrimination of northern bluefin tuna from nursery areas in the Pacific Ocean using otolith chemistry. *Marine Ecology Progress Series* 218:275–82.

Rooker, J.R., D.H. Secor, V.S. Zdanowicz, L.O. Relini, N. Santamaria, M. Deflorio, G. Palandri, and M. Relini. 2002. Otolith elemental fingerprints of Atlantic bluefin tuna from eastern and western nurseries. *ICCAT Collective Volume of Scientific Papers* 54: 498–506.

Rooker, J.R., D.H. Secor, V.S. Zdanowicz, G. DeMetrio, and L.O. Relini. 2003. Identification of Atlantic bluefin tuna stocks from putative nurseries using otolith chemistry. *Fisheries Oceanography* 12:75–84.

Rooker, J.R., J.R. Alvarado Bremer, B.A. Block, H. Dewar, G. De Metrio, A. Corriero, R.T. Kraus, et al. 2007. Life history and stock structure of Atlantic bluefin tuna (*Thunnus thynnus*). *Reviews in Fisheries Science* 15 (4): 265–310.

Rooker, J.R., D.H. Secor, G. DeMetrio, R. Schloesser, B.A. Block, and J.D. Neilson. 2008a. Natal homing and connectivity in Atlantic bluefin tuna populations. *Science* 322: 742–44.

Rooker, J.R., D.H. Secor, G. DeMetrio, J.A. Kaufman, A. Belmonte Rios, and A. Ticina. 2008b. Evidence of trans-Atlantic mixing and natal homing of bluefin tuna. *Marine Ecology Progress Series* 368:231–39.

Rooker, J.R., H. Arrizabalaga, I. Fraile, D.H. Secor, D.L. Dettman, N. Abid, P. Addis, et al. 2014. Crossing the line: Migratory and homing behaviors of Atlantic bluefin tuna, *Marine Ecology Progress Series* 504:265–76.

Rooker, J.R., R.J.D. Wells, D.G. Itano, S.R. Thorrold, and J.M. Lee. 2016. Natal origin and population connectivity of bigeye and yellowfin tuna in the Pacific Ocean. *Fisheries Oceanography* 25:277–91.

Schloesser, R.W., J.D. Neilson, D.H. Secor, and J.R. Rooker. 2010. Natal origin of Atlantic bluefin tuna (*Thunnus thynnus*) from Canadian waters based on otolith $\delta^{13}C$ and $\delta^{18}O$. *Canadian Journal of Fisheries and Aquatic Sciences* 67:563–69.

SCRS (Standing Committee on Research and Statistics). 2014. Report of the 2014 Atlantic bluefin tuna stock assessment session, ICCAT, September 22–27, Madrid, Spain. https://www.iccat.int/Documents/Meetings/Docs/2014_BFT_ASSESS-ENG.pdf.

SCRS (Standing Committee on Research and Statistics). 2016. Report of the 2016 bluefin tuna species group intersessional meeting, ICCAT, July 25–29, Madrid, Spain. https://www.iccat.int/Documents/2052-16_ENG.pdf.

Secor, D.H. 2015. *Migration Ecology of Marine Fishes*. Baltimore: Johns Hopkins University Press.

Secor, D.H., and E.J. Chesney. 1998. Summary of a workshop: Otolith microconstituent analysis of Atlantic bluefin tuna. *ICCAT Collective Volume of Scientific Papers* 48 (1): 51–58.

Secor, D.H., S.E. Campana, V.S. Zdanowicz, J.W.H. Lam, J.W. McLaren, and J.R. Rooker. 2002. Inter-laboratory comparison of Atlantic and Mediterranean bluefin tuna otolith microconstituents. *ICES Journal of Marine Science* 59:1294–1304.

Secor, D.H., J.R. Rooker, and A. Allman. 2014. Natal homing by Gulf of Mexico adult Atlantic bluefin tuna, 1976–2012. International Commission for the Conservation of Atlantic Tunas. *ICCAT Collective Volume of Scientific Papers* 69 (2): 372–74.

Secor, D.H., J.R. Rooker, B. Gahagan, M. Siskey, and R. Wingate. 2015. Depressed resilience of bluefin tuna in the western Atlantic Ocean associated with age truncation. *Conservation Biology* 29:400–408.

Siskey, M.R., M.J. Wilberg, R.J. Allman, B.K. Barnett, and D.H. Secor. 2016. Forty years of fishing: Changes in the age structure and stock mixing in northwestern Atlantic bluefin tuna (*Thunnus thynnus*) associated with size-selective and long-term exploitation. *ICES Journal of Marine Science* 73 (10): 2518–28. doi:10.1093/icesjms/fsw115.

Taylor, N.G., M.K. McAllister, G.L. Lawson, T. Carruthers, and B.A. Block. 2011. Atlantic bluefin tuna: A novel multistock spatial model for assessing population biomass. *PLOS ONE* 6 (12): e27693.

Teo, S.L.H., and A.M. Boustany. 2015. Movements and habitat use of Atlantic bluefin tuna. In *Biology and Ecology of Bluefin Tuna,* edited by T. Kitagawa and S. Kimura, 137–88. Boca Raton, FL CRC Press.

Thorrold, S.R., C. Latkoczy, and P.K. Swart. 2001. Natal homing in a marine fish metapopulation. *Science* 291:297–99.

Wells, R.J.D., J.R. Rooker, and D.G. Itano. 2012. Nursery origin of yellowfin tuna in the Hawaiian Islands. *Marine Ecology Progress Series* 461:187–96.

Wilson, S.G., I.D. Jonsen, R.J. Schallert, J.E. Ganong, M.R. Castleton, A.D. Spares, A.M. Boustany, M.J.W. Stokesbury, and B.A. Block. 2015. Tracking the fidelity of Atlantic bluefin tuna released in Canadian waters to the Gulf of Mexico spawning ground. *Canadian Journal of Fisheries and Aquatic Sciences* 72 (11): 1700–1717.

CHAPTER 3

# Life History and Migrations of Mediterranean Bluefin Tuna

Haritz Arrizabalaga, Igor Arregui, Antonio Medina,
Naiara Rodríguez-Ezpeleta, Jean-Marc Fromentin, and Igaratza Fraile

## Introduction

For successful fisheries management, stocks (management units) must correspond to either populations that are reproductively isolated, or contingents, that is, groups of individuals that are from the same population but have sufficiently different migratory behaviors (Secor 1999, Begg et al. 1999, Reiss et al. 2009). Otherwise, most productive subpopulations could be exploited at suboptimal levels, and more importantly, less productive subpopulations could collapse, causing a reduction in genetic diversity and in resilience of the species to future changes (Hauser and Carvalho 2008). Considering this a fundamental requisite of science-based management, it is essential to deliver accurate scientific knowledge on population structure and mixing to managers.

The Atlantic Bluefin Tuna (ABT), *Thunnus thynnus*, is managed by the International Commission for the Conservation of Atlantic Tunas (ICCAT) with the assumption that two stocks are separated by the 45°W meridian. However, ABT life history and migrations are complex (Block et al. 2005), and it is acknowledged that ABT population structure and mixing are among the greatest uncertainties affecting stock assessment and management (Leach et al. 2014, Fromentin et al. 2014). Recognizing the high variability in dispersal patterns for ABT (Galuardi et al. 2010), it is important to identify the origin (Gulf of Mexico or Mediterranean) of the ABT individuals that are caught throughout the Atlantic in different fisheries. Likewise, biological parameters (e.g., growth curves or maturity schedules, etc.) need to be population specific, but some of the current studies might have inadvertently used samples from both populations. In this context, the ICCAT Standing Committee on Research and Statistics (SCRS) has started to make efforts to use assessment models that consider information on movements, origin, and population-specific age-length keys (Kerr et al. 2010, Taylor et al. 2011, Carruthers et al. 2016, Anonymous 2016).

However, to date, none of these efforts consider alternative population structures within the Mediterranean Sea (MED). But new evidence suggests that the whole East Atlantic and Mediterranean stock might not correspond to a panmictic population; alternative hypotheses exist about both genetic subpopulations and migratory contingents that might require adaptation of the management regime (Fromentin and Powers 2005, Anonymous 2014). Failing to do so might cause future overexploitation of less productive subpopulations and loss of genetic and/or behavioral diversity. In fact, this could have already happened in the past during periods of heavy exploitation of Mediterranean ABT, although Riccioni et al. (2010) suggest this is not the case, probably because more extreme population decreases are needed to observe loss of genetic diversity (see Ruzzante et al. 2001).

In this chapter, we focus on the subpopulation structure of Atlantic bluefin tuna of Mediterranean origin (referred to as Mediterranean ABT), revising recent knowledge on life history and migrations generated over the past decade using a range of scientific disciplines. The review focuses, to the extent possible, on studies that can provide information about ABT that originated in different spawning areas (i.e., potential subpopulations or contingents), namely electronic tagging and genetic studies that can be associated with specific spawning areas, and reproductive studies. Old population structure hypotheses are updated, including hypotheses about the relationship between the Mediterranean ABT subpopulations or contingents and the Atlantic Ocean. The objective is to assess the complexity of the intra-Mediterranean population structure, to identify the most critical uncertainties, to discuss the main management implications, and to identify future research needs to allow for a science-based management of Mediterranean ABT.

## Population Structure and Mixing

### Hypotheses

In 2013 the ICCAT Bluefin Tuna Working Group reviewed the existing biological information and discussed alternative population structure hypotheses for ABT in the Atlantic (Figure 3.1; Anonymous 2014). Regarding the Mediterranean, three main alternative hypotheses were considered: a sin-

by Tensek et al. (2017), with two individuals tagged in Moroccan traps. They both visited the southern Tyrrenian and subsequently returned to the Atlantic.

On the contrary, Fromentin and Lopuszanski (2013) and Cermeño et al. (2015) found resident individuals occupying both the western and the central Mediterranean. Fromentin and Lopuszanski (2013) documented site fidelity to a specific feeding area in the WMED (a behavior also demonstrated on juvenile ABT in the Bay of Biscay; see below). A tagged fish remained offshore in the Gulf of Lions from September 2011 to February 2012 and went back to it one year later (in August–September 2012). The release and pop-off locations were separated by only 40 nautical miles (nm), whereas the fish had traveled 1,250 nm away from the release location a few weeks before pop-off and covered a distance of at least 8,460 nm during the year at large. The fidelity of ABT to this specific site was further supported by the back and forth of several fish. Although Fromentin and Lopuszanski (2013) could not identify putative spawning activity because of insufficient data on vertical movements from those tags, they observed migration and visits to WMED (Tyrrhenian) and CMED (Gulf of Sidra) spawning areas during the spawning period by ABT that had been feeding and residing in the WMED.

This behavior was confirmed by Cermeño et al. (2015), who characterized putative spawning events using detailed vertical movement data sets. They found one individual spawning in the CMED that visited both the CMED and the WMED. Three other individuals remained in the areas where they spawned (one in the CMED and two in the WMED), even with substantial times at liberty (up to 304 days), which supports the idea of residency not only at the scale of the whole Mediterranean, but also at the scale of Mediterranean subareas (see also Rooker et al. 2007). Subsequently, Quílez-Badia et al. (2015) expanded part of the data set in Cermeño et al. (2015) and identified the spawning activity of individuals, revealing that ABT coming from the Atlantic into the Mediterranean spawn in both the WMED and the CMED. Likewise, resident ABT also spawned in those two areas, and they mixed between the WMED and the CMED for feeding throughout the year.

The link between EMED spawners and the Atlantic Ocean was not probed until the ICCAT Atlantic-wide research program for bluefin tuna (ICCAT-GBYP) (Di Natale et al. 2016) tagged ABT in spawning aggregations in the Levantine Sea. Out of 30 ABT tagged, only 2 migrated to the Atlantic after the spawning season, a few went to the CMED, and most stayed in the EMED. However, times at liberty were generally very short (e.g., 13 tags popped off within two weeks), and thus it is possible that the real proportion of EMED origin fish using the Atlantic is higher than that inferred from these results, although a subsequent analysis by Tensek et al. (2017) considering longer times at liberty did not identify additional EMED-origin individuals exiting the MED.

One individual tagged off Ireland by Stokesbury et al. (2007) was caught with purse seine gear in a spawning aggregation during the spawning season south of Malta, suggesting that this individual of likely CMED origin used the Northeast Atlantic feeding grounds. ABT tagged in Moroccan traps that subsequently entered the Mediterranean also linked the Atlantic mostly to the WMED and CMED but also, to some extent, to the EMED (Di Natale et al. 2016). However, in this case, no detailed archival data were available to identify spawning activity, and assuming origin based on horizontal trajectories alone is difficult. Similarly, Walli et al. (2009) linked the West Atlantic feeding area (off North Carolina) mostly to the WMED and CMED, where 28 out of 29 archival tagging recoveries occurred, with a single recovery observed in the EMED. However, the authors did not identify spawning activity; the recovery rates might differ among Mediterranean subareas, affecting the interpretation of these results.

*Juveniles*

Unfortunately, far less knowledge was accumulated on juvenile ABT migrations using e-tags, compared with adults, over the past decade. Origin of juveniles is more difficult to assign using e-tag data; it requires other techniques (e.g., otolith chemistry or genetics), except when times at liberty are long enough to reach maturity. In those cases, individuals visit spawning areas, and the same criteria as for adults (i.e., identification of spawning behavior, visitation of and/or fidelity to spawning areas, or recapture on a spawning aggregation) can be used to assign origin (Table 3.1).

Arregui et al. (2018) is the most comprehensive e-tagging study on ABT juveniles (ages 1–4) of most probably Mediterranean origin (according to Fraile et al. 2015) tagged in the Bay of Biscay. To our knowledge, this study includes the longest juvenile time at liberty and the only archival tag implanted in the East Atlantic and the Mediterranean to record migrations during the transition from the early juvenile to adult stage. This individual was tagged at age 1.1 (65 cm fork length) and stayed at liberty for 7.4 years, and

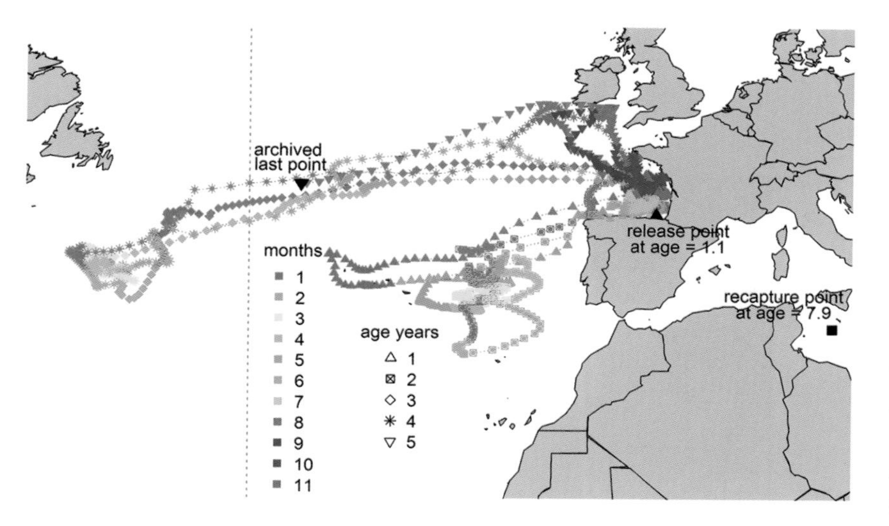


**Figure 3.2.** Long-term migration of a one-year-old Atlantic bluefin tuna tagged in the Bay of Biscay that retained the tag through the adult phase. The tag recorded information for 4.3 years, and the symbols represent the migration of the individual at different ages. The fish was caught in June 2014 in a spawning aggregation south of Malta (*black square*).

the tag recorded 4.3 years of data (Figure 3.2). At the end of the recorded track, with an age of 5.4 years (fully mature according to the currently assumed maturity schedule), the fish had not entered into the Mediterranean yet, but it did some time later, as it was caught in June 2014 by a purse seiner in a spawning aggregation south of Malta at age 7.95 years, before spending the last six months of life in a cage. Thus, assuming purse seiners in June catch aggregations that are ready to spawn, this individual was likely of CMED origin (although we cannot completely rule out the possibility of WMED or EMED origin). It's also likely that this individual spent the entirety of its juvenile stage in the Atlantic, feeding in the Bay of Biscay every summer and migrating to the area around Madeira and the Azores during the first two winters and to south of Newfoundland during the third and fourth winter, suggesting that the extent of transoceanic migration can vary among years (see Arregui et al. 2018).

The fidelity to the Bay of Biscay feeding area is confirmed when analyzing the information from all the juveniles tagged in the Bay of Biscay (Figure 3.3). Out of 20 individuals recaptured, 14 individuals provided information on 19 overwintering migrations (i.e., covering at least the whole winter period), which suggests a very high dispersal of these juveniles of likely Mediterranean origin and an extremely high fidelity to the Bay of Biscay, highlighting the importance of this feeding area in the Atlantic for juvenile ABT. Only one individual migrated to the US coast, an area also inhabited by juveniles tagged in the western Atlantic by Galuardi and Lutcavage (2012). Unfortunately there is no information about the origin of this individual; it

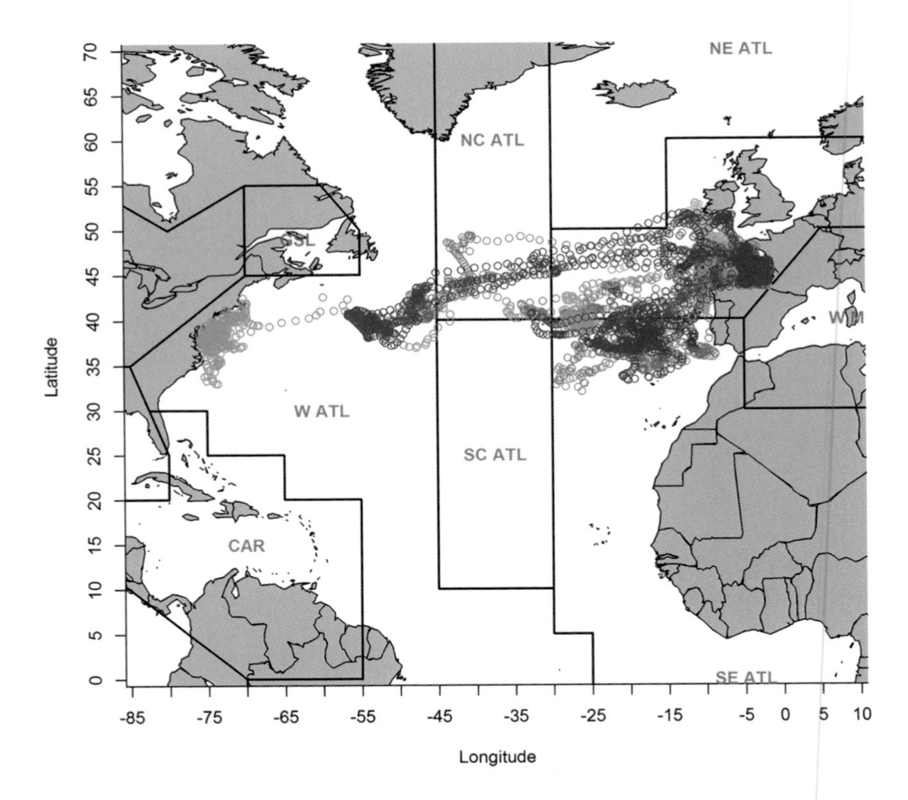


**Figure 3.3.** Tracks for all Atlantic bluefin tuna e-tagged in the Bay of Biscay (n = 20). Each color denotes a different individual. Boxes represent International Commission for the Conservation of Atlantic Tunas geographical areas. CAR = Caribbean; GSL = Gulf of St. Lawrence; NC ATL = North-central Atlantic; NE ATL = northeastern Atlantic; SE ATL = southeastern Atlantic; SC ATL = South-central Atlantic; W ATL = West Atlantic; W MED = western Mediterranean.

may have taken this migratory route because it originated in the Gulf of Mexico (as supported by Fraile et al. 2015 and Graves et al. 2015) or because it was exhibiting the behavioral plasticity of Mediterranean-origin fish, meaning that the fidelity to the Bay of Biscay might be somewhat variable (see Arregui et al. 2018).

None of the juveniles tagged in the Atlantic entered the Mediterranean during the recorded period; this suggests that Mediterranean juveniles exiting to the Atlantic live there until they need to reproduce, which is consistent with the assumed life cycle for bluefin born in the Balearics and with conventional tagging experiments that report migrations from the Bay of Biscay to the Balearics only after age 5 (Rodríguez Marín et al. 2005, Rooker et al. 2007). There are also examples from internal archival tags deployed in the western Atlantic. Those individuals remained there for several years before migrating into the Mediterranean spawning areas only after age 8. However, the two juveniles tagged by Cermeño et al. (2015) never went out of the Mediterranean, supporting the suggestion that some fraction of the juvenile population resides within the Mediterranean.

### Summary of Population Structure

The knowledge gained by e-tagging experiments in recent years has shed light on some key aspects of ABT population structure. On one hand, a large degree of connectivity for both juveniles and adults throughout the Atlantic and the Mediterranean is confirmed (consistent with Block et al. 2005, Rooker et al. 2008, and Graves et al. 2015). On the other hand, not every ABT individual performs long-distance migrations. In fact, some tracks show quite restricted spatial distribution, especially among the fish residing in the Mediterranean. The occurrence of these different behaviors supports the existence of alternative subpopulation structure or contingent hypotheses.

Recent findings suggest that the different hypotheses represented in Figure 3.1 could be refined (see Figure 3.4). Observations indicate that the Atlantic is not used solely by ABT of WMED origin, but also by ABT of CMED and EMED origin. Moreover, the proportion of Atlantic migrants might be highest among the WMED-origin ABT and lowest among the EMED-origin ABT. However, the observed pattern is probably influenced by the short times at liberty, especially among the EMED-origin adults; new experiments with longer times at liberty are needed to confirm this. Based on this information, the WMED, CMED, and EMED could constitute different contingents. However, so far, no single fish has been observed to spawn in more than one of those subareas, so there is no evidence of genetic flux between them. Thus, if natal homing was developed at the scale of Mediterranean subareas (as it is at the scale of the Atlantic, which is consistent with the observations by Block et al. [2005] for the WMED), it could be that each subarea represents a different subpopulation (Figure 3.4c). Thus, the following five hypotheses are possible:

1. A single panmictic population throughout the Mediterranean with no substructure. All individuals mix at random; they have the same chance to reproduce with any other individual, and there is no behavioral heterogeneity.
2. A single genetic population with several contingents. Each contingent uses a different spawning ground, although some genetic flux can exist—enough to homogenize genotypes between contingents. Individuals from all the contingents exit the Mediterranean to feed in the Atlantic after the spawning season.

3. A single genetic population with several contingents. Each contingent uses a different spawning ground, although some genetic flux can exist—enough to homogenize genotypes between contingents. Individuals from all the contingents exit the Mediterranean to feed in the Atlantic after the spawning season, but the westernmost contingent uses the Atlantic more than the easternmost contingent, which tends to reside more within the Mediterranean year round.
4. Several genetically isolated populations within the Mediterranean. Each population uses a different spawning ground. Individuals from all the populations exit the Mediterranean to feed in the Atlantic after the spawning season.
5. Several genetically isolated populations within the Mediterranean. Each population uses a different spawning ground. Individuals from all the populations exit the Mediterranean to feed in the Atlantic after the spawning season, but the westernmost population uses the Atlantic more than the easternmost population, which tends to reside more within the Mediterranean year round.

Discerning among the most likely hypotheses requires solving two fundamental issues that provide essential and complementary information: (1) whether different genetic groups exist within the Mediterranean, and (2) whether individuals spawning in different areas behave differently. Lack of genetic differentiation does not forcedly imply lack of subpopulation structure, as genetic connectivity could be the result of a handful of individuals mixing between groups essentially behaving as independent units and thus requiring separate management (i.e., with different population sizes, population dynamics, life history parameters, and behaviors). Thus, in the case of genetic homogeneity, additional (e.g., behavioral and life history) studies would be required to determine whether potential units are discrete enough to require independent management and thus distinguish between hypotheses 1 (single population) or 2 and 3 (single population with several contingents). Likewise, in the case of genetic differentiation among different areas of the Mediterranean, behavioral studies are still needed to understand the relative distribution of each subpopulation (e.g., to discern between hypotheses 4 and 5), as a prerequisite to allow the implementation of such separate management of the subpopulations that mix throughout the Atlantic and Mediterranean.

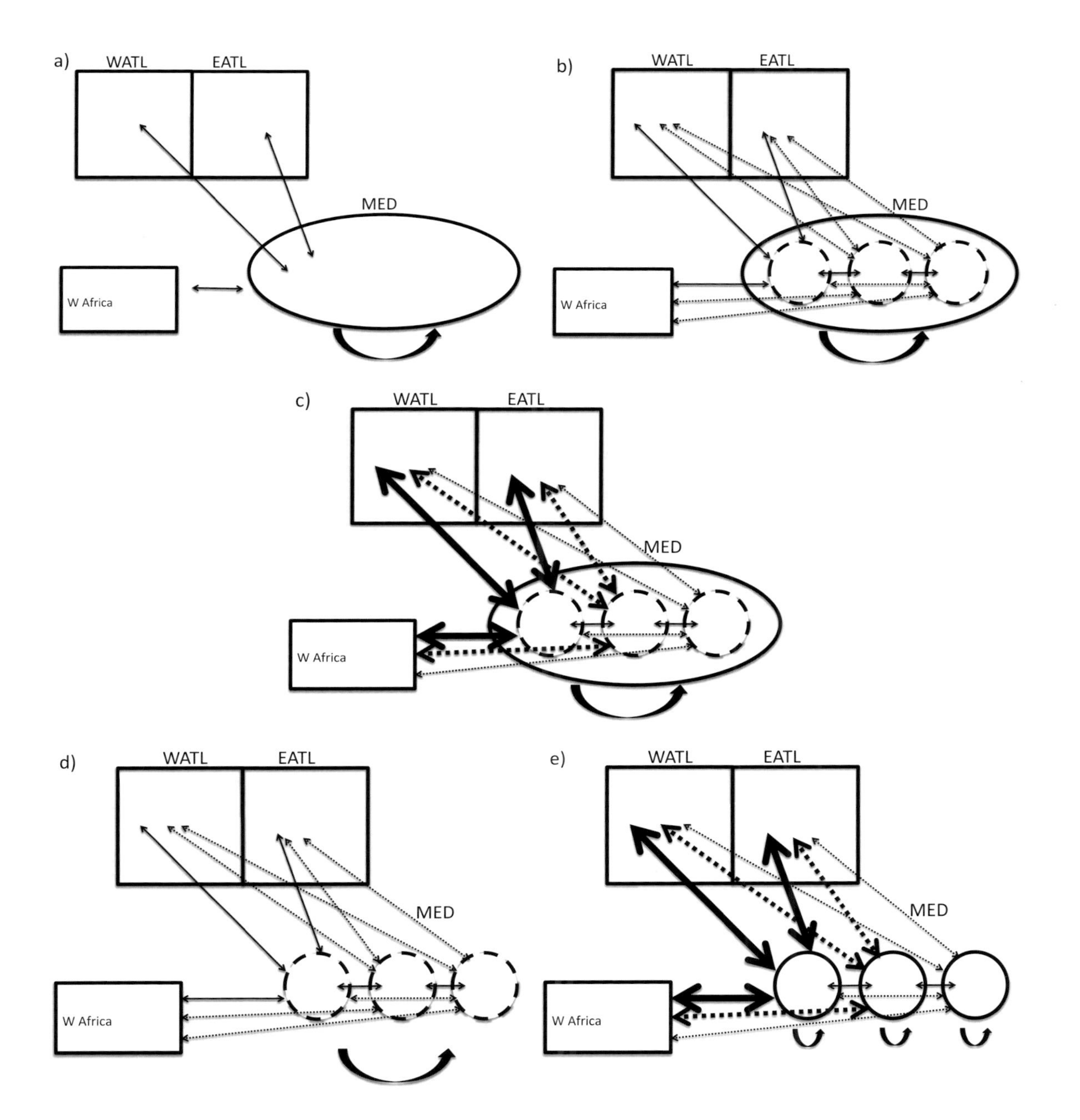


**Figure 3.4.** Updated schematic representations of potential Mediterranean Atlantic bluefin tuna population structures, based on the new knowledge (*dotted lines*) generated over past years. (*a*) A panmictic population throughout the Mediterranean. (*b*) Different contingents exist in the Mediterranean, but all of them use the Atlantic Ocean for feeding. (*c*) Different contingents exist in the Mediterranean; the Atlantic is used mostly by the western and central Mediterranean contingents, while the proportion of Mediterranean residents is highest in the eastern Mediterranean contingent. (*d*) Different genetic populations exist in the Mediterranean, but all of them use the Atlantic for feeding. (*e*) Different genetic populations exist in the Mediterranean; the Atlantic is used mostly by the western and central Mediterranean populations, while the eastern Mediterranean population is composed of mostly resident fish. WATL = West Atlantic; EATL = East Atlantic; MED = Mediterranean.

Although studies focused on the existence of behavioral differences are scarce, a handful of genetic studies aiming to decipher connectivity among Mediterranean subareas have been performed. Despite these efforts, whether there are genetic subpopulations within the Mediterranean remains a controversial issue because of the contradictory results (see Table 3.2). Most of the studies performed are not based on reference samples, that is, those known to be at or close to their spawning or birth place (larvae, young-of-the-year [YOY], or spawning adults), and thus interpreting observed genetic differences as existence of genetic subpopulations (e.g., in Riccioni et al. 2010, 2013) is not appropriate. Only four studies are based on reference samples. From those, Carlsson et al. (2004, 2007) and Boustany et al. (2008) reported heterogeneity between the WMED and CMED, whereas Viñas et al. (2011) did not find any genetic differentiation between more distant samples (WMED and EMED).

Interestingly, the analyses reporting differentiation are based on a prior grouping of individuals by location, and often the differing location was the one with the smallest number of individuals; this suggests that observed differentiation might be the result of sample allele frequencies not representative of population allele frequencies (Holsinger and Weir 2009). This is a phenomenon observed in the study of Riccioni et al. (2010), which shows differentiation when groups are a priori grouped by location, but not when no prior assumptions are considered. Yet, the most recent study focused on deciphering population structure within the Mediterranean does not find differentiation when using allele frequency differences among locations or analyses based on no prior assumptions (Viñas et al. 2011).

The abovementioned analyses were all based on mtDNA or a few microsatellite markers, which, despite being the markers of choice for population genetics for the past two decades, have drawbacks and are now being replaced by single nucleotide polymorphisms (SNPs). Despite being less polymorphic, SNPs are more informative, as they can be gathered in hundreds or thousands, they are more stable, their analysis is easier to automate and interpret, they are less sensitive to small sample sizes (Willing et al. 2012), and they have been shown to find structure where microsatellites failed (e.g., Benestan et al. 2015). Recent unpublished studies based on SNPs and using as reference samples both larvae and YOY failed to detect any structure within the Mediterranean (Puncher et al. unpublished, Rodríguez-Ezpeleta et al. unpublished). Considering the low levels of differentiation between the Gulf of Mexico and the Mediterranean, detecting

**Table 3.2.** Summary of recent studies focusing on deciphering genetic population structure within the Mediterranean

| Reference | Marker | Samples: Number | Samples: Locations | Samples: Age class | Differentiation within MED | Observations |
|---|---|---|---|---|---|---|
| Broughton and Gold 1997 | 6 microsatellites | 20 | WMED, CMED | N/A | Yes | |
| Alvarado Bremer et al. 1999 | CR mtDNA | 73 | WMED, CMED, EMED | N/A | No | No information about age class. |
| Ely et al. 2002 | CR mtDNA IdhA mtDNA | 138 | WMED, EMED | Mixture (YOY and larger) | No | Not clear where samples were collected within the Mediterranean. Some YOY from 1998 (n = 38) were used, but it is unclear which location they belong to. |
| Pujolar et al. 2003 | 37 allozymes | 734 | WMED, CMED, EMED | juveniles (0–1) and adults | No | Not clear how many reference samples (larvae or YOY) were used. |
| Carlsson et al. 2004 | CR mtDNA 9 microsatellites | 280 | WMED, CMED | YOY | Yes | Only the smallest sample (from the Ionian Sea) shows differentiation. |
| Carlsson et al. 2007 | CR mtDNA 8 microsatellites | 280 | WMED, CMED | YOY | yes | Same samples as Carlsson et al. 2004. With mtDNA, WMED showed differentiation with Gulf of Mexico, but CMED did not. |
| Boustany et al. 2008 | CR mtDNA | 109 | WMED, CMED | YOY and other | Yes | Differentiation only when using reference samples. From each location, not clear how many are YOY. |
| Riccioni et al. 2010 | 8 microsatellites | 355 | WMED, CMED | juveniles (1) and adults (2 to 18) | Yes | Some pairs do not show differentiation. Analyses based on no prior assumptions of groups do not provide evidence of structure. |
| Riccioni et al. 2013 | 7 microsatellites | 361 | WMED, CMED, EMED | 1–24 | Yes | Most samples are from Riccioni et al. 2010. Some pairs do not show differentiation. |
| Viñas et al. 2011 | 7 microsatellites | 96 | WMED, EMED | spawning adults | No | Based on reference samples (spawning adults). Some analyses were based on no prior assumption of groups. |

*Notes*: MED = Mediterranean; WMED = western Mediterranean; CMED = central Mediterranean; EMED = eastern Mediterranean; YOY = young-of-the-year; CR = control region. The sampling size refers to the number of Mediterranean samples finally analyzed. The areas sampled refer to those in Figure 3.1d.

differentiation within the Mediterranean must be challenging. Still, Pujolar et al. (2003) suggested that this lack of heterogeneity is consistent with the existence of two groups that exchange a few individuals. In this situation, Viñas et al. (2011) recommended an independent management for each Mediterranean basin to avoid the risk of possibly losing regional subpopulations.

## Reproduction

Fish population reproductive traits play a key role in determining productivity and, therefore, resiliency to exploitation by fisheries or to human-induced environmental perturbations (Morgan 2008). Although "a very limited amount of scientifically useful information is available on the reproductive biology for most tunas" (Schaefer 2001), recent work has improved our knowledge of eastern ABT reproductive traits (Karakulak et al. 2004; Corriero et al. 2005; Heinisch et al. 2008; Gordoa et al. 2009; Aranda et al. 2013a,b; Addis et al. 2016). Yet, many questions remain regarding interrelations between spawning stocks, reproductive potential, and recruitment. The reproductive performance of fishes is influenced by various intrinsic factors, many of which are poorly studied and/or understood for ABT, as summarized below.

### Age at First Maturity

Thus far, the two ABT populations have been considered to display distinct reproductive features, which are reflected in an earlier maturation age of the East Atlantic–Mediterranean population, among others. Such differences may result in a higher spawning rate, and hence a higher productivity of the eastern population. Age at first maturity has been consistently established at about age 3 for the eastern ABT population (e.g., Rodríguez-Roda 1967, Corriero et al. 2005), although Anonymous (2014) suggests that the proportion of mature individuals might be biased because samples outside the Mediterranean were not included in these analyses. In fact, among the individuals tagged in the Atlantic by Block et al. (2005), only those larger than 200 cm curved fork length (approximately 8 years of age) migrated into the Mediterranean during the spawning season. However, fish migrating from the Atlantic to the Mediterranean during the spawning migration and caught by the Spanish traps at the entrance in the Strait of Gibraltar used to range from 130 cm to 280 cm (approximately 4 to 20 years of age), indicating a much wider range than that documented by electronic tagging (Fromentin 2009).

Estimates of age of 50% maturity for western ABT are more variable, ranging from 8 to 15.8 years (e.g., Baglin 1982, Diaz and Turner 2007, Diaz 2011). In fact, determining the age at first maturity for the western population has been challenging and has become an issue of controversy in recent years. A comparative histological study has revealed similarities in reproductive characteristics between eastern and western ABT (Knapp et al. 2014). Further research based on analysis of sex hormone profiles suggests that western ABT mature at smaller sizes and at a younger age than currently assumed (Heinisch et al. 2014). Whatever the age of maturity may be in either population, assuming that some mature fish are likely to skip spawning (see below), one critical question arises: What proportion of the mature ABT does actually spawn during the reproductive season?

## Spawning Frequency, Periodicity, and Duration

In fishes with indeterminate fecundity, the annual fecundity should be calculated from data of batch fecundity, spawning frequency, and duration of the individual spawning period (Murua et al. 2003). Based on the presence and abundance of postovulatory follicles (follicle remnants left after release of mature oocytes) in histological sections of ovarian tissue, the proportion of females spawning per day in the western Mediterranean spawning ground (spawning fraction or frequency) has been estimated at ~84%, which corresponds to a spawning periodicity of 1.2 days (d) (Medina et al. 2002, Aranda et al. 2013b). Analysis of seasonal movements and diving behavior of ABT tracked with electronic tags can also provide indirect measurements of the extent of the individual spawning period (Gunn and Block 2001, Teo et al. 2007, Aranda et al. 2013a). The mean spawning duration estimated in the western Mediterranean was 23.9 d (range: 19–31 d), and the estimated mean number of spawns was 18.3 (range: 14–25); this represents a spawning fraction of 0.83 and an interspawning interval (spawning periodicity) of 1.28 d (Aranda et al. 2013a). These estimates are in good agreement with those reached through histological analysis (as shown above) and are also consistent with the results of mtDNA analysis of the eggs produced by spawners monitored in floating cages, which suggest that the spawning period of ABT females may extend over 34 days (Gordoa et al. 2015).

## Sex Ratio

Female-male proportions by size class show similar variation in different Mediterranean subareas (Fenech et al. 2003, El Tawil et al. 2004, Aranda et al. 2013b, Mèlich 2013). Overall, as in other tuna species, females are predominant in midsize classes (approximately 170–220 cm fork length); the fraction of females is up to ~0.7, but they become less frequent than males at larger sizes. Differences in growth rate between sexes as well as differential natural mortality caused by higher reproductive costs in females have been viewed as the major reasons for the size-related variation in the sex ratio (Schaefer 2001, Rooker et al. 2007, Shimose and Farley 2016).

## Skipped Spawning

Annual omission of spawning is a widespread phenomenon among long-lived teleost species. It has primarily been linked with poor feeding conditions and insufficient energy accumulation (Rideout et al. 2005, Rideout and Tomkiewicz 2011). Electronic tagging studies with multiple years of data after the initial trip to the spawning areas have not provided evidence of skipped spawning, as the tagged individuals showed repeated, annual movements into the spawning regions subsequent to the first observed migration to those areas (Block et al. 2005, Teo et al. 2007). However, the occurrence of nonreproductive individuals in nonspawning grounds in the MED during the reproductive season (Zupa et al. 2009, Fromentin and Lopuszanski 2013) suggests that some adult bluefin tuna pertaining to the eastern population might also skip spawning. Similarly, the presence of adult ABT outside of the MED (Arregui et al. 2018) or the Gulf of Mexico (Block et al. 2005, Galuardi et al. 2010) during the spawning period indicates that a substantial portion of the individuals of both populations may skip spawning. Nevertheless, the recent finding of a western Atlantic spawning location in the Slope Sea, well away from the Gulf of Mexico, suggests that some of those apparently nonspawning fish might actually be using alternate areas to breed, with larger individuals spawning in the Gulf of Mexico and smaller individuals spawning in the Slope Sea (Richardson et al. 2016).

A high prevalence of skipped spawning will have a significant impact on the reproductive output of the population (Rideout et al. 2005, Rideout and Tomkiewicz 2011). However, unfortunately, as for many fishes, long-term data on the percentage of nonreproductive individuals are extremely difficult to obtain for ABT, rendering it unfeasible to quantify the impact of skipped spawning on reproductive potential. Secor (2007), based on a general relationship between skipped spawning and longevity by Carey and Judge (2002), predicted 40% of skipped spawning for ABT. He further assessed the effects of skipped spawning modeled under various types of spawning omission in western ABT and found that skipped spawning had relatively minor impact on egg production per recruit reference point, except when juvenile fishery selectivity was considered.

### Reproduction Summary

Besides considerable uncertainty on some of the key reproductive (as well as other life history) parameters, there is scarce knowledge specific to the main spawning grounds that could allow comparison between potential subpopulations or contingents, so as to inform the different population structure hypotheses. A comprehensive study (Heinisch et al. 2008) showed that there is a gradual shift in the ABT reproductive timing across the Mediterranean basin concomitant with rising water temperatures, so that the spawning peak occurs earliest (May) in the Levantine Sea (EMED) and latest (July) in the Balearic Sea (WMED). This paper also reported a higher gonadosomatic index in the EMED than in the CMED and WMED spawning grounds. However, the observed variation of the sex ratio with fork length seems to be similar among the WMED, CMED, and EMED. Further hypothetical spatial variations in the reproductive performance of spawners throughout the MED cannot be assessed yet, since quantitative estimations (e.g., fecundity, spawning periodicity, and individual duration of spawning) have been conducted only in the WMED. Therefore, the present knowledge of the ABT reproductive biology is not very useful to a discussion of potential subpopulation structure hypotheses; further area-specific estimates and comparative analyses should be pursued.

Contrary to the current knowledge, the ICCAT stock assessment ignores the existing sex ratio and further assumes that all mature ABT actually spawn every year. Compared with shorter-term PSAT attachments, where tagging procedures could affect short-term behavior (Abascal et al. 2016) and bias perception about skipped spawning, long-term archival tags provide a safer (though more expensive) way to assess this issue. The track of the individual of presumably CMED origin shown in Figure 3.2 suggests that the use of the Atlantic for feeding might be longer than currently assumed, as spawning in the Mediterranean did not take place until the fish was at least 6 years old. This is also in agreement with Block et al. (2005), where only individuals older than 8 years old were observed to migrate into the MED. However, compared with e-tag observations, size data from ABT caught at traps revealed a wider range of sizes migrating from the Atlantic to the MED. Long multiyear tracks of Atlantic juveniles and adolescents would provide a good basis to contrast the current maturity/spawning assumptions.

## Conclusions and Future Directions

In spite of the substantial amount of research conducted over past decades, some key questions remain unresolved. In fact, with new experimental observations, we've gained a better understanding of the complexity of the life history and migration patterns, which in turn has increased the number of population-structure hypotheses considered (see section "Summary of Population Structure"). Current knowledge does not clearly favor any of the five hypotheses considered. It seems more likely that ABT originated in the WMED use the Atlantic more intensively than those originated in the EMED. However, this perception is based on a limited number of tracks that could be assigned to origin within the Mediterranean, as well as short times at liberty. Genetic results are also misleading, mostly owing to lack of proper experimental designs (i.e., high number of reference samples that adequately represent the ABT originated in the different spawning grounds), but also owing to the different markers used and analyses conducted. So in general, more research is needed to clarify whether different genetic populations exist within the Mediterranean and whether individuals spawning in different areas behave differently.

The current management approach could be supported in the future unless lack of panmixia is confirmed; for this, scientists need to confirm the differential behavior between ABT born in different spawning areas and show evidence for genetic flux between these areas. Multiyear tracks obtained with internal archival tags on adults are required to determine whether ABT individuals always use the same spawning areas, which would confirm natal homing at the Mediterranean subarea level. Block et al. (2005) observed two individuals visiting the WMED spawning ground during three and four consecutive years, respectively, but unfortunately most studies during the past decade have used PSAT tags with too-short times at liberty, and a much larger sample size is needed. We recommend conducting global, multiyear e-tag experiments, balanced between subpopulations/contingents and life stages (juveniles and adults). To be able to assign movement information to given subpopulations/contingents, it is important to tag spawning aggregations and/or to take biological samples at tagging and recovery, especially for juveniles (Cuéllar Pinzón et al. 2016).

In spite of the shorter tracks provided by PSATs, they can also be useful when physically recaptured, since detailed data sets can identify spawning activity and thus origin. More archival tags on Atlantic juveniles providing movement information throughout the adolescent and adult stages (like the one provided in Figure 3.2) will provide the information necessary to decipher whether the Atlantic Ocean is used equally by all populations/contingents and to discern between age at maturity and age of actual spawning, as well as add to the body of knowledge on skipped spawning. Finally, a curtain of acoustic receivers in the Strait of Gibraltar, as suggested by Canals et al. (2015), would prove very informative on how the different subpopulations/contingents use the Atlantic and whether Atlantic migrants go back to the Mediterranean every year to spawn.

If panmixia is not supported, then the different subpopulations or contingents would need to be managed separately. However, to allow for differential management, appropriate tools to identify the origin (within the Mediterranean) of ABT individuals caught throughout the Atlantic and the Mediterranean need to be developed. If genetic substructure is confirmed, genetic markers could be developed for this purpose. However, in the case of contingents that are genetically connected but behave differently, genetic markers will not be useful and assignment of origin might need to rely on other biological markers. In fact, some markers based on otolith chemistry have already shown potential for this (Fraile et al. 2017). Differential management would also require an estimate of subpopulation/contingent-specific biological parameters (e.g., maturity, growth, etc.) to enable the monitoring of the status of the different groups through time.

## Acknowledgments

This work was carried out under the provision of the ICCAT Atlantic-wide research program for bluefin tuna (GBYP), funded by the European Community (Grant SI2/542789), Canada, Croatia, Japan, Norway, Turkey, the United States (NMFS NA11NMF4720107), Chinese Taipei, and the ICCAT Secretariat. The contents of the chapter do not necessarily reflect the point of view of ICCAT or of the other funders. The authors thank Antonio Di

Natale and Gemma Quílez-Badia for constructive discussion on their e-tag results, and three anonymous reviewers for their constructive comments.

## REFERENCES

Abascal, F.J., A. Medina, J.M. de La Serna, D. Godoy, and G. Aranda. 2016. Tracking bluefin tuna reproductive migration into the Mediterranean Sea with electronic pop-up satellite archival tags using two tagging procedures. *Fisheries Oceanography* 25:54–66.

Addis, P., M. Secci, C. Biancacci, D. Loddo, D. Cuccu, F. Palmas, and A. Sabatini. 2016. Reproductive status of Atlantic bluefin tuna, *Thunnus thynnus*, during migration off the coast of Sardinia (western Mediterranean). *Fisheries Research* 181:137–47.

Alvarado Bremer, J., I. Naseri, and B. Ely. 1999. Heterogeneity of northern bluefin tuna populations. *ICCAT Collective Volume of Scientific Papers* 49:127–29.

Alvarado Bremer, J.R., J. Viñas, J. Mejuto, B. Ely, and C. Pla. 2005. Comparative phylogeography of Atlantic bluefin tuna and swordfish: The combined effects of vicariance, secondary contact, introgression, and population expansion on the regional phylogenies of two highly migratory pelagic fishes. *Molecular Phylogenetics and Evolution* 36:169–87.

Anonymous. 2014. Report of the 2013 bluefin meeting on biological parameters review (Tenerife, Spain—May 7–13, 2013). *ICCAT Collective Volume of Scientific Papers* 0 (1): 1–159.

Anonymous. 2016. Report of the 2015 ICCAT Bluefin Data Preparatory Meeting (Madrid, Spain—March 2–6, 2015). *ICCAT Collective Volume of Scientific Papers* 72: 1233–1349.

Aranda, G., F.J. Abascal, J.L. Varela, and A. Medina. 2013a. Spawning behaviour and post-spawning migration patterns of Atlantic bluefin tuna (*Thunnus thynnus*) ascertained from satellite archival tags. *PLOS ONE* 8:e76445.

Aranda, G., A. Medina, A. Santos, F.J. Abascal, and T. Galaz. 2013b. Evaluation of Atlantic bluefin tuna reproductive potential in the western Mediterranean Sea. *Journal of Sea Research* 76:154–60.

Arregui, I., B. Galuardi, N. Goñi, C.H. Lam, I. Fraile, J. Santiago, M. Lutcavage, et al. 2018. Movements and geographic distribution of juvenile bluefin tuna in the Northeast Atlantic, described through internal and satellite archival tags. *ICES Journal of Marine Science*. doi.org/10.1093/icesjms/fsy056.

Baglin, R.E., Jr. 1982. Reproductive biology of western Atlantic bluefin tuna. *Fishery Bulletin* 80:121–34.

Begg, G.A., K.D. Friedland, and J.B. Pearce. 1999. Stock identification and its role in stock assessment and fisheries management: An overview. *Fisheries Research* 43:1–8.

Benestan, L., T. Gosselin, C. Perrier, B. Sainte-Marie, R. Rochette, and L. Bernatchez. 2015. RAD-genotyping reveals fine-scale genetic structuring and provides powerful population assignment in a widely distributed marine species; the American lobster (*Homarus americanus*). *Molecular Ecology* 24 (13): 3299–3315.

Block, B.A., S.L.H. Teo, A. Walli, A. Boustany, M.J.W. Stokesbury, C.J. Farwell, K.C. Weng, H. Dewar, and T.D. Williams. 2005. Electronic tagging and population structure of Atlantic bluefin tuna. *Nature* 434:1121–27.

Boustany, A.M., C.A. Reeb, and B.A. Block. 2008. Mitochondrial DNA and electronic tracking reveal population structure of Atlantic bluefin tuna (*Thunnus thynnus*). *Marine Biology* 156:13–24.

Broughton, R.E., and J.R. Gold. 1997. Microsatellite development and survey of variation in northern bluefin tuna (*Thunnus thynnus*). *Molecular Marine Biology And Biotechnology* 6:308–14.

Canals, M., E. Balguerías, M. Stokesbury, F. Whoriskey, A. Sánchez, A. Medina, F.J. Abascal, and G. Aranda. 2015. An acoustic telemetry curtain across the Strait of Gibraltar? Working paper, SCRS/2015/056:1–3.

Carey, J., and D. Judge. 2002. *Longevity records: Life spans of mammals, birds, amphibians, reptiles, and fish*. Monographs on Population Aging, 8. Odense, Denmark: Odense University Press.

Carlsson, J., J.R. McDowell, P. Diaz-Jaimes, J.E.L. Carlsson, S.B. Boles, J.R. Gold, and J.E. Graves. 2004. Microsatellite and mitochondrial DNA analyses of Atlantic bluefin tuna (*Thunnus thynnus thynnus*) population structure in the Mediterranean Sea. *Molecular Ecology* 13:3345–56.

Carlsson, J., J.R. McDowell, J.E.L. Carlsson, and J.E. Graves. 2007. Genetic identity of YOY bluefin tuna from the eastern and western Atlantic spawning areas. *Journal of Heredity* 98 (1): 23–28.

Carruthers, T., A. Kimoto, J.E. Powers, L. Kell, D.S. Butterworth, M.V. Lauretta, and T. Kitakado. 2016. Structure and estimation framework for Atlantic bluefin tuna operating models. *ICCAT Collective Volume of Scientific Papers* 72:1782–95.

Cermeño, P., G. Quílez-Badia, A. Ospina-Alvarez, S. Sainz-Trápaga, A.M. Boustany, A.C. Seitz, S. Tudela, and B.A. Block. 2015. Electronic tagging of Atlantic bluefin tuna (*Thunnus Thynnus*, L.) reveals habitat use and behaviors in the Mediterranean Sea. *PLOS ONE* 10:e0116638.

Corriero, A., S. Karakulak, N. Santamaria, M. Deflorio, D. Spedicato, P. Addis, S. Desantis, et al. 2005. Size and age at sexual maturity of female bluefin tuna (*Thunnus thynnus* L. 1758) from the Mediterranean Sea. *Journal of Applied Ichthyology* 21:483–86.

Cuéllar-Pinzón, J., P. Presa, S.J. Hawkins, and A. Pita. 2016. Genetic markers in marine fisheries: Types, tasks and trends. *Fisheries Research* 173:194–205.

De Metrio, G, G.P. Arnold, J.M. de la Serna, B.A. Block, P. Megalofonou, M. Lutcavage, I. Oray, and M. Deflorio. 2005. Movements of bluefin tuna (*Thunnus thynnus* L.) tagged in the Mediterranean Sea with pop-up satellite tags. *ICCAT Collective Volume of Scientific Papers* 58:1337–40.

Di Natale, A., S. Tensek, and A.P. García. 2016. Preliminary information about the ICCAT GBYP tagging activities in Phase 5. *ICCAT Collective Volume of Scientific Papers* 72: 1589–1613.

Diaz, G.A. 2011. A revision of western Atlantic bluefin tuna age of maturity derived from size samples collected by the Japanese longline fleet in the Gulf of Mexico (1975–1980). *ICCAT Collective Volume of Scientific Papers* 66:1216–26.

Diaz, G.A., and S.C. Turner. 2007. Size frequency distribution analysis, age composition, and maturity of western bluefin tuna in the Gulf of Mexico from the US (1981–2005) and Japanese (1975–1981) longline fleets. *ICCAT Collective Volume of Scientific Papers* 6:1160–70.

El Tawil, M., N. El Kabir, J. Ortiz-de-Urbina-Gutiérrez, J. Valeiras, and E. Abad. 2004. Analysis of sex-ratio by length-class for bluefin tuna (*Thunnus thynnus* L.) caught from the Lybian trap fishery. *ICCAT Collective Volume of Scientific Papers* 56 (3): 1189–91.

Ely, B., D.S. Stoner, J.R. Alvarado Bremer, J.M. Dean, P. Addis, A. Cau, E.J. Thelen, et al. 2002. Analyses of nuclear ldhA gene and mtDNA control region sequences of Atlantic northern bluefin tuna populations. *Marine Biotechnology* 4 (6): 583–88.

Fenech, A., J. de la Serna, and J. Ortiz. 2003. Sex-ratio by length-class of bluefin tuna (*Thunnus thynnus* L.) caught by Maltese longliners. *ICCAT Collective Volume of Scientific Papers* 55:1145–47.

Fraile, I., H. Arrizabalaga, and J.R. Rooker. 2015. Origin of Atlantic bluefin tuna (*Thunnus thynnus*) in the Bay of Biscay. *ICES Journal of Marine Science* 72:625–34.

Fraile, I., H. Arrizabalaga, D. Macías, M. Valastro, P. Addis, I. Oray, and J.R. Rooker. 2017. First insights into the Atlantic bluefin tuna stock structure within the Mediterranean Sea. SCRS/P/2017/001.

Fromentin, J.M. 2009. Lessons from the past: Investigating historical data from bluefin tuna fisheries. *Fish and Fisheries* 10:197–216.

Fromentin, J.-M., and D. Lopuszanski. 2013. Migration, residency, and homing of bluefin tuna in the western Mediterranean Sea. *ICES Journal of Marine Science* 71:510–18.

Fromentin, J.M., and J.E. Powers. 2005. Atlantic bluefin tuna: Population dynamics, ecology, fisheries and management. *Fish and Fisheries* 6 (4): 281–306.

Fromentin, J.-M., S. Bonhommeau, H. Arrizabalaga, and L.T. Kell. 2014. The spectre of uncertainty in management of exploited fish stocks: The illustrative case of Atlantic bluefin tuna. *Marine Policy* 47:8–14.

Galuardi, B., and M. Lutcavage. 2012. Dispersal routes and habitat utilization of juvenile Atlantic bluefin tuna, *Thunnus thynnus*, tracked with mini PSAT and archival tags. *PLOS ONE* 7 (5): e37829.

Galuardi, B., F. Royer, W. Golet, J. Logan, J. Neilson, and M. Lutcavage. 2010. Complex migration routes of Atlantic bluefin tuna (*Thunnus thynnus*) question current population structure paradigm. *Canadian Journal of Fisheries and Aquatic Sciences* 67: 966–76.

Gordoa, A., M.P. Olivar, R. Arevalo, J. Viñas, B. Molí, and X. Illas. 2009. Determination of Atlantic bluefin tuna (*Thunnus thynnus*) spawning time within a transport cage in the western Mediterranean. *ICES Journal of Marine Science* 66:2205–10.

Gordoa, A., N. Sanz, and J. Viñas. 2015. Individual spawning duration of captive Atlantic bluefin tuna (*Thunnus thynnus*) revealed by mitochondrial DNA analysis of eggs. *PLOS ONE* 10:e0136733.

Graves, J.E., A.S. Wozniak, R.M. Dickhut, M.A. Cochran, E.H. MacDonald, E. Bush, H. Arrizabalaga, and N. Goñi. 2015. Transatlantic movements of juvenile Atlantic bluefin tuna inferred from analyses of organochlorine tracers. *Canadian Journal of Fisheries and Aquatic Sciences* 72:625–33.

Gunn, J., and B. Block. 2001. Advances in acoustic, archival, and satellite tagging of tunas. In *Tuna: Physiology, Ecology, and Evolution*, edited by B.A. Block and E.D. Stevens, 167–224. San Diego, CA: Academic Press.

Hauser, L., and G.R. Carvalho. 2008. Paradigm shifts in marine fisheries genetics: Ugly hypotheses slain by beautiful facts. *Fish and Fisheries* 9:333–62.

populations; together they have demonstrated that two discrete eastern and western ABT management units exist, move to separate spawning grounds, and mix extensively on North Atlantic foraging grounds (Block et al. 2005; Carlsson et al. 2007; Rooker et al. 2008, 2014; Boustany et al. 2008; Schloesser et al. 2010; Siskey et al. 2016; Hanke et al. 2016a, 2018; Puncher et al. 2018). Data obtained from electronic tags improve modeling of key parameters of life history traits related to maturation, natural mortality, and fisheries mortality (Kurota et al. 2009, Taylor et al. 2011). Bluefin tuna population models rely on biological assumptions. Reducing the uncertainty in these assumptions with electronic tagging data is vital to improving new assessments and to developing operating models that are built on realistic life history parameters of the species being managed (Taylor et al. 2011, Kerr et al. 2016). Despite the rapid advances of the past decade, key questions remain about the level of population mixing, recruitment dynamics, maturity schedules, and the number and size of the independent populations.

ABT are currently managed by the International Commission for the Conservation of Atlantic Tunas (ICCAT) as two populations separated by the 45° meridian. Both populations are considered overfished, and current total allowable catches are facilitating rebuilding (ICCAT 2014, 2017). An essential part of the ICCAT stock assessment process is focused on determining the current spawning population size and status in comparison to the historical spawning stock biomass (Taylor et al. 2011). This provides a measure of the biomass decline over time, or mature stock biomass relative to the maximum sustainable yield. Key to conducting appropriate assessments is accurately accounting for new biological information when constructing management models. This is essential for tracking the progress of rebuilding models and assessing bluefin tuna biomass when the catches begin to increase. To date, this has been difficult because ABT populations are known to be mixed on the North Atlantic foraging (and fishing) grounds. Tagging and biological markers (Block et al. 2005, Boustany et al. 2008, Rooker et al. 2008, Dickhut et al. 2009, Hanke et al. 2016a,b) demonstrate that eastern and western ABT stocks are extensively mixed.

Conventional and electronic tagging provides evidence of large-scale migrations, where juveniles, adolescent, and mature ABT from eastern and western spawning populations move across the ICCAT stock boundary at the 45° meridian (Block et al. 2001, 2005; Stokesbury et al. 2007). More recently, tagging has identified a significant variability in dispersal patterns displayed by individual ABT (Block et al. 2005, Galuardi et al. 2010, Cermeño et al. 2015, Wilson et al. 2015). Key to the future of Atlantic bluefin management is understanding the life history, dispersal patterns, ontogenetic changes in migration patterns, and reproductive strategies of each spawning population to account for the fact that fisheries on one side of the ocean can impact population biomass on either side of ICCAT's stock management boundary. Likewise, biological parameters specific to each stock (e.g., growth curves or maturity schedules) need to be deciphered to accurately assess spawning stock biomass and status for all bluefin populations in the Atlantic (Hanke et al. 2016a,b, 2018). These population movements indicate that the way single-stock models capture standing stock biomass declines is not entirely correct, because catch removals have not always been attributed to the proper stock of origin in the past (Taylor et al. 2011, Kerr et al. 2016). Movement toward new operational models that integrate the mixing of populations informed annually by biological inputs should improve ABT assessment accuracy.

## Electronic Tagging in the West Atlantic

In the North Atlantic, the Tag-A-Giant (TAG) research program of Stanford University and the Monterey Bay Aquarium has electronically tagged ABT since 1996. To date, 1,350 electronic tags (internal archival, pop-up archival satellite [PAT], and acoustic) have been deployed on adolescent and mature fish, primarily in the western Atlantic Ocean along the eastern seaboard of North America. Additionally, more than 75 tags have been put out in the MED (Cermeño et al. 2015) and in the eastern Atlantic along the coasts of Morocco and Ireland (Stokesbury et al. 2007). This electronic tagging effort has created a detailed time series for ABT from 4 to 25 years of age, with more than 39,000 tracking days from western Atlantic releases (Figures 4.1 and 4.2). Population assignments of a proportion of the western Atlantic track data have been possible using genetics of fin clips taken during tagging (Boustany et al. 2008); assignments are also possible when mature electronically tagged fish visit a known spawning site (Block et al. 2005, Wilson et al. 2015). Models incorporating these data have improved the capacity of scientists to predict spatial habitat use of each population (western and eastern), to estimate mortality (Kurota et al. 2009), and to project biomass trajectories (Taylor et al. 2011). Recent efforts to conduct more complicated stock assessments incorporating the spatial and tempo-

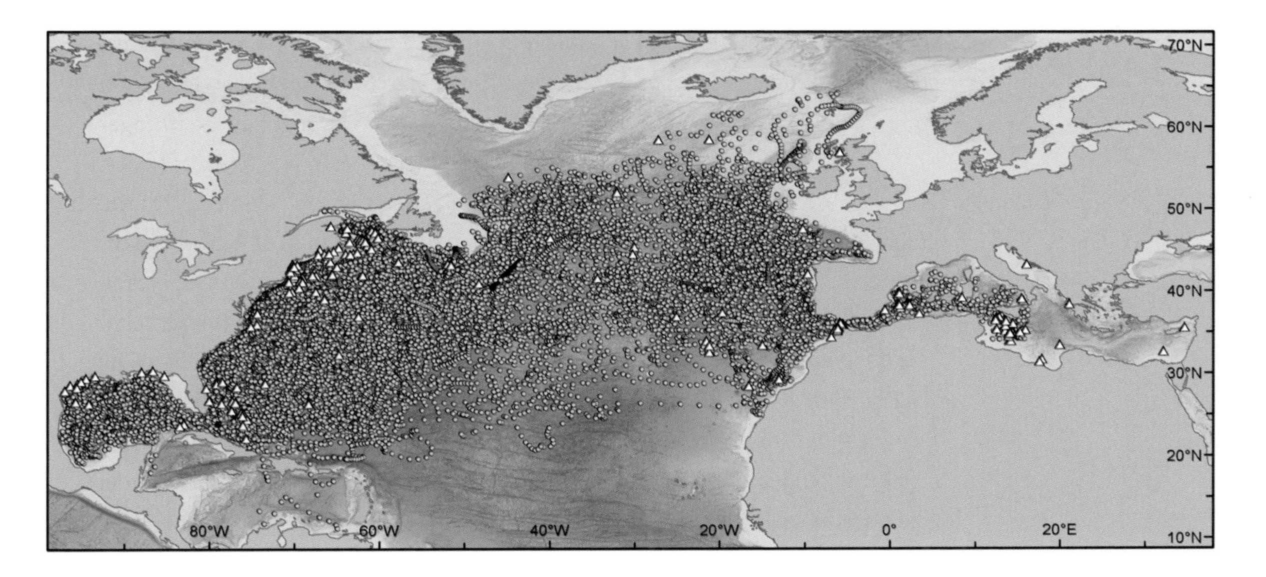


**Figure 4.1.** Atlantic bluefin tuna electronic tag tracks (n = 378) revealing the geolocation positions of archival- and satellite-tagged bluefin tuna from western releases. Triangles indicate recovered tag locations from archival and satellite tags. The minimum length of a fish measured at tagging was 112 cm curved fork length (CFL), and the maximum size was 313 cm CFL.

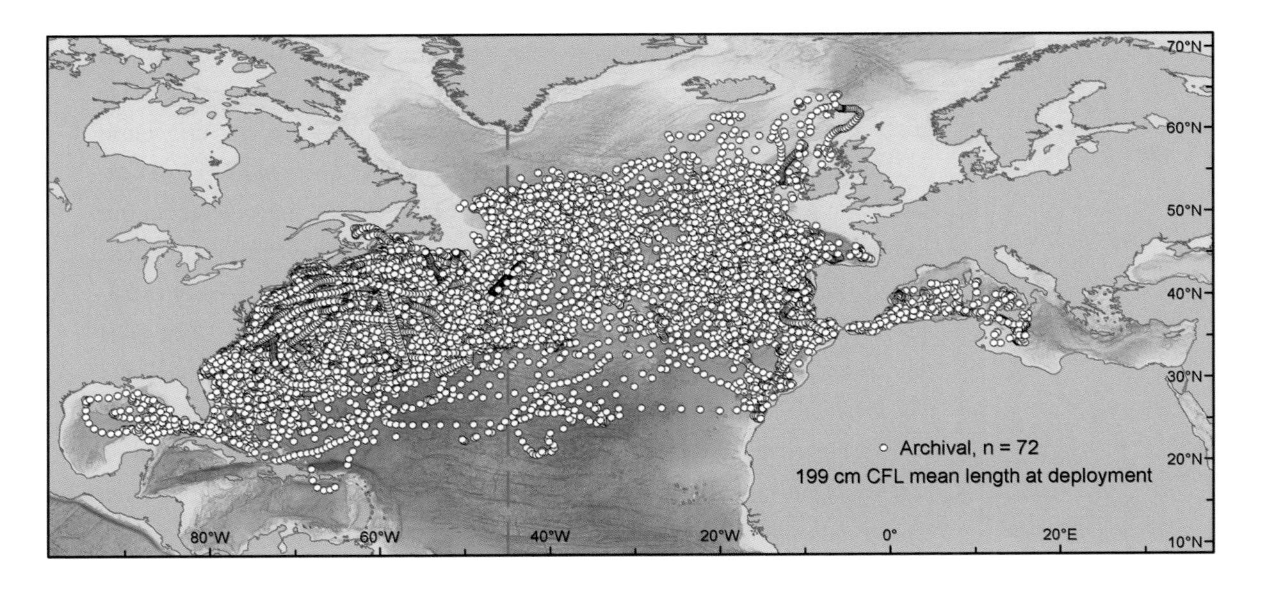


**Figure 4.2.** Positions (geolocations) of archival-tagged Atlantic bluefin tuna (ABT) in the western Atlantic that were deployed with tags that recorded 30 days or more of tracking data (recovered from ABT, 1997–2012). The tags were deployed on fish with a mean measured length of 199 cm curved fork length (CFL) from North Carolina waters. Few archival tag tracks released in the western Atlantic recorded visitations to Gulf of Mexico waters, but many recorded visitations to the western and central Mediterranean.

ral use of the North Atlantic and the GOM are vital to sorting out the extent of mixing within the immature and mature fish year classes.

## Archival Tagging

Electronic tags provide valuable information on movements of ABT but have important limitations by tag type (archival, pop-up satellite archival, acoustic). Implantable archival tags (n = 727) were deployed in ABT (mean measured length 210.7 cm ±20.3 curved fork length [CFL]) off North Carolina from 1996 to 2012. To date, ~21.7% of the archival tags (n = 158) have been recovered from fisheries. Archival tag tracks have recorded the fidelity of ABT movements in successive years to North Atlantic foraging grounds, back to the GOM breeding grounds for 3 consecutive years, and to the western Mediterranean for periods of up to 4.2 years (Block et al. 2005, Teo et al. 2007a, Wilson et al. 2015). Multiyear fidelity to a spawning ground in consecutive years has been seen in five individual western tagged and released fish that moved into the MED. Two fish tagged with archival tags have revealed consecutive breeding years and fidelity to the GOM (Figure 4.3). Archival tags have provided a critical long-term and multiyear data set; however, keeping stalked tags functioning and batteries operational has presented engineering challenges, and the total expenses of maintaining a long-term archival tagging research program have been high. Surgery on large mature fish of the size that visit the GOM is difficult at sea, so most archival tags are implanted in subadult fish in foraging hot spots (e.g., North Carolina waters) where fish are aggregated. This improves the capacity to get archival tags out in large numbers, which is required to ensure recoveries.

Archival tags are the only tags to date that provide the capacity to follow a fish through its ontogeny from an immature to a mature ABT. The importance of this capacity is underappreciated, as it's one of the few direct ways we can learn exactly when ABT of different populations move into a spawning area to potentially spawn. Archival tags provide the most detailed tracks, often taking data every 1–60 seconds over the longest continuous durations (up to five years). While it often takes many years to accumulate a large data set with annual migrations, diving records, and oceanographic information spanning multiple years, these data remain the most valuable and difficult data to obtain. The archival tag tracks are extremely important, but the ability of national and international teams to deploy tags surgically, recover the tags, and download and process data sets requires a com-

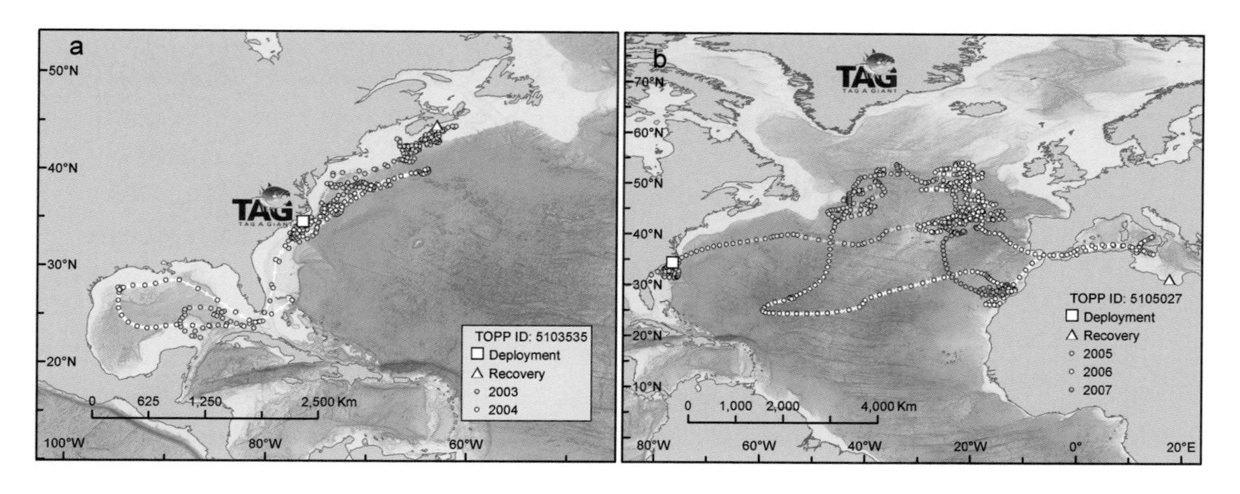


**Figure 4.3.** Archival tags use geolocation to reveal multiyear movements and spawning ground migrations. (*a*) A multiyear track of an Atlantic bluefin tuna that went to the Gulf of Mexico in two consecutive years after release in North Carolina waters, and (*b*) a multiyear track showing movements from the western foraging grounds off North Carolina waters (2005) into the North Atlantic, followed by a migration into the Mediterranean spawning grounds (2006) and out again to the eastern Atlantic Ocean (2007). TOPP = Tagging of Pacific Predators, an international collaboration of scientists using new technologies to track pelagic animal migrations. All bluefin tuna data are archived with a TOPP ID.

mitment to long-term funding and cooperation across fishing fleets. The latest models of electronic tags from the two main manufacturers of light-based geolocation tags (Wildlife Computers and Lotek Wireless Inc.) now have models of archival tags that are three-to-five-year instruments.

To date, the archival-tagged ABT deployed in the western Atlantic Ocean have demonstrated movements into the GOM at a minimal size of 8 years of age (Block et al. 2005). Younger fish, archivally tagged off North Carolina, did not often move into the GOM spawning ground. However, many of the fish tagged in the waters off the mid-Atlantic winter fishery did return to the MED, and in some cases these fish spent a year in western North Atlantic waters prior to returning to the MED spawning ground (Block et al. 2005). This was a significant finding (nonvisitation to a spawning area the year post tagging) and is consistent with more recent deployments (Wilson et al. 2015, Hazen et al. 2016) in the past decade. The limitation of the archival tags in the earlier years was that the tag's memory log often filled up (two years post tagging) before a fish matured and moved into a spawning area. This provided ample behavioral data on North Atlantic foraging migrations, diving patterns, and core habitat preferences (Walli et al. 2009). A small proportion of these fish may be Slope Sea spawners; however, many never visited waters with temperatures indicative of spawning conditions (>23°C), and only a few individual bluefin recorded the ambient temperatures consistent with spawning (Richardson et al. 2016, Block pers. com.). In addition, multiple types of tag failures (light-stalk breaks and battery failure) limited the ability to recover the electronic tag data. Given the capacity of the recent generations of archival tags to provide unparalleled data for long durations, ICCAT should invest in this type of electronic tagging. Building capacity across countries will require workshops and multinational collaborations and sustained funding for training, deployments, tag recovery, rewards, and data analyses.

The major advantage of the TAG archival tag program is that this type of electronic tag technology is the only one to date that provides reliable multiyear tracks revealing fidelity to spawning grounds (Figures 4.2 and 4.3) and ontogenetic shifts in core habitat utilization. Western Atlantic archival-tagged ABT released from mid-Atlantic locations rarely recorded trips to the GOM and most often moved from coastal US waters to the MED spawning area (Figure 4.2), raising questions about the origin of the fish off the coast of North Carolina. The archival tags demonstrated that adolescent ABT traverse back and forth across the North Atlantic prior to maturing and spawning in the MED, most likely seeking optimal foraging opportunities. Archival tracks provide valuable information about North Atlantic core habitat use and preferences (Walli et al. 2009, Lawson et al. 2010), and these tracks have produced rich behavioral data sets, complete with body temperature, that have become "training data" for understanding pop-up satellite archival data from tags recaptured years later (Hazen et al. 2016). To gather data on spawning in the GOM (primarily from tags on fish > 10–14 years old) or the MED (Cermeño et al. 2015), the challenge is to keep the archival tags functioning and recording data as the fish matures (usually 1–4 years after surgery).

## Pop-Up Satellite Archival Tags

Pop-up satellite archival tags (commonly known as PAT tags or mini-PAT tags) are attached externally on adolescent and mature ABT and are typically retained for 12 months or less. Satellite tags were first placed on tunas in 1997 (Block et al. 1998) in the winter fishery in the waters off North

Carolina, and the following summer they were deployed off New England waters (n = 288, to fish that measured 210.7 ± 20.3 CFL). Because of the size of these fish at tagging, the timing of deployments early in the year (winter in North Carolina, late fall in New England), and the short durations of these early deployments, which were due in part to premature release of the external tags from the fish (< 7 months), the number of tracks to spawning grounds obtained was approximately 15% of the US tag deployments. The satellite and archival tagging revealed that the US Atlantic fisheries were significantly mixed, with a number of ABT of Mediterranean origin and subadult GOM fish foraging together in western North Atlantic waters. Additionally, in these US experiments, PAT tagging did not, in the short 4–7 months of tag attachment, often reveal spawning due to short attachment times.

The Canadian foraging aggregation in the Gulf of St. Lawrence (GSL) offered a new opportunity to apply the lessons learned in the US winter and fall fisheries to larger ABT, which were presumably more mature. The international team first obtained permits for electronic tagging by US and Canadian scientists in 2005 to test the concept of tagging very large adult fish in Nova Scotian waters. The experiments revealed that only PAT tagging would work, and surgery (at the time) was deemed too challenging on fish over 400 kg. To date, TAG scientists have placed 158 satellite tags on these Canadian-caught fish (measured mean size 265.4 cm CFL ± 21.2 cm). This assemblage contains the largest fish electronically tagged (313 cm CFL). The early years in Canadian waters were focused on establishing proper methods for capture and tagging techniques that permitted successful handling and satellite tagging for these large, mature fish. In all cases, these ABT are brought aboard the boat, where the tags are carefully placed prior to release. Mortality studies were also conducted on the tag-and-release fishery to ensure that large fish survived the stress of heavy tackle line capture and release (Stokesbury et al. 2011). In addition to satellite tagging, acoustic mark-recapture experiments have been ongoing since 2008, with 128 tags deployed externally. In Canadian deployments, as in other operations, electronic tagging occurs on board the vessel and all fish are accurately measured. Of the tags deployed in Canadian GSL waters, 113 satellite tags remained on the fish for 14 days or more. Of the tags that successfully transmitted data, 54 individuals were shown to have migrated into the GOM spawning area (Figure 4.4). Seven additional tags popped up at the entrance to the GOM, where oscillatory diving to deep depths (800–1000 m) is common and may cause

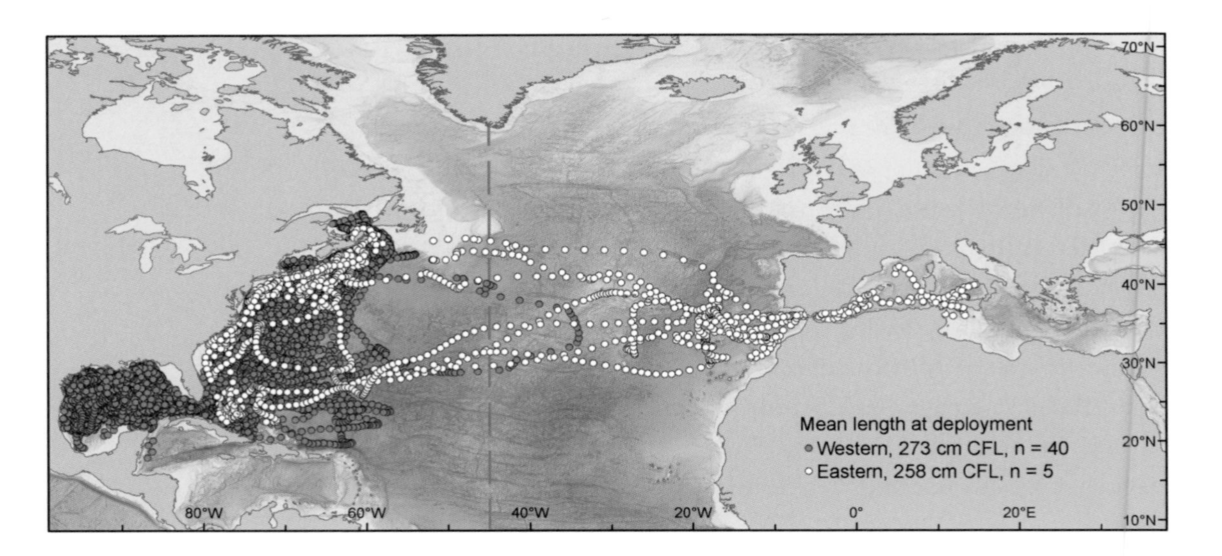


**Figure 4.4.** Geolocations from pop-up satellite archival tags (n = 45) deployed in Canadian waters that remained on fish for six months or more. Circles represent state-space model outputs from geolocation models from light-based and sea surface temperature–based position estimations, and colors indicate tracks of satellite-tagged Atlantic bluefin tuna that are coded by their movements onto known spawning grounds (Gulf of Mexico, *orange;* Mediterranean, *white*). Adapted from Wilson et al. 2015. CFL = curved fork length.

premature tag release. To date, five satellite-tagged individuals have moved from the GSL foraging area into the MED or have been at the entrance of Gibraltar, indicating that mixing is occurring in the GSL. In the past five years, a focus on satellite tag recovery has enabled 34 satellite archival GOM records to be obtained, thus providing archival records of ABT on the spawning grounds.

These new tagging data sets from the GSL serve two purposes. Satellite archival tag attachments have improved, and tracks initiating in the Canadian waters provide 7–12 months of data, with detailed archival time series and oceanography of the period when ABT occupy the GOM and tags are recovered. By selecting Canada's GSL as the deployment location (a western population location with an assemblage of mature fish), the majority of the electronic tags deployed are on the GOM spawners, which represent the smaller population of ABT. In addition, the satellite archival tags provide detailed time series data sets on their behaviors (foraging, reproductive, and oceanographic preferences), and the satellite tags record temporal data on the journey from the GSL foraging ground to the GOM, the exit date from the GSL, the entrance date to the GOM, the spawning date location, and the initiation and termination of spawning behaviors in relation to local GOM oceanography. These data are vital for modeling the spawning area, for understanding the oceanographic preferences of adults, and for enabling the designation of closed areas protecting spawning habitat for ABT (Teo et al. 2007a,b; Wilson and Block 2009; Wilson et al. 2015; Hazen et al. 2016).

The recent tagging work in the GSL of Canada and the rapidly growing time series and tracking data sets from the joint US and Canadian scientific campaigns have shown that in the years of tagging, GSL bluefin tuna were mostly but not exclusively fish of GOM origin. Previously, otolith analyses had shown 100% of GSL fish to be from the GOM (Rooker et al. 2008), and as a result, assessment models that evaluated mixing attributed pure GOM populations of ABT to the GSL fisheries (Taylor et al. 2011). But electronic tagging demonstrates this is not the case—to date, at least five fish tagged in the GSL have shown migrations to and in some cases back from the MED. In addition, underwater acoustic receivers detected a single acoustic-tagged fish when it passed through the Strait of Gibraltar along the northern regions of the MED. This negates the hypothesis that GSL fish are exclusively western, and it informs fisheries scientists that the catch per unit effort index in the GSL should be managed with caution as an index of the status of the GOM population. Of the GSL tags deployed from 2007 to 2012 that transmitted in January–August, 54% (44) recorded visits to the GOM (Figures 4.3 and 4.4), supporting the hypothesis that to date most of these fish are western spawners. A small number of fish did not visit any known Atlantic spawning ground during breeding season and spent the track duration in the North Atlantic. Many of the fish outside the GOM (Figure 4.3) have been classified as borderline spawners, using the standard 8–14 age range for minimum spawning size in the GOM. Additionally, a proportion of these tags popped prematurely, thus making it challenging to assign any spawning area origin to the tracks.

The satellite tag experiments in Canada provide data on movements of GSL fish into the GOM spawning area (Figure 4.4). Moreover, fidelity to foraging grounds enabled a recovery of more than 34 tags after pop-up and drift (Wilson et al. 2015, Hazen et al. 2016). These tags have provided complete round-trip archival tracks from mature fish (Figure 4.7). The positions and time series data sets provide remarkable new information on mature GOM bluefin tuna habitat utilization in the GOM and in the North

Atlantic, as archival data recorded detailed high-resolution temperature and diving profiles. Therefore, the location and behavior of the oldest, most fecund size classes of this population can be accessed, habitat preferences assigned, and the information included in a variety of models.

The recent advancement in satellite tagging techniques in GSL waters provides a glimpse of the full annual cycle for mature ABT of GOM origin (Figures 4.4, 4.5, 4.6, and 4.7). From these new data, key information has been obtained that has led to the following important discoveries: (a) Entrance into the GOM is occurring over a longer seasonal time period (November–June) than previously recorded from implantable archival tags on the North Carolina aggregation of GOM fish. (b) Known geographic areas occupied in the GOM are increasing as larger fish are tagged, and these fish utilize new areas in Mexican and international waters. (c) Spawning takes place primarily in March, April, and May, as previously reported, and the area of primary occupation during the spawning duration gets reduced as fish aggregate near the US continental shelf waters across the GOM. (d) Residence times and exit times vary among fish, and a trend in body size appears to play a role (Wilson et al. 2015). Bluefin tuna exiting the GOM spawning ground can transit in less than 17 days to New England waters (passing the southeast US coast within two weeks). ABT can transit in 30 days from the warmth of the GOM (25°C–31°C) to the cold Canadian foraging grounds (9°C–13°C) where they were initially tagged.

Importantly, tagging data from the GSL foraging areas revealed that the positions of the majority of the GSL fish tracked show a western bias (Figures 4.4, 4.6, and 4.7), with only a few of the fish crossing the 45° meridian. This western residency of the mature GSL/GOM fish post visitation to the foraging ground is important and may represent restricted spatial habitat preferences post maturity and spawning. The electronic tagging data can be incorporated into a variety of traditional (virtual population analysis) and new mixing models that incorporate the spatial and temporal movements to influence the assessment of the western and eastern biomass of ABT (Taylor et al. 2011, Kerr et al. 2016).

Electronic tagging also informs fisheries managers about the age at which ABT are moving into the GOM, presumably to spawn. Only tagging provides the exact time an individual fish moves into the spawning area. Plotting the ABT age cumulatively over the entire period when bluefin were tracked with electronic tags provides information on when and at what age tuna enter the GOM (November through June). The youngest tagged fish

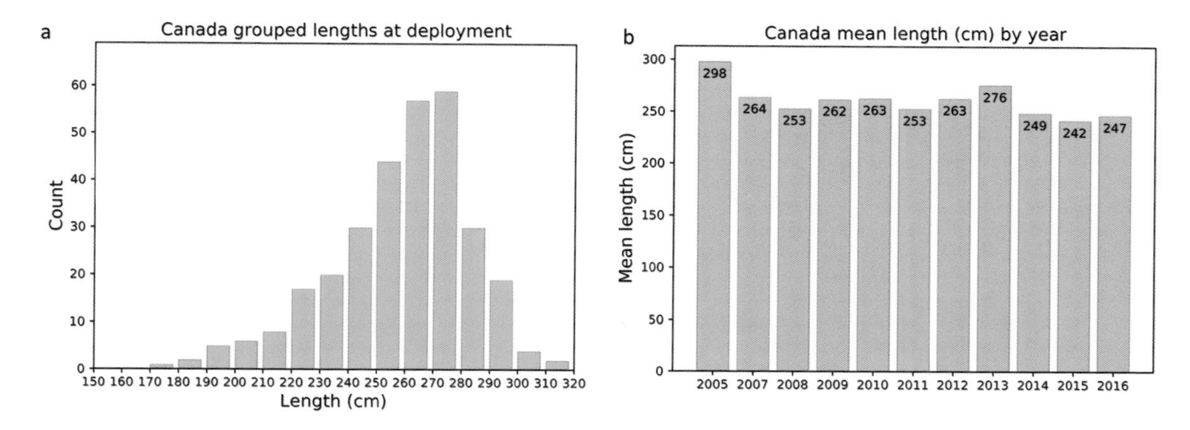


**Figure 4.5.** The length and frequency of Atlantic bluefin tuna (ABT) tagged over the course of the Canadian electronic tagging program. The availability of a relatively large fish—almost always mature (the mean curved fork length [CFL] of the ABT was 258 cm)—makes tagging in Canada of high interest. (*a*) Bluefin tuna satellite tagged in the GSL until 2016 (n = 146). (*b*) Combined data for measured CFL from satellite- and acoustic-tagged fish by year (n = 245 individuals). The year 2005 was an experimental year in which only three fish were tagged.

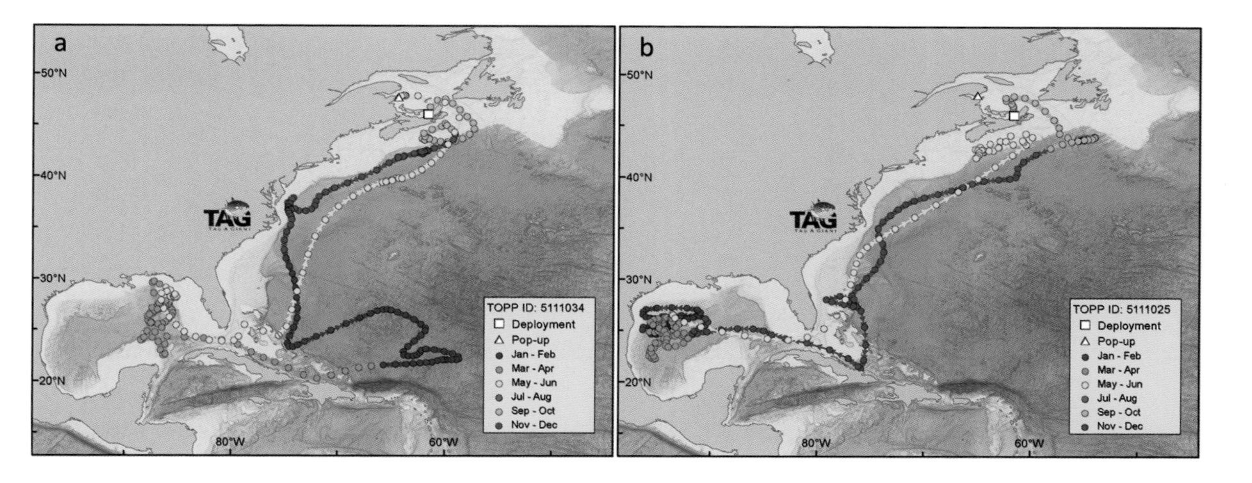


**Figure 4.6.** Examples of the state-space modeled pop-up satellite archival tag tracks obtained from tagging Canadian-released Atlantic bluefin tuna with pop-up satellite archival tags. Tracks show trips from the foraging ground to spawning areas in the Gulf of Mexico (GOM) and demonstrate use of (*a*) eastern GOM areas and (*b*) western GOM waters. Squares represent deployment of tagged fish, and triangles indicate pop-up release point. TOPP = Tagging of Pacific Predators, an international collaboration of scientists using new technologies to track pelagic animal migrations. All bluefin tuna data are archived with a TOPP ID.

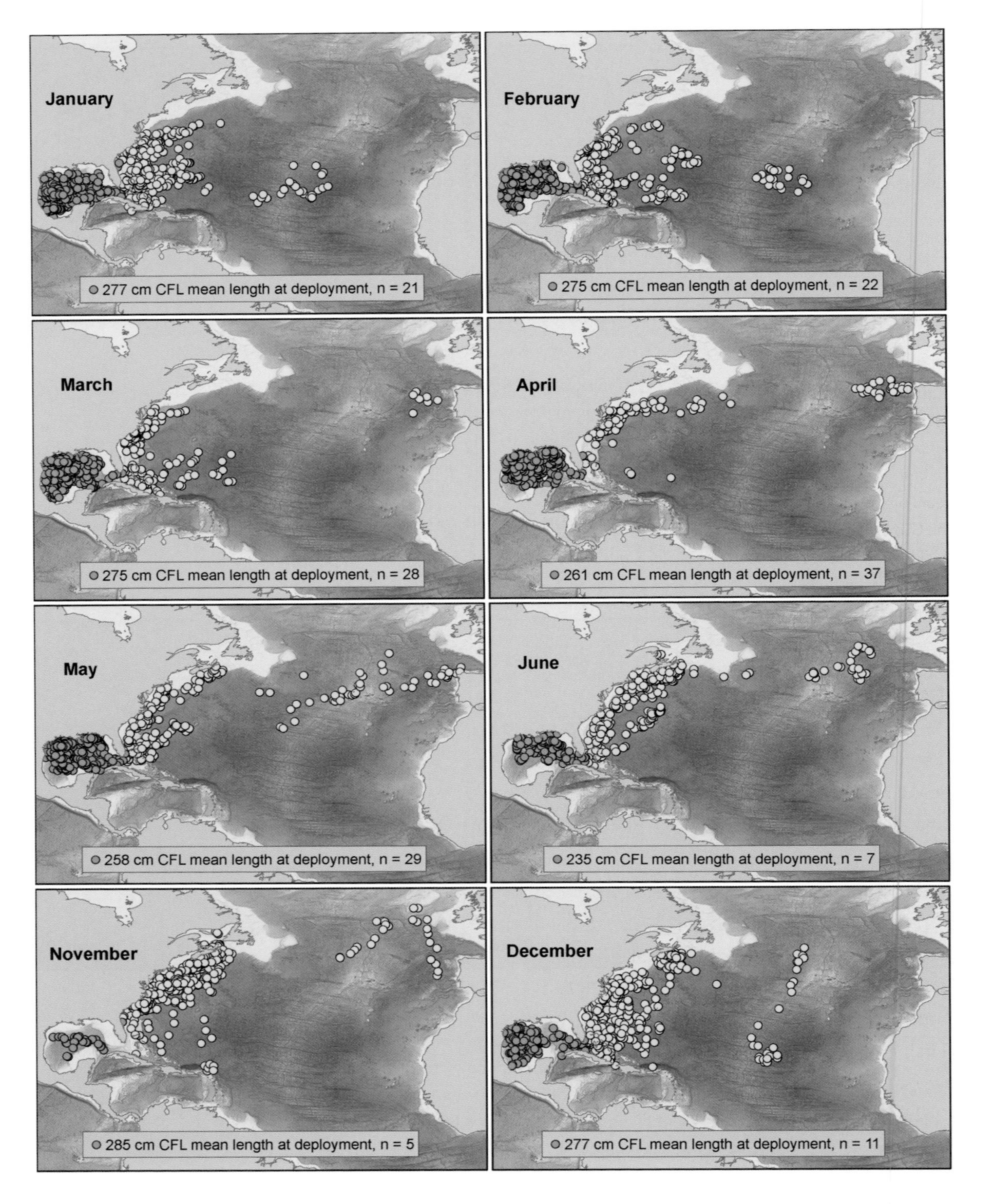


**Figure 4.7.** Pooled monthly geolocations from a sample of electronic tags that showed visitation to the Gulf of Mexico (GOM) spawning site. Orange circles indicate position estimations in the GOM. Yellow circles indicate geolocations from these fish that visited the GOM while in the North Atlantic. All lengths are in curved fork length (CFL).

MacKenzie, B.R., M. Payne, J. Boje, J.L. Højer, and H. Siegstad. 2014. A cascade of warming impacts brings bluefin tuna to Greenland waters. *Global Change Biology* 20 (8): 2484–91. doi:10.1111/gcb.12597.

Mather, F.J., J.M. Mason Jr., and A.C. Jones. 1995. Historical document: Life history and fisheries of Atlantic bluefin tuna. NOAA Technical Memorandum NMFS-SEFSC-370.

Muhling, B.A., S.-K. Lee, J.T. Lamkin, and Y. Liu. 2011. Predicting the effects of climate change on bluefin tuna (*Thunnus thynnus*) spawning habitat in the Gulf of Mexico. *ICES Journal of Marine Science* 68 (6): 1051–62. doi:10.1093/icesjms/fsr008.

Neilson, J.D., and S.E. Campana. 2008 A validated description of age and growth of western Atlantic bluefin tuna (*Thunnus thynnus*). *Canadian Journal of Fisheries and Aquatic Sciences* 65:1523–27.

Puncher, G.N., A. Cariani, G.E. Maes, J. Van Houdt, K. Herten, R. Cannas, N. Rodriguez-Ezpeleta, et al. 2018. Spatial dynamics and mixing in the Atlantic Ocean and Mediterranean Sea revealed using next-generation sequencing. *Molecular Ecology Resources* 18(3):620–38. doi:10.1111/1755-0998.12764.

Restrepo, V.R., G.A. Diaz, J.F. Walter, J.D. Neilson, S.E. Campana, D. Secor, and R.L. Wingate. 2010. Updated estimate of the growth curve of Western Atlantic bluefin Tuna. *Aquatic Living Resources* 23:335–42.

Richardson, D.E, K.E. Marancik, J.R. Guyon, M.E. Lutcavage, B. Galuardi, C.H. Lam, H.J. Walsh, S. Wildes, D.A. Yates, and J.A. Hare. 2016. Discovery of a spawning ground reveals diverse migration strategies in Atlantic bluefin tuna (*Thunnus thynnus*). *Proceedings of the National Academy of Sciences* 113 (12): 3299–3304.

Rooker, J.R., D.H. Secor, G. DeMetrio, R. Schloesser, B.A. Block, and J.D. Neilson. 2008. Natal homing and connectivity in Atlantic bluefin tuna populations. *Science* 322:742–44.

Rooker, J.R, H. Arrizabalaga, I. Fraile, D.H. Secoe, D.L. Dettman, N. Abid, P. Addis, et al. 2014. Crossing the line: Migratory and homing behaviors of Atlantic bluefin tuna. *Marine Ecology Progress Series* 504:265–76.

Royer, F., J.-M. Fromentin, and P. Gaspar. 2005. A state-space model to derive bluefin tuna movement and habitat from archival tags. *Oikos* 109 (3): 473–84.

Schlosser, R.W., J.D. Neilson, D.H. Secor, and J.R. Rooker. 2010. Natal origin of Atlantic bluefin tuna (*Thunnus thynnus*) from Canadian waters based on otolith $\delta^{13}C$ and $\delta^{18}O$. *Canadian Journal of Fisheries and Aquatic Sciences* 67:563–69.

Siskey, M.R., M.J. Wilberg, R.J. Allman, B.K. Barnett, and D.H. Secor. 2016. Forty years of fishing: Changes in age structure and stock mixing in the northwestern Atlantic bluefin tuna (*Thunnus thynnus*) associated with size-selective and long-term exploitation. *ICES Journal of Marine Science* 73 (10): 2518–28. doi:10.1093/fsw115.

Stokesbury, M.J.W., R. Cosgrove, S.L.H. Teo, D. Browne, R.K. O'Dor, and B.A. Block. 2007. Movement of Atlantic bluefin tuna from the eastern Atlantic Ocean to the western Atlantic Ocean as determined with pop-up satellite archival tags. *Hydrobiologia* 582:91–97.

Stokesbury, M.J.W., J.D. Neilson, E. Susko, and S.J. Cooke. 2011. Estimating mortality of Atlantic bluefin tuna (*Thunnus thynnus*) in an experimental recreational catch-and-release fishery. *Biological Conservation* 144:2684–91.

Taylor, N.G., M.K. McAllister, G.L. Lawson, T. Carruthers, and B.A. Block. 2011. Atlantic bluefin tuna: A novel multistock spatial model for assessing population biomass. *PLOS ONE* 6 (12): e27693. doi:10.1371/journal.pone.0027693.

Teo, S.L.H., and B.A. Block. 2010. Comparative influence of ocean conditions on yellowfin and Atlantic bluefin tuna catch from longlines in the Gulf of Mexico. *PLOS ONE* 5:e10756. doi:10.1371/journal.pone.0010756.

Teo, S.L.H., A. Boustany, H. Dewar, M.J.W. Stokesbury, K.C. Weng, S. Beemer, A.C. Seitz, C.J. Farwell, E.D. Prince, and B. A. Block. 2007a. Annual migrations, diving behavior, and thermal biology of Atlantic bluefin tuna, *Thunnus thynnus*, on their Gulf of Mexico breeding grounds. *Marine Biology* 151:1–18.

Teo, S.L.H., A. Boustany, and B.A. Block. 2007b. Oceanographic preferences of Atlantic bluefin tuna, *Thunnus thynnus*, on their Gulf of Mexico breeding grounds. *Marine Biology* 152:1105–19.

Walli A., S.L.H Teo, A. Boustany, C.J. Farwell, T. Williams, H. Dewar, E. Prince, and B.A. Block. 2009. Seasonal movements, aggregations and diving behavior of Atlantic bluefin tuna (*Thunnus thynnus*) revealed with archival tags. *PLOS ONE* 4:1–18.

Wilson, S.G., and B.A. Block. 2009. Habitat use in Atlantic bluefin tuna *Thunnus thynnus* inferred from diving behavior. *Endangered Species Research* 10:355–67. doi:10.3354/esr00240.

Wilson, S., I. Jonsen, R. Schallert, J.E. Ganong, M.R. Castleton, A.D. Spares, A.M. Boustany, et al. 2015. Tracking the fidelity of Atlantic bluefin tuna released in Canadian waters to the Gulf of Mexico spawning grounds. *Canadian Journal of Fisheries and Aquatic Sciences* 72 (11): 1700–1717.doi:10.1139/cjfas-2015-0110.

CHAPTER 5

# Spatial Mixing Models for Atlantic Bluefin Tuna

Nathan Taylor

## Introduction

Among the bluefin tunas discussed at the Bluefin Futures Symposium, Atlantic bluefin tuna (ABT) pose some unique stock assessment challenges. All the bluefin tuna populations discussed at the meeting exhibit long-range migrations, but ABT are unique in that the fishery is comprised of mixed stock, with at least two (Boustany et al. 2008), and possibly three (Riccioni et al. 2010), populations mixing across the range. The mixed-stock spatial dynamics of ABT give rise to a specific stock-assessment modeling challenge, namely, how to simultaneously address both movement and mixed-stock population dynamics.

By 2008 there was clear evidence for the mixed-stock movement dynamics of ABT. Movements had been observed within the Mediterranean stock and between the Mediterranean and the Atlantic Ocean stocks since the 1920s (Sella 1929), and transatlantic and transequatorial migrations had been observed using tagging data since the 1950s and 1960s (Mather 1995). The development of new technology for tagging (Block et al. 1998) led to improved understanding and representation of the scale of ABT movement. Similarly, the development of otolith microchemistry techniques (Rooker et al. 2001) led to the ability to assign observed fish to their spawning area of origin and provided clear depictions of the relative stock composition of western versus eastern ABT in the Atlantic Ocean (Rooker et al. 2008). In addition to tagging and otolith microchemistry data, genetic techniques that differentiate between eastern and western stocks were also developed (Boustany et al. 2008).

There have been several models developed to capture aspects of the spatial dynamics apparent in the ABT data, using an assortment of approaches.

Punt and Butterworth (1995) developed a method to use tagging data in the existing virtual population analysis (VPA) framework to estimate migration rates of bluefin tuna across the Atlantic. Porch et al. (2001) developed the VPA-2BOX model that used tagging data in combination with virtual population analyses of ABT with alternative models of transatlantic migration. Kurota et al. (2009) developed another approach to sequentially analyze the conventional, pop-up satellite archival, and archival tag data to estimate area and age-based fishing mortality rates that account for movement rates by age. These models generally ignored genetic stock structure in assigning tagged fish to a stock of origin, because the stock of origin was for the most part unknown. Output from the VPA-2BOX is consistently applied to ABT stock assessments as part of the regular process, but the single-stock VPA remains the base-case assessment model for assessing ABT today. By 2008 the stock assessment challenge was how to use all these data to make some inferences about stock movement, population size, and productivity.

The Multistock Age-Structured Tag-integrated model (MAST) (Taylor et al. 2011) for ABT was built to address this challenge. MAST provided a model to reconstruct the stocks' biomass time series to determine stock status, to estimate the mean productivity of the stock, to quantify movement rates between major areas, and to predict the performance of alternative catch policies. The model was updated with catch and catch-per-unit-effort (CPUE) data through 2010 and was used by Kerr et al. (2016) to parameterize a set of simulation models that could be used as operating models. However, as a statistical model, the MAST approach encountered some challenges that remain unresolved today. This chapter reviews some of these challenges and suggests methods by which they may be overcome in future use of mixed-stock spatial models for ABT stock assessment and management.

## Review of Results

MAST is a statistical catch-at-age model. The model is described in more detail in Taylor et al. (2011); only a cursory description of the model formulation follows herein. The model runs at quarterly time steps within five areas: the Gulf of Mexico, assumed to be the western stock spawning area; the Gulf of St. Lawrence, assumed to be primarily fish of western stock; the western and eastern Atlantic Ocean, assumed to be of mixed stock; and the Mediterranean Sea, assumed to contain exclusively fish of eastern stock. Data inputs include abundance indices, age proportions, and otolith microchemistry, as well as conventional and electronic mark-recapture data (Taylor et al. 2011). The modeling approach involves first the initialization of the model based on a steady-state condition given the model's "leading" parameters—in this case, the maximum sustainable yield (MSY) and the fishing mortality consistent with achieving MSY, $F_{MSY}$ (Martell et al. 2008). Next, numbers and biomass at age were updated in each area for each time step according to parameters that included movement and natural and fishing mortality. The parameters describing these dynamics were estimated using nonlinear minimization techniques.

One key step, prior to the statistical fitting procedure required for parameter estimation, was to assign tagged fish to spawning stock on the basis of their recapture history. To do this, all the mark-recapture histories were reviewed, and if a tagged fish had ever been observed in either the Gulf of Mexico or the Canadian Gulf of St. Laurence (Rooker et al. 2008), then it was considered a fish of Gulf of Mexico origin. Tags were assembled into cohorts of fish that were released in the same area at the same time. The vast majority of the mark-recapture data used in MAST could not be attributed to stock or origin. For example, when this method was used for the conventional tag data used in Taylor et al. (2011), it yielded the following tag assignments: 125 cohorts determined to be Gulf of Mexico stock, 142 cohorts determined to be from the Mediterranean stock, and 1,465 conventional tag cohorts for which the stock of origin was unknown.

The spatial dynamics were parameterized in MAST in two forms. For each, the probability of fish moving from one area to another was defined in terms of a movement matrix $M$. Each movement matrix consisted of rows representing area of origin and columns representing the destination area. Each row element of $M$ therefore was the probability of fish moving from area $j$ (rows) to area $j'$ (columns), and each row represented a probability vector that summed to unity. One parameterization estimated a single propensity of fish to stay in a given area—that is, "gravity" (the diagonal elements of $M$)—which was assumed to capture the attractiveness of that area relative to the areas associated with the off-diagonal elements. The second "bulk transfer" parameterization estimated the full matrix of movement probabilities, so allowed for more complicated movement to be modeled. While the bulk transfer parameterization could capture more complicated spatial dynamics, using it came at the cost of estimating many more parameters, with 66 versus 132 movement parameters for estimating gravity

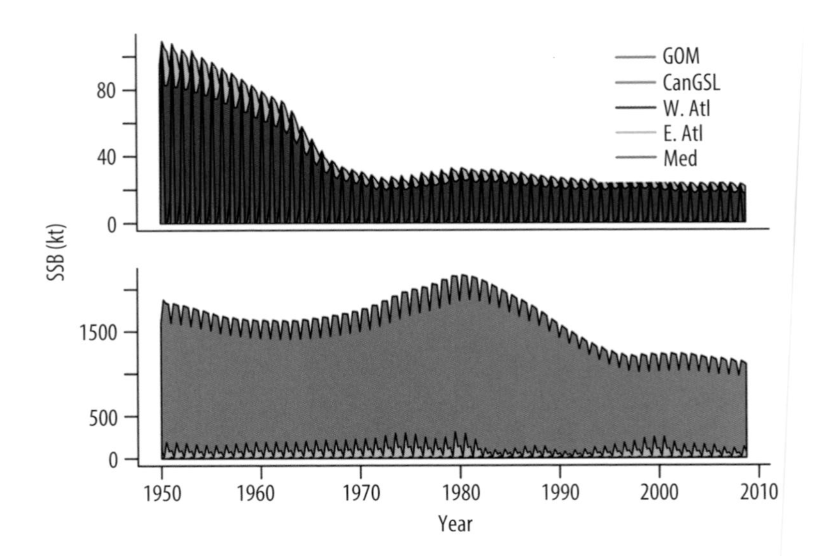


**Figure 5.1.** Example of multistock age-structured tag-integrated model spawning stock biomass (SSB) estimates by year. The model results are from the Taylor et al. 2011 base case. The upper panel represents the western SSB and the lower panel represents the eastern SSB. The Gulf of Mexico (GOM) is represented in red, the Gulf of St. Lawrence (CanGSL) in green, the western Atlantic (W. Atl) in dark blue, the eastern Atlantic (E. Atl) in light blue, and the Mediterranean Sea (Med) in pink.

and bulk transfer approaches, respectively (Taylor et al. 2011). Operationally, the model behaves like a standard, single-species, single-area, statistical catch-at-age model, but the biomass at each time step is partitioned among the areas. In the model's population dynamics, fish are born once a year, then at each quarterly time step, they grow and survive though natural mortality and fishing mortality and are then distributed among the model's areas according to the movement transition matrices. This gives rise to a sawtoothlike pattern in the biomass as new recruits enter the population (Figure 5.1) and as fish are distributed at each time step to other areas.

To validate the MAST predictions, we used a Bayesian statistical fitting procedure to maximize the probability of the parameters given the data. The objective function that defined the probability of the parameters, given the data, used the conventional tagging data, the archival tagging data, the pop-up satellite archival tagging data, the otolith microchemistry data, the CPUE indices, and the proportions of age in catch (derived from the catch-at-age matrix that was input into the VPA) (Taylor et al. 2011). The hope had been that using this "integrated approach," where there was simultaneous estimation of the full set of parameters that determined the state dynamics, would result in a more complete set of uncertainty represented in the model's predictions, and this uncertainty would help with decision-making. Unfortunately, the simplifying assumptions needed to get the model to converge resulted in a representation of uncertainty that was less than complete.

## Challenges

Fitting MAST to the data was not without significant challenges. Fitting the model to the data was computationally burdensome; at the time the model was built, a single function call to run the state dynamics and to calculate the objective function took approximately one minute, so estimating the model's parameter could take as long as one hour. During the development period, it was consistently difficult to establish model convergence, because the model included more complicated state dynamics than were ultimately published. For example, early incarnations of the model included growth-type groups with different asymptotic sizes to model changes in the distribution of size at age, as well as multiple fleets. However, the computational burden of fitting the model to the data made modeling these dynamics prohibitively slow; thus, the model was progressively simplified until it operated at more practical speeds and converged. Even with the simplifications, the version of the model published in Taylor et al. (2011) required three weeks for the Markov-Chain Monte Carlo simulation (MCMC) to converge. Essentially, the problem was that, as a statistical model, it was parameterized at the limits of parsimony so that only a limited number of parameters could be estimated and therefore only a limited range of state dynamics could be considered.

Tagged fish of unknown stock origin contributed to the model's slowness and to convergence difficulties. As described above, when MAST was developed, the majority of the tagging data of all types were not assigned to stock of origin. However, the model was estimating stock-specific movement parameters. For fish of unknown origin, the statistical likelihood was computed twice; that is, using movement probability matrices from western and eastern stocks with likelihood weights given by the ratio of vulnerable numbers of each stock to total vulnerable numbers in that area at that time (Taylor et al. 2011). This ratio was determined with considerable uncertainty (Figure 5.2). Thus, considerable computational burden was added by computing the statistical likelihood twice for these data, and there was a very large uncertainty propagated to the estimates of movement parameters, i.e., coefficients of variation in the order of one, because of the lack of knowledge regarding the stock of origin.

Several additional major simplifying assumptions were needed to reduce the number of estimated parameters and to get the model to converge. First, MAST forced movement during the following season so that the probability

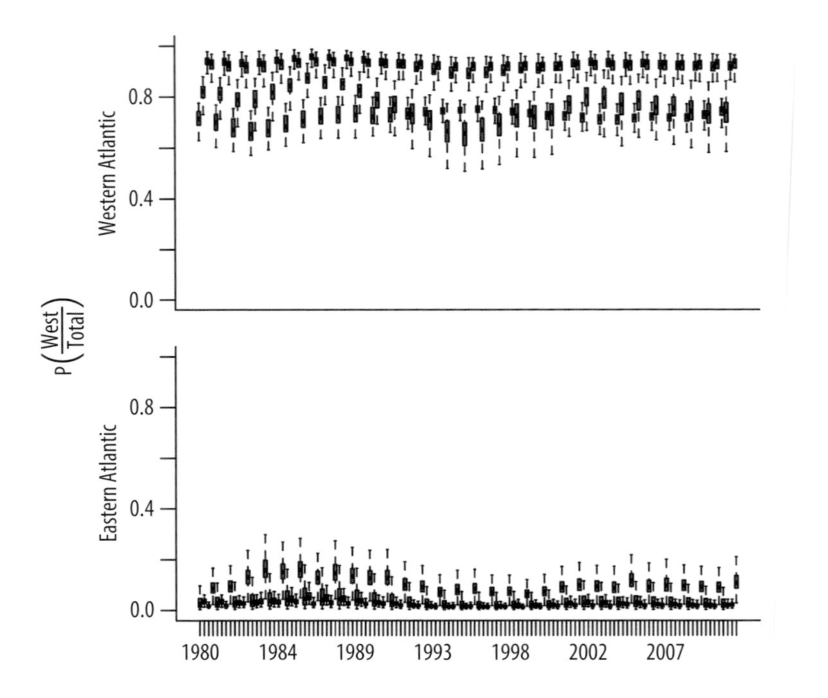


**Figure 5.2.** Example of multistock age-structured tag-integrated model posterior samples for the estimated ratio of western stock fish to the total numbers over time estimated using the Taylor et al. 2011 base case.

of going from any area to the stock-specific spawning area was given by the maturity ogive; this was a key structural assumption in that exploring alternative maturity ogives meant alternative hypotheses about how the stock was distributed during spawning seasons. Another simplification was to assume that there was a single, time-invariant reporting rate for the conventional mark-recapture data. While this approach followed other analyses done for bluefin tuna (Kurota et al. 2009), it is not likely to be a valid assumption. It is more likely that reporting rates vary with time, fleet, and area. However, without making a time-invariant reporting-rate assumption, the model would not converge. Similarly, a second simplifying assumption was to assume that selectivity was constant over time, which is also not a realistic assumption, given how targeting behavior in response to market demands has changed over the course of the fishery.

While the inability to estimate time-varying or area-specific reporting rates for conventional tagging data was an issue for model development, it points to a bigger problem with the use of these data at all. There is a broad suite of problems with reporting rates, including the potential for time-varying rates as rewards and the changing public perception about the state of the stock. When the reporting points potentially vary with time, fleet, and area, any observed recaptures of conventional tags may contain little about either the movement or the mortality parameters that they are intended to inform. Therefore, the conventional tagging data have limited utility.

## Discussion: What Is the Future of Mixed-Stock Spatial Models for Atlantic Bluefin Tuna?

From a stock-assessment modeling perspective, the MAST project illustrates a situation in which there are enough data to show that there are mixed-stock spatial dynamics, but not enough data to reliably estimate several key management parameters. For example, of the five sensitivity scenarios explored in Taylor et al. (2011), the predicted status of the western stock at fixed catch scenarios (1,750 and 12,900 metric tons for eastern and western stocks, respectively) ranged from 43% to 160% of the biomass that would produce MSY. Nevertheless, because the linkage of fisheries through stock mixing exists in practice, harvest policies applied in any given area affect biomass and fishing opportunities in other areas, so policy concerns demand that mixing be considered for ABT management; i.e., in spite of large uncertainties introduced by the mixed-stock spatial dynamics, decision-making must still occur. Accordingly, the key problem remains how best to capture and represent these population dynamics to inform the decision-making process?

One solution is that mixed-stock movement models like MAST may be better for forming the basis of operating models for management strategy evaluation (MSE) than for direct application to quota setting in a given year. There are several reasons for this, but they basically sum to an argument for the superiority of an MSE approach over the so-called traditional approach (TA) (Butterworth 2007 and elsewhere). In the case of ABT, the added complexities of a mixed-stock spatial model can make the deficiencies of the TA even more pronounced. When the International Commission for the Conservation of Atlantic Tunas (ICCAT) met in 2009 to consider the status of ABT populations with respect to the Biological Listing Criteria of the Convention on International Trade in Endangered Species of Wild Fauna and Flora (CITES) (ICCAT 2009), for the eastern stock alone there were 48 alternative rebuilding scenarios provided, assuming perfect implementation, but with a mixed-stock spatial model there could have been many more. By their very nature, mixed-stock spatial models have more data choices and a more diverse universe of parameterization options (imagine, for example, a set of time-varying movement times time-varying productivity scenarios), resulting in potentially endless haggling over the best data and model options. Moreover, there are practical limitations to running a model like MAST for the purpose of decision-making; the model version published by Taylor et al. (2011) took three weeks for the MCMC to converge; thus, even if solutions to the problems identified in the data and more compact parameterizations were developed, the data for a stock assessment at ICCAT would need to be compiled and models run in the course of two weeks. For practical reasons alone, there is no option but to consider complex movement models as operating models for ABT management rather than as stock-assessment models. But management procedures must be designed that are robust to the apparent mixed-stock spatial dynamics; part of that task is to show that any stock-assessment model selected as part of the management procedure has reasonable performance, whether it is explicitly spatial or not. Closed-loop simulations done in an MSE context would help to provide an analysis of this performance. To this end, the estimates of movement rates from MAST have already been used for operating-model development (Kerr et al. 2016).

But it is important to note that even in an MSE context, the problem of how to defensibly parameterize a set of alternative operating models for spatially structured, mixed-stock fisheries remains. While the output from MAST has been used to parameterize operating models for ABT, there are several remedies to the challenges encountered during MAST development that can be overcome to further advance operating-model development for the ABT MSE.

First, in mixed-stock situations it is essential that the tagged fish used for estimating movement parameters be genotyped to a stock of origin; for some of the existing tagged fish, these data exist, and for future samples it is essential. Once the tagged fish are genotyped, the electronic tagging data will provide a much clearer picture of the movement dynamics. I am not convinced that conventional tag data is among the remedies to improving spatial mixed-stock models for ABT; given that the conventional tagging data suffered from both time- and possibly fleet-varying reporting rates, it is difficult to disentangle the effects of time-varying reporting rates from fishing mortality and from movement based on this class of data. Conventional tagging data might provide a more misleading rather than an informative picture of ABT population dynamics.

In addition, there are better alternatives to parameterizing the movement than the estimated movement matrices approach that was employed for MAST. The animal movement dynamics underlying an operating model could be parameterized using the telemetry data alone (Sibert et al. 1999) or using an increasingly rich set of state-space methods (Patterson et al. 2008, McClintock et al. 2012). The movement dynamics of ABT could be characterized using such methods by analyzing the electronic data to define the movement dynamics outside the stock-assessment model used to estimate the biomass.

There are several advantages to this approach. First, in the simultaneous estimation procedure used for MAST, the movement parameters are estimated by minimizing an objective function that includes fitting all the other data sources. Movement parameters are adjusted in the fitting procedure to maximize the fit to data sources that may not necessarily represent spatial population dynamics; keeping the estimates of movement separate from the other data sources would solve this problem. Second, separating movement into five areas according to the MAST formulation is a very coarse approximation of the population and fishing dynamics: the areas are large enough that within each, fishing mortality and the fish themselves are not homogeneously distributed. Using the electronic tagging data alone, a more diverse set of movement models (Ver Hoef et al. 2018) could be considered. For example, conditional auto-regressive models could be used to define a three-dimensional blob whose shape changes in response to seasonal movement and potentially time-varying changes in water column temperature and nutrient profiles. The fine-scale movement dynamics output from these analyses could underlie a simulation model that partitions the biomass into coarser time steps (i.e., quarters, years, or statistical areas) that a stock-assessment model would use in the closed-loop simulation component of MSE and in practice for regular quota setting. In this way, the more complicated and realistic spatial dynamics could be considered, which would in turn allow a richer set of hypotheses to be considered (e.g., the potential distributional changes caused by climate change). The flexibility to consider such hypotheses about movement for the ABT MSE is a key consideration for ABT futures.

## REFERENCES


Block, B.A., H. Dewar, C. Farwell, and E.D. Prince. 1998. A new satellite technology for tracking the movements of Atlantic bluefin tuna. *Proceedings of the National Academy of Sciences USA* 95:9384–89.

Boustany, A.M., C.A. Reeb, and B.A. Block. 2008. Mitochondrial DNA and electronic tracking reveal population structure of Atlantic bluefin tuna (*Thunnus thynnus*). *Marine Biology* 156:13–24.

Butterworth, D.S. 2007. Why a management procedure approach? Some positives and negatives. *ICES Journal of Marine Science* 64 (4): 613–17. doi:10.1093/icesjms/fsm003.

ICCAT (International Commission for the Conservation of Atlantic Tunas). 2009. Extension of the 2009 SCRS meeting to consider the status of Atlantic bluefin tuna populations with respect to CITES Biological Listing Criteria. International Commission for the Conservation of Atlantic Tunas, October 21–23, Madrid, Spain. http://iccat.int/Documents/Meetings/Docs/PA2-604%20ENG.pdf.

Kerr, L.A., S.X. Cadrin, D.H. Secor, and N.G. Taylor. 2016. Modeling the implications of stock mixing and life history uncertainty of Atlantic bluefin tuna. *Canadian Journal of Fisheries and Aquatic Sciences,* doi.org/10.1139/cjfas-2016-0067.

Kurota, H., M.K. McAllister, G.L. Lawson, J.I. Nogueira, S.L.H. Teo, and B.A. Block. 2009. A sequential Bayesian methodology to estimate movement and exploitation rates using electronic and conventional tag data: Application to Atlantic bluefin tuna (*Thunnus thynnus*). *Canadian Journal of Fisheries and Aquatic Sciences* 66:321–42.

Martell, S.J.D., W.E. Pine, and C.J. Walters. 2008. Parameterizing age-structured models from a fisheries management perspective. *Canadian Journal of Fisheries and Aquatic Sciences* 65:1586–1600.

Mather, F.J. 1995. Historical document: Life history and fisheries of Atlantic bluefin tuna. NOAA Technical Memorandum NMFS-SEFSC 370:1–165. doi.org/10.5962/bhl.title.4783.

McClintock, B.T., R. King, L. Thomas, J. Matthiopoulos, B.J. McConnell, and J.M. Morales. 2012. A general discrete-time modeling framework for animal movement using multistate random walks. *Ecological Monographs* 82:335–49.

Patterson, T.A., L. Thomas, C. Wilcox, O. Ovaskainen, and J. Matthiopoulos. 2008. State–space models of individual animal movement. *Trends in Ecology and Evolution* 23: 87–94.

Porch, C.E., S.C. Turner, and J.E. Powers. 2001. Virtual population analyses of Atlantic bluefin tuna with alternative models of transatlantic migration: 1970–1997. *ICCAT Collective Volume of Scientific Papers* 52:1022–45.

Punt, A.E., and D.S. Butterworth. 1995. Use of tagging data within a VPA formalism to estimate migration rates of bluefin tuna across the North Atlantic. *ICCAT Collective Volume of Scientific Papers* 44:166–82.

Riccioni, G., M. Landi, G. Ferrara, I. Milano, A. Cariani, L. Zane, M. Sella, G. Barbujani, and F. Tinti. 2010. Spatio-temporal population structuring and genetic diversity retention in depleted Atlantic bluefin tuna of the Mediterranean Sea. *Proceedings of the National Academy of Sciences USA* 107:2102–7.

Rooker, J.R., D.H. Secor, V.S. Zdanowicz, and T. Itoh. 2001. Discrimination of northern bluefin tuna from nursery areas in the Pacific Ocean using otolith chemistry. *Marine Ecology Progress Series* 218:275–82.

Rooker, J.R., D.H. Secor, G. De Metrio, R. Schloesser, B.A. Block, and J.D. Neilson. 2008. Natal homing and connectivity in Atlantic bluefin tuna populations. *Science* 322: 742–44.

Sella, M. 1929. Migrations and habitat of the tuna (*Thunnus thynnus* L.), studied by the method of the hooks, with observations on growth, on the operation of fisheries, etc. Translated by W.G. Van Campen, 1952, in *Special Scientific Report of the U. S. Fish and Wildlife Service* 76:20.

Sibert, J.R., J. Hampton, D.A. Fournier, and P.J. Bills. 1999. An advection-diffusion-reaction model for the estimation of fish movement parameters from tagging data, with application to skipjack tuna (*Katsuwonus pelamis*). *Canadian Journal of Fisheries and Aquatic Sciences* 56 (6): 925–38. doi.org/10.1139/f99-017.

Taylor, N.G., M.K. McAllister, G.L. Lawson, T. Carruthers, and B.A. Block. 2011. Atlantic bluefin tuna: A novel multistock spatial model for assessing population biomass. *PLOS ONE* 6:e27693.

Ver Hoef, J.M., E.E. Peterson, M.B. Hooten, E.M. Hanks, and M. Fortin. 2018. Spatial autoregressive models for statistical inference from ecological data. *Ecological Monographs* 88:36–59. doi: 10.1002/ecm.1283.

# PACIFIC

CHAPTER 6

# Life History of Pacific Bluefin Tuna, *Thunnus orientalis*

Tamaki Shimose

## Introduction

Life history studies provide basic biological information for a species, which can be synthesized over time to better understand overall population ecology. Basic life history parameters (e.g., growth rates, spawning age, fecundity) are also required for adequate stock assessment and fisheries management. Major components of life history studies include habitat, diet, growth, and reproduction, as well as seasonal cycle(s) and changes over ontogeny. These components are closely interconnected and interdependent. For example, seasonal and ontogenetic habitat changes can relate to both feeding success and reproduction; feeding success will influence reproductive output and growth rate; and growth rate will in turn influence maturation and spawning. Therefore, all of these parameters must be investigated and described to adequately understand the basic biology and proper management approaches for a given species.

On the one hand, Pacific bluefin tuna (*Thunnus orientalis*) is a large pelagic fish, and its size and habitat make it difficult to sample efficiently by research vessels. On the other hand, there are various kinds of commercial fisheries that target and land Pacific bluefin, from juveniles to adults. Since the first age determination was attempted in the 1930s (Aikawa and Kato 1938), commercial fisheries have been instrumental to Pacific bluefin life history studies by providing a platform for tagging live bluefin and by providing samples of harvested bluefin for age/growth, genetic, isotopic, and reproductive analyses. This chapter presents a general picture of our current understanding of the life history of Pacific bluefin, focusing specifically on the above components that have been based on fishery-dependent data and samples.

## Age and Growth

Analyses of fish life stages describe the morphological and physiological development of a given species with increasing body size and age. Age determination across size classes is necessary to calculate age-length relationships, which are fundamental to describing overall life history. Age determination and growth estimation studies of Pacific bluefin have been conducted for decades, using various methods (Shimose and Farley 2015). Age determination requires a sufficient sampling effort, which often is available only from commercial landings. Supportive information for growth studies, such as tag-recapture growth rate(s) and modal progression analyses, also has required assistance from commercial fisheries. Currently, age-length relationships at different life stages are estimated using multiple methods.

Otolith micro-increment (daily growth increment) has been used in published studies to estimate daily age of young Pacific bluefin up to 420 days (Itoh 2009) or ~5.5 years (Foreman 1996). Pacific bluefin grow rapidly in their first 180 days, attaining ~50 cm fork length (FL). Growth rate then decreases, with Pacific bluefin reaching 57–60 cm at 1 year (Itoh 2009). These rapid and slow growth periods correspond to summer and winter seasons in the western North Pacific Ocean (Yukinawa and Yabuta 1967, Itoh 2009). In previous studies using otolith micro-increments, growth of Pacific bluefin could be estimated up to 1,607–1,992 days (4.4–5.5 years) old and 131–158 cm (Foreman 1996). Estimates from these studies are supported by tag-recapture data (Bayliff et al. 1991) and modal progression analysis (Yukinawa and Yabuta 1967).

Daily growth increments become difficult to identify in otoliths of older fish; thus, annual growth increments in sectioned otoliths have been used for age determination in larger fish > 1 year old (Shimose et al. 2009). The commonly used growth model, the von Bertalanffy growth function (Shimose et al. 2009), suggests that Pacific bluefin grow to ~80 cm (~10 kg) in 2 years, 150 cm (50–70 kg) in 5 years, and 210 cm (150–200 kg) in 10 years, on average (Figure 6.1). Growth rate then decreases, and asymptotic length is estimated to be ~250 cm (250–350 kg). Ages of some larger individuals are estimated to be more than 20 years (Ishihara et al. 2017, Shiao et al. 2017). Sexual dimorphism is also recognized, with males reaching slightly larger size than females (Shimose et al. 2009). Some records of maximum size/age suggest ~290 cm FL (Hsu et al. 2000), 555 kg in body

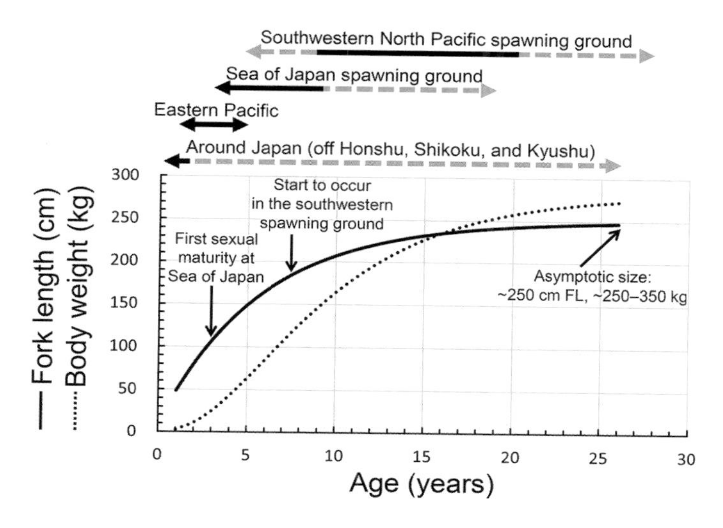


**Figure 6.1.** Regional occurrence at age and growth curve of Pacific bluefin tuna (*Thunnus orientalis*) in fork length (FL, cm) and gilled/gutted body weight (kg).

weight (Foreman and Ishizuka 1990), and 28 years of age (Shiao et al. 2017). Growth of Pacific bluefin is faster than that of southern bluefin *T. maccoyii* (Gunn et al. 2008) but slower than that of Atlantic bluefin *T. thynnus* (Restrepo et al. 2010). Maximum longevity is lower than in both Atlantic (>30 years; Neilson and Campana 2008) and southern (>40 years; Gunn et al. 2008) bluefin. The age-length relationships and associated age determination techniques in the studies above have enabled age-based description of other life history characters (e.g., habitat, diet) in Pacific bluefin.

## Ontogenetic Habitat and Diet Shifts

The distribution and migration patterns of Pacific bluefin have been summarized in previous reviews (e.g., Collette and Nauen 1983, Bayliff 1994). Pacific bluefin occur across an extensive area of the North Pacific and are mainly caught by fisheries in the western Pacific (waters around Japan to the Philippines) and in the eastern Pacific (waters off the United States and Mexico). Distribution has been estimated by collecting fish with associated information (e.g., date, time, location, and depth of capture) and sometimes abundance estimates for each size class. Habitat use over time can be estimated by spatiotemporal differences in fisheries' catches for each life stage. Seasonal habitat shifts reflect underlying migration patterns, though fishery-dependent data do not show high-resolution migration patterns of individuals (as has been accomplished with electronic tagging studies). Furthermore, fisheries-based sampling efforts can be biased spatially (horizontal and vertical) and temporally (diurnal and seasonal), necessitating caution when using commercial catch data to reconstruct habitat. In spite of these limitations, fisheries can provide some data with higher efficiency and lower cost than the data from dedicated research vessels. Sample specimens have been made available from fisheries-captured individuals, potentiating approaches such as traditional stomach content analysis and gonadal maturation assessments. Ontogenetic habitat shifts and associated stomach content information, based on fishery-dependent data/samples and supplemented by tagging and research vessel–associated studies, are summarized here in Figure 6.2 and Table 6.1.

Though life starts at the fertilized egg, there is no information on egg abundance/distribution for Pacific bluefin in the wild. Larvae (< 10 mm standard length [SL], < 19 days) and juveniles (> 10 mm SL, > 19 days) (Sabate et al. 2010) of Pacific bluefin have been collected by research vessels since the 1950s (e.g., Yabe et al. 1966, Nishikawa et al. 1985). Most of these samples are less than 12 mm SL, corresponding to < 20 days of age (Sabate et al. 2010), and collection of these newly hatched juveniles indirectly indicates spawning ground and season. Currently, two main spawning grounds are known from larval collections: the southwestern North Pacific (around the Ryukyu Islands to Philippine waters) and the southern Sea of Japan (Tanaka and Suzuki 2015). Larvae have been collected from early May to early July at the southwestern spawning grounds (Yabe et al. 1966) and from early to mid-August at the Sea of Japan spawning ground (Okiyama 1974, Kitagawa et al. 1995). Stomach content information of larvae, available only from the southwestern North Pacific, shows that larvae feed on small crustacean zooplankton (e.g., copepod; Uotani et al. 1990). Larvae and juveniles (size unknown) have been collected off the Pacific coast of southern Japan (<35°N; Yabe et al. 1966, Nishikawa et al. 1985), suggesting that the described southwestern spawning ground possibly extends northeastward when oceanographic conditions are preferable.

Small juveniles ranging from 108 to 280 mm FL have been collected by midwater trawls on research vessels in the Sea of Japan (Tanaka et al. 2007), and individuals < 100 mm have been rare (T. Tanabe, Seikai National Fisheries Research Institute, unpublished data). Juveniles of ~17 cm FL start to occur near coastal areas of southern Japan in mid-July and are caught by spawners near the surface in the Sea of Japan, while longline can capture bluefin in the southwestern North Pacific in any spawning state. Spawning activity is known to increase throughout the spawning season and also around the new moon period in the southwestern spawning ground (Shimose et al. 2018). Unfortunately, spawning duration has not been estimated for Pacific bluefin in the wild.

Sea surface temperatures near collection of spawning females are 26°C–29°C off eastern Taiwan (Chen et al. 2006), >25°C around the Yaeyama Islands (Ashida et al. 2015), and 19.3°C–27.7°C (mean 23.2°C in 2011, 23.4°C in 2012), with a majority of measurements >22°C in the Sea of Japan (Okochi et al. 2016). Lower latitudes in the southwestern (18°N–25°N) versus southern Sea of Japan (35°N–40°N) spawning grounds cause warmer water temperatures in the former area earlier in the year. Spawning season is indeed earlier, and observed sea surface temperature in spawning season is higher, in the southwestern ground. In net cages used for aquaculture, Pacific bluefin spawn at water temperatures of 23.5°C–29.6°C at Amami Station (28°08′N; Masuma et al. 2006) and 21.6°C–29.2°C at Oshima Station (33°28′N; Miyashita et al. 2000).

As mentioned above, main size and age compositions are different between the two spawning groups—i.e., 180–250 cm (8–21 years old) in the southwestern (Shimose et al. 2016) and 110–200 cm FL (3–9 years old) in the Sea of Japan (Okochi et al. 2016) spawning grounds. The size/age compositions and spawning conditions (latitude, spawning season, spawning output) at the two North Pacific spawning grounds are similar to those of Atlantic bluefin (Baglin 1982, Corriero et al. 2005, Heinisch et al. 2008). Smaller/younger adults spawn later in the season in higher-latitude areas (Pacific bluefin in the Sea of Japan, Okochi et al. 2016; Atlantic bluefin in Mediterranean spawning ground, Corriero et al. 2005 and Heinisch et al. 2008). However, larger/older adults spawn earlier in the season in lower-latitude areas (Pacific bluefin in the southwestern North Pacific, Ashida et al. 2015 and Shimose et al. 2018; Atlantic bluefin in the Gulf of Mexico, Baglin 1982). The differences in age/length composition between two differential latitudinal spawning grounds merit further study. Although no evidence currently exists, there is a possibility that young adults first spawn in the Sea of Japan, then change to southwestern spawning grounds at ~8–9 years old.

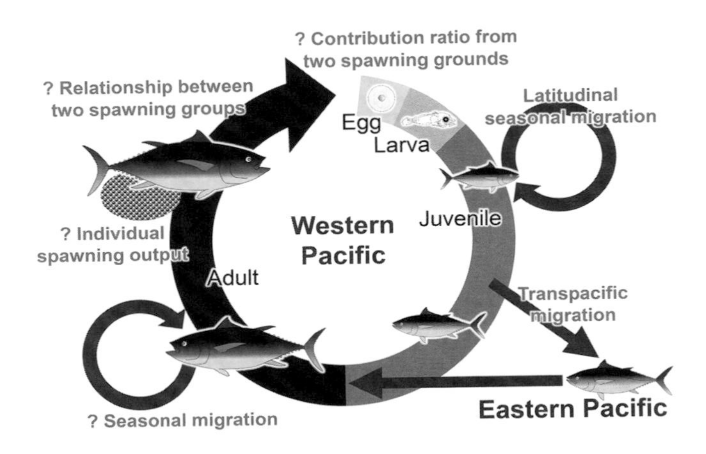


**Figure 6.3.** Schematic drawing of the life cycle of Pacific bluefin tuna (*Thunnus orientalis*), with comments on known and unknown aspects of its life history.

## Conclusion

Basic life history information of Pacific bluefin has been studied since the 1930s, owing largely to the commercial importance of the species. Commercial fisheries capture a large number of fish across seasons and life stages, helping to accumulate valuable data for science. Recently developed electronic tagging technology revealed seasonal migration routes of juveniles, supporting historical understanding of migration patterns. Age-determination techniques, using otolith microstructure and annual increments, have clarified growth rate and lifespan estimates of Pacific bluefin. But some issues remain poorly understood (Figure 6.3). Migration routes of adult Pacific bluefin during the nonspawning season are not well described. Quantitative information on reproductive output for individual fish, as well as seasonal spawning duration, has not been studied. The connectivity between spawning groups at two spawning grounds is unclear, and the contribution ratio of these two spawning groups to the Pacific bluefin stock is not certain. These issues should be clarified in future studies to understand the entire life history of the species and to properly manage this valuable fishery.

## Acknowledgments

The author thanks D. Madigan and two anonymous reviewers for reviewing, grammatical editing, and improving this chapter. The author's

CHAPTER 7

# Migrations of Pacific Bluefin Tuna Tagged in the Western Pacific Ocean

Takashi Kitagawa, Ko Fujioka, and Nobuaki Suzuki

## Introduction

Pacific bluefin tuna (PBT; *Thunnus orientalis*) constitute one of the most important fishery resources in the Pacific Ocean because of the economic value of their high-quality fish meat. In addition, with increase in market globalization, the value of this species has been increasing rapidly during the past 10 years. In 2013, at the year's first auction at the Tsukiji market in Tokyo, a 222-kg PBT sold for a record price of $1.76 million (155.4 million yen), which was three times as high as the previous record price. This kind of enthusiastic consumption leads to overfishing and necessitates strict stock management for the species (Kitagawa 2013).

An understanding of PBT migratory ecology is important for stock management. Spawning grounds of the single PBT population are located between the Philippines and the Nansei Islands of Japan, where spawning occurs from April to June, and in the Sea of Japan, where spawning occurs in August (Figure 7.1; e.g., Okiyama 1974, Bayliff 1994, Chen et al. 2006, Tanaka et al. 2007). After migrating to Japanese coastal areas (e.g., by the Kuroshio Current and the Tsushima Warm Current) at 60–90 days after hatching (Kitagawa et al. 2010), the juveniles appear in coastal areas off Kochi and Nagasaki prefectures during the summer of their first year (Figure 7.1). While some juvenile PBT remain in coastal waters around Japan (Bayliff 1994), others migrate during the latter half of their first or second year from the Kuroshio-Oyashio transition region to the eastern Pacific, a distance of approximately 8,000 km in what is referred to as the transpacific migration (TPM) (Orange and Fink 1963, Clemens and Flittner 1969, Bayliff et al. 1991, Bayliff 1994). (Before the confirmation of the TPM, PBT western and eastern stocks were considered different species: *Thunnus*

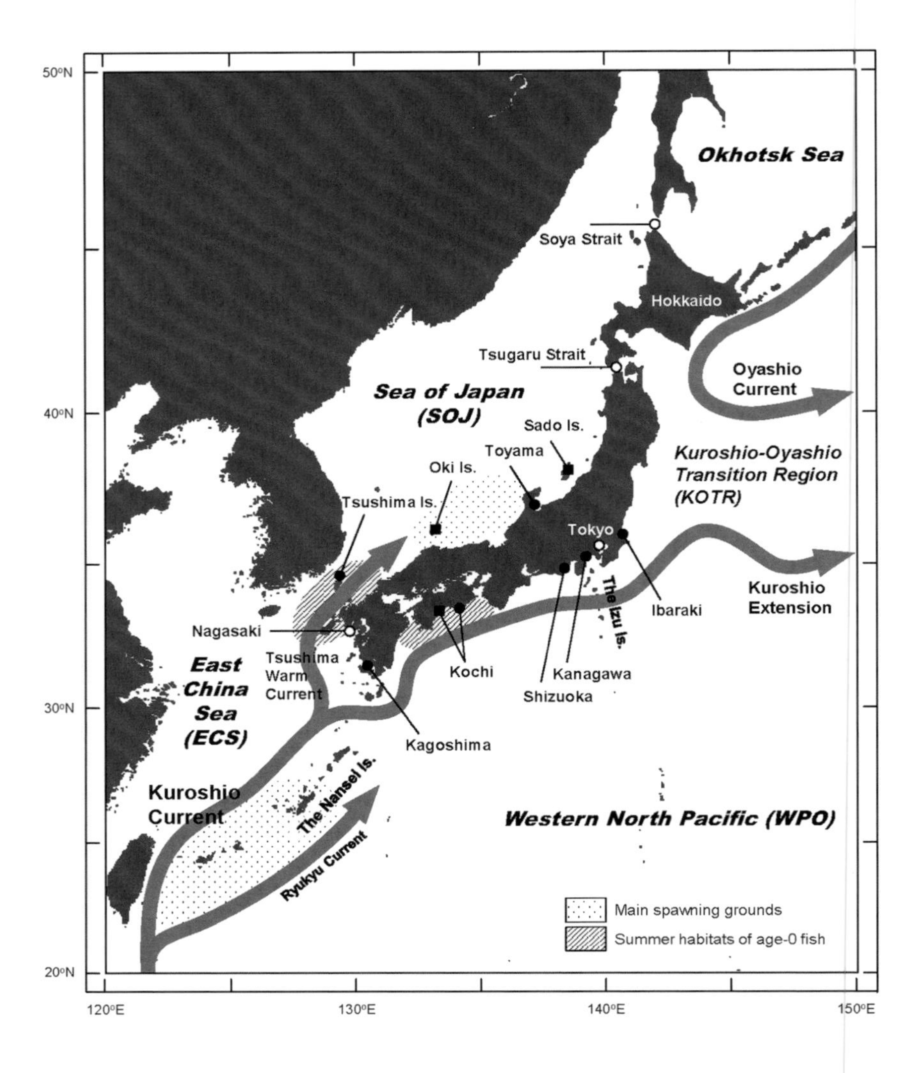


**Figure 7.1.** Two main spawning grounds of Pacific bluefin tuna in the area between the Philippines and the Ryukyu (Nansei) Islands and distribution areas of age-0 PBT in summer in the northwestern Pacific Ocean and in the Sea of Japan. Major currents around Japan are also shown. Locations of tagging sites (solid squares are ongoing sites), islands, and straits of Japan are marked for reference.

*saliens* was adopted for PBT in the eastern Pacific [Bell 1963], but those PBT are now grouped with *T. orientalis*.) After several years in the eastern Pacific, bluefin return to the spawning area (Bayliff 1994).

Combining fisheries data and the results of conventional tagging experiments, Bayliff (1980) drew up a migration diagram. This diagram simply and effectively represents the dynamism of PBT's migration. Later, however, to bridge the gap between information obtained for various fishery grounds or seasons, detailed data on migration, independent from the fisheries, were required. As PBT breeding sites are located in the western Pacific close to Japan, international stock management is urgently required to collect data on PBT migration in the western Pacific Ocean.

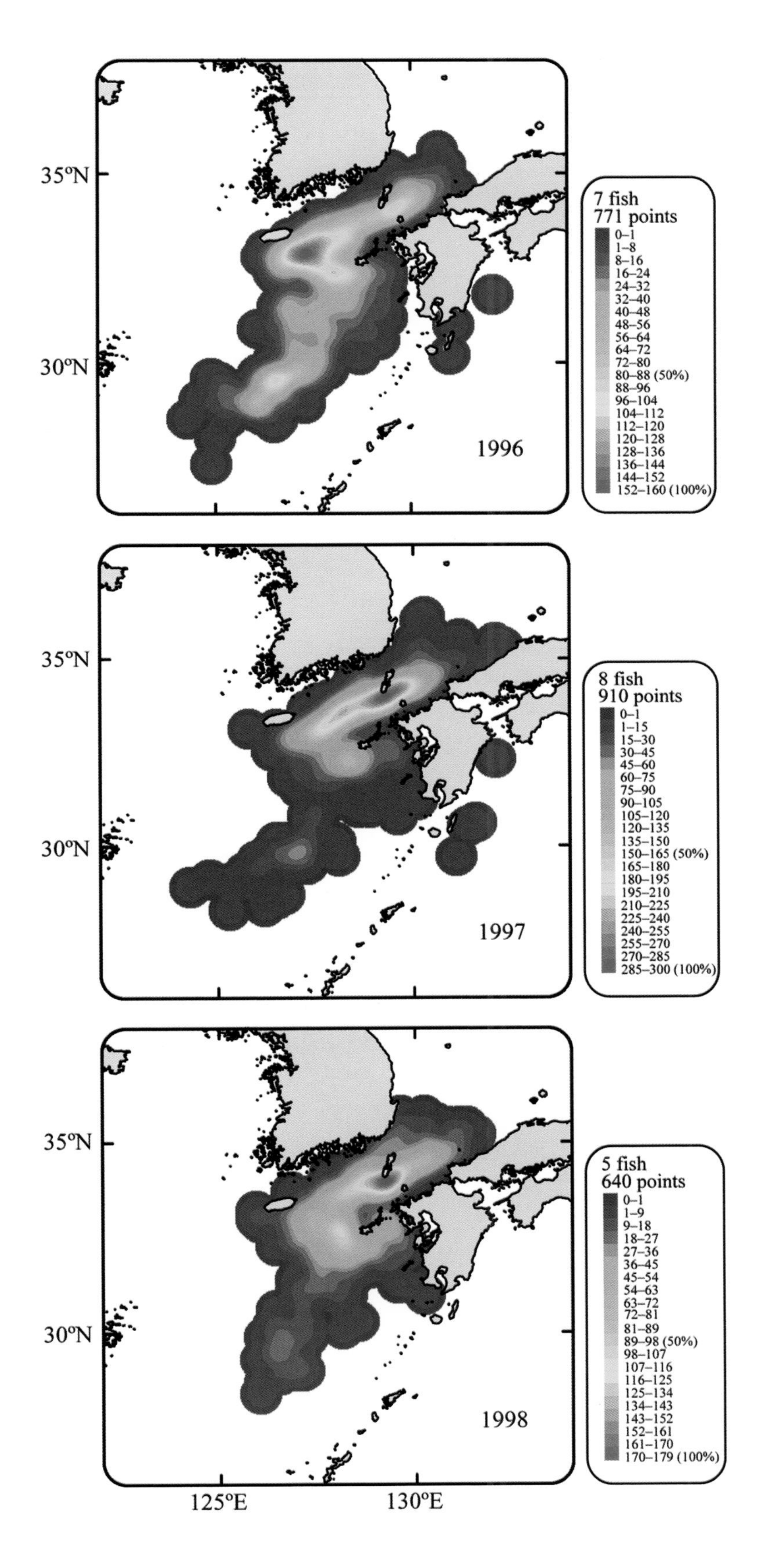


**Figure 7.3.** Yearly distribution of tagged Pacific bluefin tuna in the East China Sea from January to June in 1996 (La Niña year, *upper panel*), 1997 (normal year, *middle panel*) and 1998 (El Niño year, *lower panel*), shown as Kernel density. *Kitagawa et al. 2006b.*

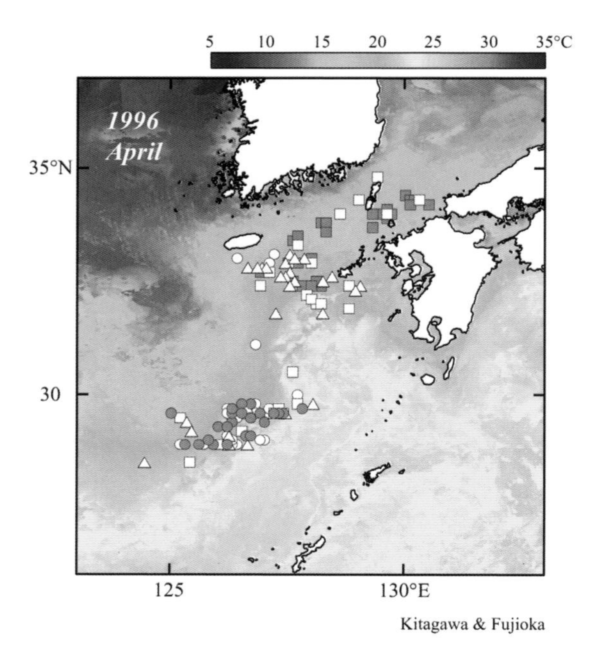


**Figure 7.4.** Monthly horizontal distributions of sea surface temperature and geolocations of tagged bluefin in the East China Sea (ECS) in April 1996 (from Kitagawa et al. 2006b), when a water mass from the Kuroshio intruded into the ECS, dividing the fish distribution, suggesting that this warm water prevented the bluefin from migrating north. Symbols code for geolocations of individual fish.

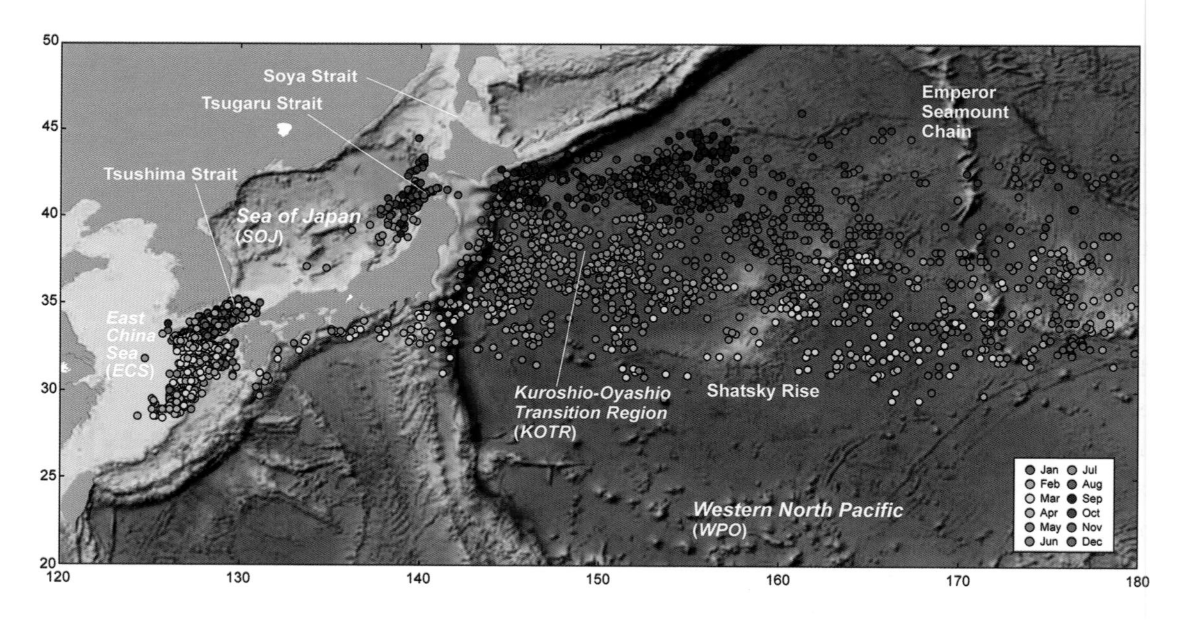


**Figure 7.5.** Daily geolocations of representative fish in the western Pacific Ocean shown in Inagake et al. 2001 and Kitagawa et al. 2004b (Pacific bluefin tuna tag numbers 177, 194, 209, and 241), with geographical information.

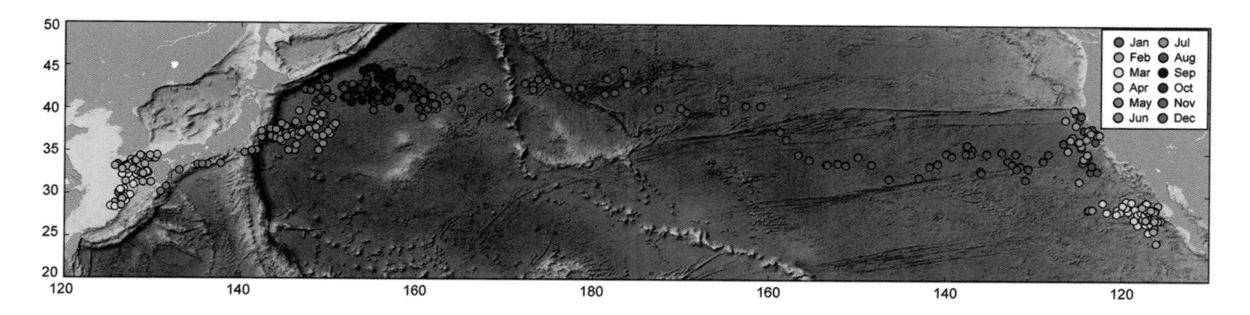


**Figure 7.6.** The migration pathway of an archival-tagged Pacific bluefin tuna. *Modified from Itoh et al. 2003a; Kitagawa et al. 2009, 2013a.*

Another fish, released in the ECS in November 1996, migrated to the KOTR (Figure 7.6) in the spring of 1997. After a period of residence in the KOTR, in November 1997 the fish initiated its TPM, moving eastward along the Subarctic Frontal Zone (Inagake 2001; Itoh et al. 2003a; Kitagawa et al. 2009, 2013). After its arrival in the eastern Pacific, the fish was recaptured in August 1998. Tagging studies revealed that PBT adopted the sit-and-go form of migration, that is, a residence mode with random movements and a migration mode with constant direction. Daily distance covered by the fish during the TPM was 100–200 km, significantly higher than during other periods (Kitagawa et al. 2009, 2013). In contrast, it is unknown what proportion of the total PBF population consists of migratory fish or even whether there are separate nonmigratory and migratory subpopulations (Fujioka et al. 2015). The proportion of PBF making TPMs appears to be variable on interannual scales and may be tied to prey availability in the western Pacific; this possibility is supported by the hypothesis that during years when sardines are abundant off Japan, a higher proportion of bluefin stay in the western Pacific than when sardines are scarce (Polovina 1996).

PBTs spend most of their time at the surface in early summer (Figure 7.7; e.g., Kitagawa et al. 2000, 2004b, 2007b). However, the young PBTs monitored in the ECS (nine fish with 56–86 cm FL when they were recovered) often dove below the thermocline during the day. This resulted in a marked change in the ambient temperature for the fish because of the existence of a seasonal thermocline. As the thermal gradient became steeper in May–June, in particular, the dive duration decreased. The peritoneal cavity (body) temperature increased and remained high in the day and then decreased at night. Fish body insulation (bulk effect) and heat production in the body were calculated based on the relationship between peritoneal

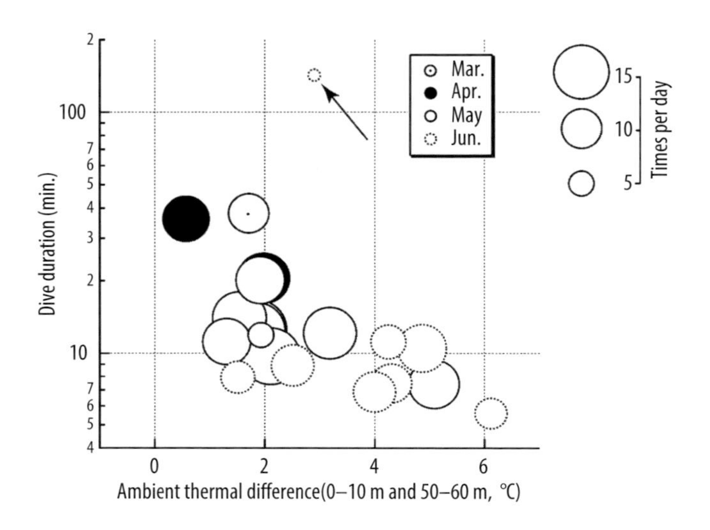


**Figure 7.7.** Relationships between the thermal gradient (0–10 m and 50–60 m) and the mean dive duration in the daytime with the dive frequency per day (from Kitagawa et al. 2007b). Only one fish (*arrow*) showed a lower dive frequency with a longer dive duration in June because a high-temperature water mass often existed at the surface.

and ambient temperatures, and it was found that because of the PBTs' greater insulation, the recovery of body temperature was slow (Kitagawa et al. 2001). Therefore, the fish may thermoregulate by spending most of its time at the surface in summer, executing multiple short-term dives that do not influence the body temperature (Kitagawa et al. 2007b). That is, such dives are adaptive behavior to seasonal changes in the vertical thermal structure of the water column. The fish compensate for the need to feed in colder depths, while maintaining a high body temperature, by undertaking multiple short-duration dives rather than fewer longer dives.

Although the daily dive frequency in the ECS was more than 10 in March–May, it decreased in June. This indicates that the dive frequency is not related to the strength of the thermal gradient (Figure 7.8; Kitagawa et al. 2007b). It is likely that food availability was greater at the surface in June than in the earlier months (Ohshimo 2004); or the increase in sea surface temperature may redirect the motivation of the PBT to migration to the Sea of Japan.

Kitagawa et al. (2004b) examined PBTs' vertical movements in relation to oceanographic conditions and the occurrence of feeding events inferred from thermal fluctuations in the peritoneal cavity. PBTs in the ECS

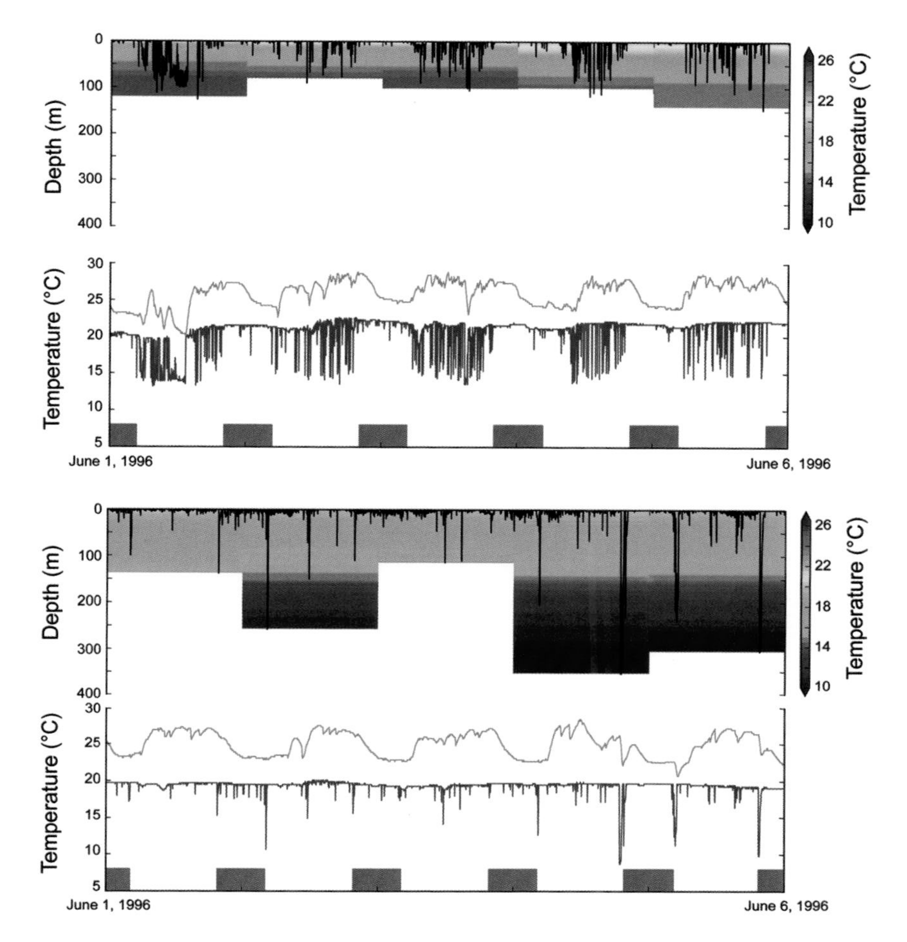


**Figure 7.8.** Time series data for five days in early summer in (*a*) the East China Sea and in (*b*) the Kuroshio-Oyashio transition region for swimming depth and vertical thermal structure (*upper*) and peritoneal cavity temperature and ambient temperature (*lower*) obtained from PBT 177 and 209, respectively (modified from Kitagawa et al. 2004b). Horizontal bars indicate nighttime. The PBTs make dives at dawn and dusk irrespective of seasons and distributed areas.

dove to depths though the thermocline more than 10 times in the daytime (Figure 7.8a), but feeding events occurred only a few times, thus suggesting that the purpose of the dive is foraging but that actual feeding is limited. However, when fish migrated to the KOTR (Figure 7.8b), they dove less frequently than in the ECS but fed much more frequently (Kitagawa et al. 2004). This suggests that food is plentiful at the surface in the KOTR, facilitating rapid growth. This may be one of the reasons for the migration of the fish to the KOTR.

Bluefin make dives at dawn and dusk in an automatic or mechanical way irrespective of seasons or distribution areas (Figure 7.7). This suggests that the dives are probably related to avoidance of a specific light level (Itoh et al. 2003b, Kitagawa et al. 2004b).

## Pacific Bluefin Tuna: Studies for the Future

As shown above, since 1995 electronic tagging has been conducted around Japan, but sporadically except for the ECS and Kochi (Table 7.1; Figure 7.2). Therefore, we must continue to conduct electronic tagging studies off the coasts of Japan, especially to monitor the movement of first-year PBTs to generate information about natural mortality for annual recruitment estimation. For the time being, as shown in Figure 7.9 (Furukawa et al. 2017), juvenile fish smaller than 18 cm FL are currently tracked and were released off Kochi (Table 7.1, Furukawa et al. 2017). As for the Sea of Japan, Kitagawa et al. (2002) described the vertical movements for a few fish; however, information is scarce. Tagging has been conducted off Shimane since 2016 (Table 7.1, Figure 7.2); therefore, more information about juvenile PBT behavior in both the Pacific and the Sea of Japan will be available.

So far, we have investigated only the first few years of the life of PBTs, which live more than 20 years (Bayliff 1994, Shimose et al. 2009). It is necessary to clarify what routes they take to the spawning areas, using electronic tags. In the Sea of Japan, in particular, one of the three branches of the Tsushima Warm Current flows along the Japanese coast in waters shallower than 200 m, but more than 90% of the water is colder than 5°C because of the Proper Water phenomenon (e.g., Yasui et al. 1967, Hase et al. 1999), and precocious PBTs mature at 3 years of age (95% maturity occurred at approximately 4 years old) (Okochi et al. 2016). In such a harsh environment, therefore, the behavior of spawners as well as juveniles needs to be measured intensively to determine the ecological importance of the area, although electronic tagging of spawners is currently conducted off Niigata (Table 7.1). In addition, the behavior of spawners in the Pacific needs to be measured; only one adult fish (with an estimated body weight of 230 kg) was tracked for 10 days (256 hours) in the spawning area in the Pacific (Ohta and Yamada 2016, Ohshimo et al. 2017). In particular, we need to clarify information regarding migrations to spawning grounds, the timing of spawning, and after-spawning behavior.

Additionally, to clarify PBT stock structures in the Pacific or the importance of TPM, more tags need to be released to help predict ocean environments over the next 50–100 years. Although it is especially important to do so, we cannot yet demonstrate how many fish are represented by a single tagged individual. To compensate for this shortcoming, electric tag-

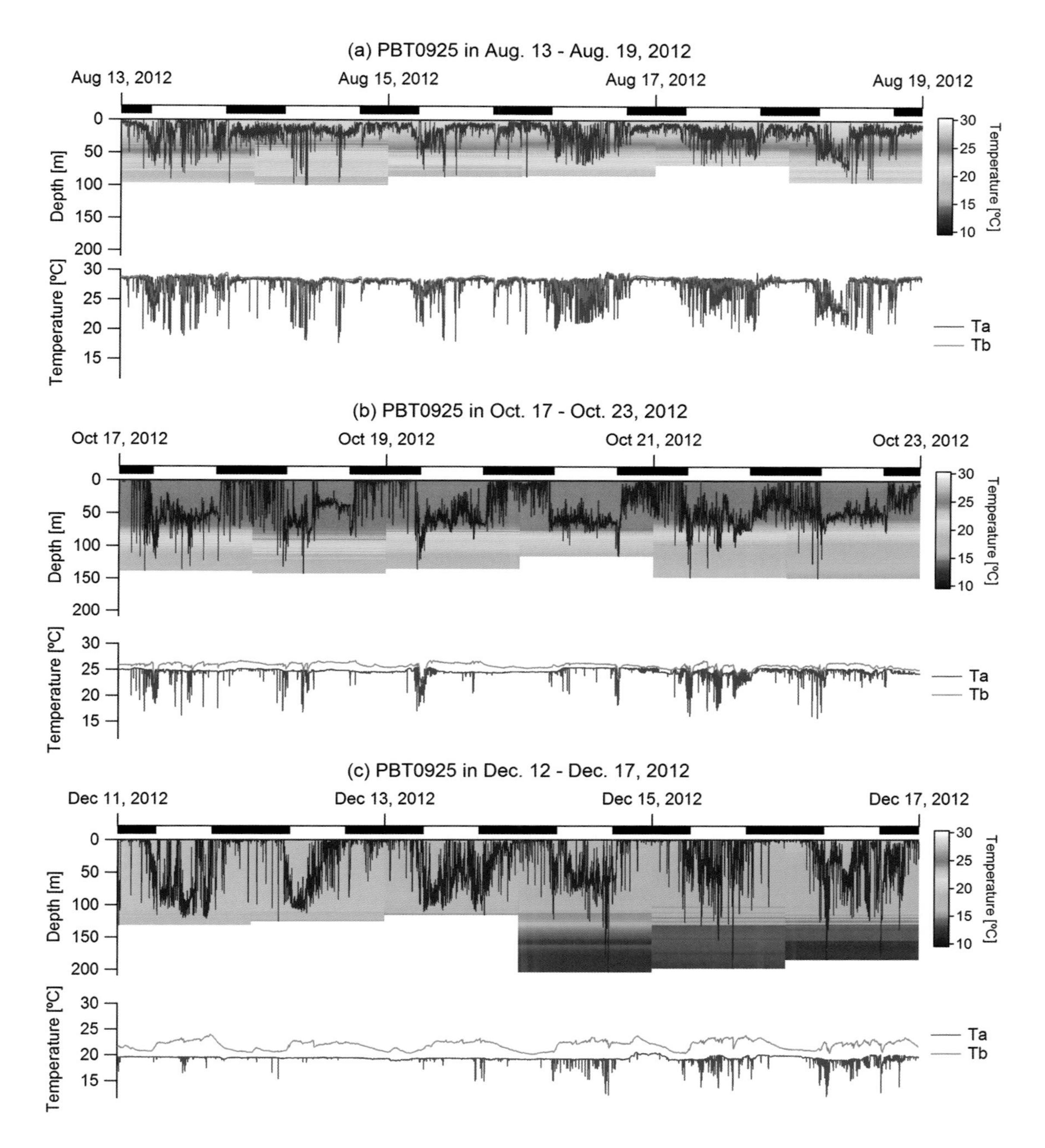


**Figure 7.9.** Weekly time series data in (*a*) August, (*b*) October, and (*c*) December of 2012 for swimming depth and vertical thermal structure (*upper*) and peritoneal cavity temperature and ambient temperature (*lower*) obtained from fish with 24 cm fork length (FL) at release off Kochi. Horizontal black bars indicate nighttime. *Furukawa et al. 2017*.

ging studies should be used in combination with other studies such as stable isotope analysis, numerical modeling, and molecular genetics such as close-kin analysis (Kitagawa et al. 2010; Madigan et al. 2016; Bravington et al. 2014, 2016), in addition to fisheries data analysis and conventional tagging experiments. It will not be long before the synthesis of all information regarding distributions, movements, genetic stock structure, and kinship illustrates the true behavior of PBT.

Noda et al. (2016) suggested that electronic tags should be further developed for measuring fish schools. In addition, new types of less expensive archival tags should be developed. Current electronic tags are very sophisticated, but the price is still very high. In addition, these tags were sourced from overseas and are therefore not designed to be user-friendly in Japan. The price is also affected by the foreign exchange rate. For these reasons, a development project for new types of electronic tags is currently running in Japan (Miyashita et al. 2014).

## Acknowledgments

The authors wish to thank the Japanese Society of Fisheries Science and Springer (License 4035171299156), which gave them permission to reproduce figures, as well as Dr. Yoshinori Aoki of the National Research Institute of Far Seas Fisheries, Japan Fisheries Research and Education Agency, and Dr. Seishiro Furukawa of Japan Sea National Fisheries Research Institute, Japan Fisheries Research and Education Agency, for their kind support. This work was supported by JSPS KAKENHI grant numbers JP2438010 and JP16H01769. The authors would also like to thank Enago for the English language review.

### REFERENCES

Bayliff, W.H. 1980. Synopsis of biological data on the northern bluefin tuna, *Thunnus thynnus* (Linnaeus, 1758), in the Pacific Ocean. *Inter-American Tropical Tuna Commission Scientific Report* 2:261–93.

Bayliff, W.H. 1994. A review of the biology and fisheries for northern bluefin tuna, *Thunnus thynnus*, in the Pacific Ocean. *Interactions of Pacific Tuna Fisheries* 336 (2): 244–94.

Bayliff, W.H., Y. Ishizuka, and R.B. Deriso. 1991. Growth, movements, and mortality of northern bluefin, *Thunnus thynnus*, in the Pacific Ocean, as determined from tagging experiments. *Bulletin of Inter-American Tropical Tuna Commission* 20:380–421.

Bell, R.R. 1963. Synopsis of biological data on California bluefin tuna *Thunnus saliens* Jordan and Evermann 1926. Species Synopsis No. 12, FAO Fisheries Biology Synopsis No. 55. *FAO Fisheries Report* 6:380–421. www.fao.org/docrep/017/ap882e/ap882e.pdf.

Block, B.A., H. Dewar, S.B. Blackwell, T. Williams, E. Prince, A.M. Boustany, C. Farwell, D.J. Dau, and A. Seitz. 2001. Archival and pop-up satellite tagging of Atlantic bluefin tuna. In *Electronic Tagging and Tracking in Marine Fisheries*, edited by J.R. Sibert and J.L. Nielsen. 65–88. Dordrecht, Netherlands: Kluwer Academic. doi:10.1007/978-94-017-1402-0_3.

Block, B.A., I.D. Jonsen, S.J. Jorgensen, A.J. Winship, S.A. Shaffer, S.J. Bograd, E.L. Hazen, et al. 2011. Tracking apex marine predator movements in a dynamic ocean. *Nature* 475:86–90. doi:10.1038/nature10082.

Boustany, A.M., D.J. Marcinek, J. Keen, H. Dewar, and B.A. Block. 2001. Movements and temperature preferences of Atlantic bluefin tuna (*Thunnus thynnus*) off North Carolina: A comparison of acoustic, archival and pop-up satellite tags. In *Electronic Tagging and Tracking in Marine Fisheries*, edited by J.R. Sibert and J.L. Nielsen, 89–108. Dordrecht, Netherlands: Kluwer Academic. doi:10.1007/978-94-017-1402-0_4.

Boustany, A.M., R. Matteson, M. Castleton, C. Farwell, and B.A. Block. 2010. Movements of Pacific bluefin tuna (*Thunnus orientalis*) in the Eastern North Pacific revealed with archival tags. *Progress in Oceanography* 86:94–104.

Bravington, M.V., P.M. Grewe, and C.R. Davies. 2014. Fishery-independent estimate of spawning biomass of Southern bluefin tuna through identification of close-kin using genetic markers. FRDC Report 2007/034, CSIRO, Australia.

Bravington, M. V., H.J. Skaug, and E.C. Anderson. 2016. Close-kin mark-recapture. *Statistical Science* 31 (2): 259–74.

Carey, F.G., and K.D. Lawson. 1973. Temperature regulation in free-swimming bluefin tuna. *Comparative Biochemistry and Physiology* 44 (2): 375–92.

Chen, K.-S., P. Crone, and C.-C. Hsu. 2006. Reproductive biology of female Pacific bluefin tuna *Thunnus orientalis* from south-western North Pacific Ocean. *Fisheries Science* 72 (5): 985–94. doi: 10.1111/j.1444-2906.2006.01247.x.

Clemens, A.E., and G.A. Flittner. 1969. Bluefin tuna migrate across the Pacific Ocean. *California Fish and Game* 55:132–35.

Ekstrom, P.A. 2004. An advance in geolocation by light. *Memoirs of the National Institute of Polar Research (Tokyo)* 58:210–26.

Fujioka, K., A.J. Hobday, R. Kawabe, K. Miyashita, K. Honda, T. Itoh, and Y. Takao. 2010a. Interannual variation in summer habitat utilization by juvenile southern bluefin tuna (*Thunnus maccoyii*) in southern Western Australia. *Fisheries Oceanography* 19:183–95.

Fujioka, K., R. Kawabe, A.J. Hobday, Y. Takao, K. Miyashita, O. Sakai, and T. Itoh. 2010b. Spatial and temporal variation in the distribution of juvenile southern bluefin tuna *Thunnus maccoyii*: Implication for precise estimation of recruitment abundance indices. *Fisheries Science* 76:403–10.

Fujioka, K., A.J. Hobday, R. Kawabe, K. Miyashita, Y. Takao, O. Sakai, and T. Itoh. 2012. Departure behavior of juvenile southern bluefin tuna (*Thunnus maccoyii*) from southern Western Australia temperate waters in relation to the Leeuwin Current. *Fisheries Oceanography* 21:269–80.

Fujioka, K., H. Fukuda, S. Okamoto, and Y. Takeuchi. 2013a. First record of the small (age-0) Pacific bluefin tuna (Thunnus orientalis) migration in the sea off Kochi revealed by archival tags. Abstract of 9th Indo-Pacific Fish Conference, June 24–28, Okinawa, Japan.

Fujioka, K., H. Fukuda, S. Okamoto, and Y. Takeuchi. 2013b. Migration patterns of juvenile (age-0) Pacific bluefin tuna (*Thunnus orientalis*) in coastal nursery areas of Japan. *Proceedings of 64th Annual Tuna Conference, Lake Arrowhead, CA, May 20–23*, p. 40. https://docs.wixstatic.com/ugd/ba25d2_6d6c9367d2332e03ebc89cb2992a86bd.pdf.

Fujioka, K., M. Masujima, A.M. Boustany, and T. Kitagawa. 2015. Horizontal movements of Pacific bluefin tuna. In *Biology and Ecology of Bluefin Tuna*, edited by T. Kitagawa and S. Kimura, 101–22. Boca Raton, FL: CRC Press.

Furukawa, S., Y. Tsuda, G.N. Nishihara, K. Fujioka, S. Ohshimo, S. Tomoe, N. Nakatsuka, et al. 2014. Vertical movements of Pacific bluefin tuna (*Thunnus orientalis*) and dolphinfish (*Coryphaena hippurus*) relative to the thermocline in the northern East China Sea. *Fisheries Research* 149:86–91.

Furukawa, S., K. Fujioka, H. Fukuda, N. Suzuki, Y. Tei, and S. Ohshimo. 2017. Archival tagging reveals swimming depth and ambient and peritoneal cavity temperature in age-0 Pacific bluefin tuna, *Thunnus orientalis*, off the southern coast of Japan. *Environmental Biology of Fishes* 100:35–48.

Gunn J., and B. Block. 2001. Advances in acoustic, archival, and satellite tagging of tunas. In *Tuna: Physiology, Ecology and Evolution*, edited by B.A. Block and E.D. Stevens, 167–224. San Diego, CA: Academic Press.

Hase, H., J.-H. Yoon, and W. Koterayama. 1999. The current structure of the Tsushima Warm Current along the Japanese Coast. *Journal of Oceanography* 55:217–35.

Inagake, D., H. Yamada, K. Segawa, M. Okazaki, A. Nitta, and T. Itoh. 2001. Migration of young bluefin tuna, *Thunnus orientalis* Temminck et Schlegel, through archival tagging experiments and its relation with oceanographic condition in the Western North Pacific. *Bulletin of the National Research Institute of Far Seas Fisheries*. 38:53–81.

Itoh, T., Tsuji, S. and Nitta, A. 2003a. Migration patterns of young Pacific bluefin tuna (*Thunnus orientalis*) determined with archival tags. *Fishery Bulletin* 101: 514–34.

Itoh, T., S. Tsuji, and A. Nitta. 2003b. Swimming depth, ambient water temperature preference, and feeding frequency of young Pacific bluefin tuna (*Thunnus orientalis*) determined with archival tags. *Fishery Bulletin* 101:535–44.

Kitagawa, T. 2013. Behavioral ecology and thermal physiology of immature Pacific bluefin tuna (*Thunnus orientalis*). In *Physiology and Ecology of Fish Migration*, edited by H. Ueda and K. Tsukamoto, 152–78. Boca Raton, FL: CRC Press.

Kitagawa, T., H. Nakata, S. Kimura, T. Itoh, S. Tsuji, and A. Nitta. 2000. Effect of ambient temperature on the vertical distribution and movement of Pacific bluefin tuna *Thunnus thynnus orientalis*. *Marine Ecology Progress Series* 206:251–60.

Kitagawa, T., H. Nakata, S. Kimura, and S. Tsuji. 2001. Thermoconservation mechanisms inferred from peritoneal cavity temperature in free-swimming Pacific bluefin tuna *Thunnus thynnus orientalis*. *Marine Ecology Progress Series* 220:253–63.

Kitagawa, T., H. Nakata, S. Kimura, T. Sugimoto, and H. Yamada. 2002. Differences in vertical distribution and movement of Pacific bluefin tuna (*Thunnus thynnus orientalis*) among areas: The East China Sea, the Sea of Japan and the western North Pacific. *Marine and Freshwater Research* 53:245–52.

Kitagawa, T., S. Kimura, H. Nakata, and H. Yamada. 2004a. Overview of the research on tuna thermo-physiology using electric tags. *Memoirs of the National Institute of Polar Research,* Special Issue, 58:69–79.

Kitagawa, T., S. Kimura, H. Nakata, and H. Yamada. 2004b. Diving behavior of immature, feeding Pacific bluefin tuna (*Thunnus thynnus orientalis*) in relation to season and area: The East China Sea and the Kuroshio-Oyashio transition region. *Fisheries Oceanography* 13:161–80.

Kitagawa, T., S. Kimura, H. Nakata, and H. Yamada. 2006a. Thermal adaptation of Pacific bluefin tuna *Thunnus orientalis* to temperate waters. *Fisheries Science* 72:149–56.

Kitagawa, T., A. Sartimbul, H. Nakata, S. Kimura, H. Yamada, and A. Nitta. 2006b. The effect of water temperature on habitat use of young Pacific bluefin tuna *Thunnus orientalis* in the East China Sea. *Fisheries Science* 72:1166–76.

Kitagawa, T., A.M. Boustany, C.J. Farwell, T.D. Williams, M.R. Castleton, and B.A. Block. 2007a. Horizontal and vertical movements of juvenile bluefin tuna (*Thunnus orientalis*) in relation to seasons and oceanographic conditions in the eastern Pacific Ocean. *Fisheries Oceanography* 16:409–21.

Kitagawa, T., S. Kimura, H. Nakata, and H. Yamada. 2007b. Why do young Pacific bluefin tuna repeatedly dive to depths through the thermocline? *Fisheries Science* 73:98–106.

Kitagawa, T., S. Kimura, H. Nakata, H. Yamada, A. Nitta, Y. Sasai, and H. Sasaki. 2009. Immature Pacific bluefin tuna, *Thunnus orientalis,* utilizes cold waters in the Subarctic Frontal Zone for trans-Pacific migration. *Environmental Biology of Fishes* 84:193–96.

Kitagawa, T., Y. Kato, M.J. Miller, Y. Sasai, H. Sasaki, and S. Kimura. 2010. The restricted spawning area and season of Pacific bluefin tuna facilitate use of nursery areas: A modeling approach to larval and juvenile dispersal processes. *Journal of Experimental Marine Biology and Ecology* 393:23–31.

Kitagawa, T., S. Kimura, H. Nakata, H. Yamada, A. Nitta, Y. Sasai, and H. Sasaki. 2013. Immature Pacific bluefin tuna, *Thunnus orientalis,* utilizes cold waters in the Subarctic Frontal Zone for trans-Pacific migration (vol. 84, pg. 193, 2009). *Environmental Biology of Fishes* 96:797–98.

Lutcavage, M.E., R.W. Brill, G.B. Skomal, B.C. Chase, J.L. Goldstein, and J. Tutein. 2000. Tracking adult North Atlantic bluefin tuna (*Thunnus thynnus*) in the northwestern Atlantic using ultrasonic telemetry. *Marine Biology* 137:347–58.

Madigan, D.J., W.C. Chiang, N.J. Wallsgrove, B.N. Popp, T. Kitagawa, C.A. Choy, J. Tallmon, N. Ahmed, N.S. Fisher, and C. Sun. 2016. Intrinsic tracers reveal recent foraging ecology of giant Pacific bluefin tuna at their primary spawning grounds. *Marine Ecology Progress Series* 553:253–66.

Marcinek, D.J., S.B. Blackwell, H. Dewar, E.V. Freund, C. Farwell, D. Dau, A.C. Seitz, et al. 2001. Depth and muscle temperature of Pacific bluefin tuna examined with acoustic and pop-up satellite archival tags. *Marine Biology* 138 (4): 869–85.

Miyashita, K., T. Kitagawa, Y. Miyamoto, N. Arai, H. Shirakawa, H. Mitamura, T. Noda, T. Sasakura, and T. Shinke. 2014. Construction of advanced biologging system to implement high data-recovery rate—a challenging study to clarify the dynamics of fish population and community. *Nippon Suisan Gakkaishi* 80:1009–15.

Musyl, M.K., R.W. Brill, D.S. Curran, J.S. Gunn, J.R. Hartog, R.D. Hill, D.W. Welch, J.P. Eveson, C.H. Boggs, and R.E. Brainard. 2001. Ability of archival tags to provide estimates of geographical position based on light intensity. In *Electronic Tagging and Tracking in Marine Fisheries Reviews: Methods and Technologies in Fish Biology and Fisheries*, edited by J.R. Sibert and J.L. Nielsen, 343–68. Dordrecht, Netherlands: Kluwer Academic.

Noda, T., K. Fujioka, H. Fukuda, H. Mitamura, K. Ichikawa, and N. Arai. 2016. The influence of body size on the intermittent locomotion of a pelagic schooling fish. *Proceedings of the Royal Society of London B: Biological Sciences* 283:20153019.

Ohshimo, S. 2004. Spatial distribution and biomass of pelagic fish in the East China Sea in summer, based on acoustic surveys from 1997 to 2001. *Fisheries Science* 70:389–400.

Ohshimo, S., A. Tawa, T. Ota, S. Nishimoto, T. Ishihara, M. Watai, K. Satoh, T. Tanabe, and O. Abe. 2017. Horizontal distribution and habitat of Pacific bluefin tuna, *Thunnus orientalis*, larvae in the waters around Japan. *Bulletin of Marine Science* 93:769–87.

Ohta, I., and H. Yamada. 2016. Formation of a Pacific bluefin tuna fishing ground on their spawning grounds around the Ryukyu Islands: Implication of a relationship with mesoscale eddies. In *Biology and Ecology of Bluefin Tuna*, edited by T. Kitagawa and S. Kimura, 123–36. Boca Raton, FL: CRC Press.

Okiyama, M. 1974. Occurrence of the postlarvae of bluefin tuna, *Thunnus thynnus*, in the Japan Sea. *Bulletin of the Japan Sea National Fisheries Research Institute. Niigata* 25:89–97.

Okochi, Y., O. Abe, S. Tanaka, Y. Ishihara, and A. Shimizu. 2016. Reproductive biology of female Pacific bluefin tuna, *Thunnus orientalis*, in the Sea of Japan. *Fisheries Research* 174:30–39.

Orange, C.J., and B.D. Fink. 1963. Migration of tagged bluefin tuna across the Pacific Ocean. *California Fish and Game* 49:307–9.

Polovina, J.J. 1996. Decadal variation in the trans-Pacific migration of northern bluefin tuna (*Thunnus thynnus*) coherent with climate-induced change in prey abundance. *Fisheries Oceanography* 5:114–19.

Shimose T., T. Tanabe, K.-S. Chen, C.-C. Hsu. 2009. Age determination and growth of Pacific bluefin tuna, *Thunnus orientalis*, off Japan and Taiwan. *Fisheries Research* 100: 134–39.

Tanaka, Y., M. Mohri, and H. Yamada. 2007. Distribution, growth and hatch date of juvenile Pacific bluefin tuna *Thunnus orientalis* in the coastal area of the Sea of Japan. *Fisheries Science* 73 (3): 534–42. doi: 10.1111/j.1444-2906.2007.01365.x.

Welch, D.W., and J.P. Eveson. 1999. An assessment of light-based geoposition estimates from archival tags. *Canadian Journal of Fisheries and Aquatic Sciences* 56:1317–27.

Whitlock, R.E., E.L. Hazen, A. Walli, C. Farwell, S.J. Bograd, D.G. Foley, M. Castleton, and B.A. Block. 2015. Direct quantification of energy intake in an apex marine predator suggests physiology is a key driver of migrations. *Science Advances* 1 (8): e1400270.

Yasui, M., T. Yasuoka, K. Tanioka, and O. Shiota. 1967. Oceanographic studies of the Japan Sea (1). *Oceanographical Magazine* 18:177–92.

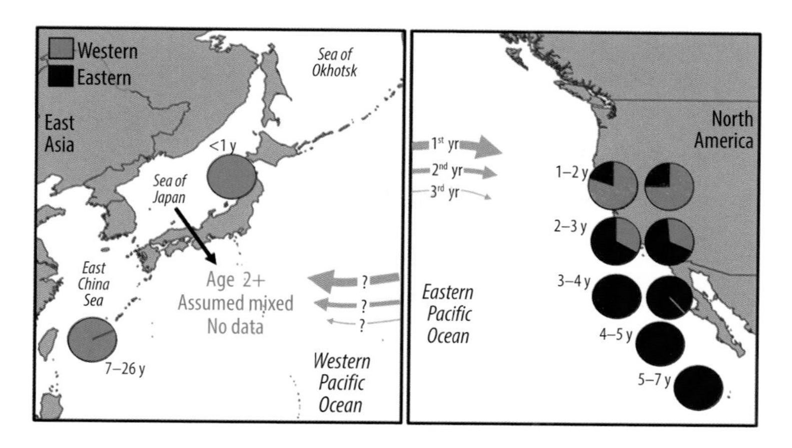


**Figure 8.5.** Recent migratory origin of Pacific bluefin tuna in the western and eastern Pacific Ocean (WPO and EPO) across age classes. Recent origin reflects the past ~1.5 years and is based on $\delta^{15}N$ of muscle tissue. Thickness of grey arrows representing migrants into the WPO and EPO reflects relative proportion of migrants represented by the age class indicated. Red text for age 2+ in the left panel indicates that the Sea of Japan is potentially a mix of WPO residents and EPO migrants, but published chemical tracer data from this region are currently unavailable.

interannual variability (Foreman and Ishizuka 1990), is not determined by larger, older bluefin migrating from the WPO to the EPO. Rather, abundance of larger bluefin in the EPO will be dictated by (1) recruitment of juveniles (ages 0–2) to the EPO; (2) survivability of juveniles for multiple years in the EPO; and (3) retention in the EPO, owing to oceanographic and/or biological variables. Consequently, high mortality on the youngest bluefin (age 0) will influence the abundance of large fish in the EPO by decreasing recruitment in early life stages (Whitlock et al. 2012). High mortality on fish ages 1–2 in the EPO can decrease the number of fish reaching larger size in the EPO and eventually returning to the WPO to spawn. Owing to the current lack of size-structured estimates of exchange between the EPO and WPO, the specific impacts of fishing mortality on EPO and WPO groups of Pacific bluefin remain difficult to quantify.

## Future Directions

Chemical tracers, and SIA in particular, have provided new insights into the movements of Pacific bluefin tuna in the North Pacific Ocean. More importantly, recent work has provided a tool to coarsely but reliably estimate the recent migratory history of Pacific bluefin in both the EPO and WPO. This approach circumvents some bias associated with traditional tag-recapture approaches and many of the financial and technological limitations of electronic tagging approaches. Pacific bluefin are still harvested in large numbers on both sides of the North Pacific Ocean, and the recent migratory history of any individual can be estimated with a small plug of muscle tissue. Researchers and management bodies thus have a wealth of continuously generated, potential data from commercial bluefin fisheries that could be used to create age- and time-structured estimates of Pacific bluefin migration patterns.

The most critical gap in the understanding of transpacific movements is the unknown migratory history of the age group in the WPO, which is likely a mix of long-term WPO residents and recent EPO migrants. Since the muscle isotope signal degrades over time (~1.5 years), it cannot be ascertained which individual older (e.g., 10+ years) Pacific bluefin in the WPO ever migrated to the EPO. Otolith microchemistry approaches circumvent this issue, as otoliths are accretionary structures that archive past movements ontogenetically. Analysis of otoliths of larger, older Pacific bluefin in the WPO following the methodologies of Baumann et al. (2015) would identify the bluefin that did and did not migrate to the EPO. However, for population-wide estimates of movement, and to capture interannual variability of transpacific movements, multiyear studies using large sample sizes across multiple age classes are necessary. This may preclude the more time- and cost-intensive tagging and otolith approaches. Large-scale and continuous sampling of Pacific bluefin throughout their range could create the sample sizes necessary for population-wide generalizations of transpacific movements that are within realistic budgets. Such an effort would be supported by higher-resolution data from otolith microchemistry and by electronic tagging from smaller numbers of bluefin. Understanding transpacific dynamics will facilitate management decision-making by clarifying the potential fisheries impacts of variable fishing pressures on both sides of the North Pacific Ocean.

## Acknowledgments

This chapter builds upon and synthesizes work by a large collective group, including H. Dewar, O. Snodgrass, N. Fisher, Z. Baumann, A. Carlisle, B. Block, B. Popp, W-C. Chiang, C. Sun, N. Wallsgrove, A. Choy, T. Kitagawa, N. Wallsgrove, J. Tallmon, N. Ahmed, and others. A. Norton, C. Farwell, L.

Rodriguez, E. Estess, D. Klinger, L. Gardner, and others from the Tuna Research and Conservation Center (TRCC) provided valuable field and lab assistance over many years of work. Sampling and analysis of Pacific bluefin in the EPO would not be possible without many researchers and assistants at NOAA's Southwest Fisheries Science Center (SWFSC).

## REFERENCES

Baumann, H., R. J. D. Wells, J. R. Rooker, S. Zhang, Z. Baumann, D. J. Madigan, H. Dewar, O. E. Snodgrass, and N. S. Fisher. 2015. Combining otolith microstructure and trace elemental analyses to infer the arrival of juvenile Pacific bluefin tuna in the California current ecosystem. *ICES Journal of Marine Science* 72 (7): 2128–38.

Bayliff, W. H. 1993. Growth and age composition of northern bluefin tuna, *Thunnus thynnus,* caught in the eastern Pacific Ocean, as estimated from length-frequency data, with comments on Trans-Pacific migrations. *Inter-American Tropical Tuna Commission Bulletin* 20 (9): 501–40.

Bayliff, W. H. 1994. A review of the biology and fisheries for northern bluefin tuna, *Thunnus thynnus,* in the Pacific Ocean. *FAO Fisheries Technical Paper* 336:244–95.

Bayliff, W. H., Y. Ishizuka, and R. Deriso. 1991. Growth, movement, and attrition of northern bluefin tuna, *Thunnus thynnus,* in the Pacific Ocean, as determined by tagging. *Inter-American Tropical Tuna Commission Bulletin* 20:3–94.

Block, B. A., I. D. Jonsen, S. J. Jorgensen, A. J. Winship, S. A. Shaffer, S. J. Bograd, E. L. Hazen, et al. 2011. Tracking apex marine predator movements in a dynamic ocean. *Nature* 475:86–90.

Boustany, A. M., R. Matteson, M. R. Castleton, C. J. Farwell, and B. A. Block. 2010. Movements of Pacific bluefin tuna (*Thunnus orientalis*) in the Eastern North Pacific revealed with archival tags. *Progress in Oceanography* 86:94–104.

Campana, S. E. 1999. Chemistry and composition of fish otoliths: Pathways, mechanisms and applications. *Marine Ecology Progress Series* 188:263–97.

Carlisle, A. B., S. L. Kim, B. X. Semmens, D. J. Madigan, S. J. Jorgensen, C. R. Perle, S. D. Anderson, T. K. Chapple, P. E. Kanive, and B. A. Block. 2012. Using stable isotope analysis to understand migration and trophic ecology of northeastern Pacific white sharks (*Carcharodon carcharias*). *PLOS ONE* 7:e30492.

Chen, K.-S., P. Crone, and C.C. Hsu. 2006. Reproductive biology of female Pacific bluefin tuna *Thunnus orientalis* from south-western North Pacific Ocean. *Fisheries Science* 72:985–94.

Das, K., G. Lepoint, V. Loizeau, V. Debacker, P. Dauby, and J. M. Bouquegneau. 2000. Tuna and dolphin associations in the North-east Atlantic: Evidence of different ecological niches from stable isotope and heavy metal measurements. *Marine Pollution Bulletin* 40:102–9.

Deshpande, A. D., R. M. Dickhut, B. W. Dockum, R. W. Brill, and C. Farrington. 2016. Polychlorinated biphenyls and organochlorine pesticides as intrinsic tracer tags of foraging grounds of bluefin tuna in the northwest Atlantic Ocean. *Marine Pollution Bulletin* 105:265–76.

Dickhut, R. M., A. D. Deshpande, A. Cincinelli, M. A. Cochran, S. Corsolini, R. W. Brill, D. H. Secor, and J. E. Graves. 2009. Atlantic bluefin tuna (*Thunnus thynnus*) population dynamics delineated by organochlorine tracers. *Environmental Science and Technology* 43:8522–27. doi:10.1021/Es901810e.

Domeier, M. L., D. Kiefer, N. Nasby-Lucas, A. Wagschal, and F. O'Brien. 2005. Tracking Pacific bluefin tuna (*Thunnus thynnus orientalis*) in the northeastern Pacific with an automated algorithm that estimates latitude by matching sea-surface-temperature data from satellites with temperature data from tags on fish. *Fishery Bulletin* 103:292–306.

Foreman, T. J., and Y. Ishizuka. 1990. Giant bluefin tuna off southern California, with a new California size record. *California Fish and Game* 76 (3): 181–86.

Fraile, I., H. Arrizabalaga, J. Santiago, N. Goñi, I. Arregi, S. Madinabeitia, R. J. D. Wells, and J. R. Rooker. 2016. Otolith chemistry as an indicator of movements of albacore (*Thunnus alalunga*) in the North Atlantic Ocean. *Marine and Freshwater Research* 67: 1002–13.

Fujioka, K., M. Masujima, A. M. Boustany, and T. Kitagawa. 2015. Horizontal movements of Pacific bluefin tuna. In *Biology and Ecology of Bluefin Tuna*, edited by T. Kitagawa and S. Kimura, 102–22. Boca Raton, FL: CRC Press.

Graves, J. E., A. S. Wozniak, R. M. Dickhut, M. A. Cochran, E. H. MacDonald, E. Bush, H. Arrizabalaga, and N. Goñi. 2015. Transatlantic movements of juvenile Atlantic bluefin tuna inferred from analyses of organochlorine tracers. *Canadian Journal of Fisheries and Aquatic Sciences* 72:625–33.

Hazen, E. L., S. Jorgensen, R. R. Rykaczewski, S. J. Bograd, D. G. Foley, I. D. Jonsen, S. A. Shaffer, J. P. Dunne, D. P. Costa, and L. B. Crowder. 2013. Predicted habitat shifts of Pacific top predators in a changing climate. *Nature Climate Change* 3:234–38.

Itoh, T., S. Tsuji, and A. Nitta. 2003a. Migration patterns of young Pacific bluefin tuna (*Thunnus orientalis*) determined with archival tags. *Fishery Bulletin* 101:514–35.

Itoh, T., S. Tsuji, and A. Nitta. 2003b. Swimming depth, ambient water temperature preference, and feeding frequency of young Pacific bluefin tuna (*Thunnus orientalis*) determined with archival tags. *Fishery Bulletin* 101:535–44.

Kawabata, T. 1955. Studies on the radiological contamination of fishes I. A consideration on the distribution and migration of contaminated fishes on the basis of the compiled data of radiological survey. *Japanese Journal of Medical Science and Biology* 8:337–46.

Kitagawa, Y., Y. Nishikawa, T. Kuboto, and M. Okiyama. 1995. Distribution of ichthyoplankton in the Japan Sea during summer, 1984, with special reference to scombroid fishes [in Japanese, with English abstract]. *Bulletin of the Japanese Society of Fisheries Oceanography* 59:107–14.

Kitagawa, T., S. Kimura, H. Nakata, H. Yamada, A. Nitta, Y. Sasai, and H. Sasaki. 2009. Immature Pacific bluefin tuna, *Thunnus orientalis*, utilizes cold waters in the Subarctic Frontal Zone for trans-Pacific migration. *Environmental Biology of Fishes* 84:193–96.

Klaassen, M., T. Piersma, H. Korthals, A. Dekinga, and M. W. Dietz. 2010. Single-point isotope measurements in blood cells and plasma to estimate the time since diet switches. *Functional Ecology* 24:796–804.

Krygier, E. E., and W. G. Pearcy. 1977. Source of cobalt-60 and migrations of albacore off the west coast of north America. National Marine Fisheries Service, NOAA, Newport, OR.

Lin, Y.-T., C.-H. Wang, C.-F. You, and W.-N. Tzeng. 2013. Ba/Ca Ratios in otoliths of southern bluefin tuna (*Thunnus maccoyii*) as a biological tracer of upwelling in the Great Australian Bight. *Journal of Marine Science and Technology* 21:733–41.

Logan, J. 2009. Tracking diet and movement of Atlantic bluefin tuna (*Thunnus thynnus*) using carbon and nitrogen stable isotopes. PhD diss., University of New Hampshire.

Lorrain, A., B. S. Graham, B. N. Popp, V. Allain, R. J. Olson, B. P. V. Hunt, M. Potier, et al. 2015. Nitrogen isotopic baselines and implications for estimating foraging habitat and trophic position of yellowfin tuna in the Indian and Pacific Oceans. *Deep Sea Research II* 113:188–98.

Luque, P. L., S. Zhang, J. R. Rooker, G. Bidegain, and E. Rodríguez-Marín. 2017. Dorsal fin spines as a non-invasive alternative calcified structure for microelemental studies in Atlantic bluefin tuna. *Journal of Experimental Marine Biology and Ecology* 486:127–33.

Macdonald, J. I., J. H. Farley, N. P. Clear, A. J. Williams, T. I. Carter, C. R. Davies, and S. J. Nicol. 2013. Insights into mixing and movement of South Pacific albacore *Thunnus alalunga* derived from trace elements in otoliths. *Fisheries Research* 148:56–63.

Madigan, D. J., Z. Baumann, and N. S. Fisher. 2012a. Pacific bluefin tuna transport Fukushima-derived radionuclides from Japan to California. *Proceedings of the National Academy of Sciences* 109:9483–86.

Madigan, D. J., S. Y. Litvin, B. N. Popp, A. B. Carlisle, C. J. Farwell, and B. A. Block. 2012b. Tissue turnover rates and isotopic trophic discrimination factors in the endothermic teleost, Pacific bluefin tuna (*Thunnus orientalis*). *PLOS ONE* 7:e49220.

Madigan, D. J., Z. Baumann, A. B. Carlisle, D. K. Hoen, B. N. Popp, H. Dewar, O. E. Snodgrass, B. A. Block, and N. S. Fisher. 2014. Reconstructing trans-oceanic migration patterns of Pacific bluefin tuna using a chemical tracer toolbox. *Ecology* 95:1674–83.

Madigan, D. J., W.-C. Chiang, N. J. Wallsgrove, B. N. Popp, T. Kitagawa, C. A. Choy, J. Tallmon, N. Ahmed, N. S. Fisher, and C.-L. Sun. 2016. Intrinsic tracers reveal recent foraging ecology of giant Pacific bluefin tuna at their primary spawning grounds. *Marine Ecology Progress Series* 553:253–66.

Madigan, D. J., Z. Baumann, A. B. Carlisle, O. Snodgrass, H. Dewar, and N. S. Fisher. 2018. Isotopic insights into migration patterns of Pacific bluefin tuna in the eastern Pacific Ocean. *Canadian Journal of Fisheries and Aquatic Sciences* 75 (2): 260–70. doi.org/10.1139/cjfas-2016-0504.

Ménard, F., A. Lorrain, M. Potier, and F. Marsac. 2007. Isotopic evidence of distinct feeding ecologies and movement patterns in two migratory predators (yellowfin tuna and swordfish) of the western Indian Ocean. *Marine Biology* 153:141–52.

Munschy, C., N. Bodin, M. Potier, K. Héas-Moisan, C. Pollono, M. Degroote, W. West, et al. 2016. Persistent organic pollutants in albacore tuna (*Thunnus alalunga*) from Reunion Island (southwest Indian Ocean) and South Africa in relation to biological and trophic characteristics. *Environmental Research* 148:196–206.

Neville, D. R., A. J. Phillips, R. D. Brodeur, and K. A. Higley. 2014. Trace levels of Fukushima disaster radionuclides in East Pacific albacore. *Environmental Science and Technology* 48:4739–43.

Ohta, I., and H. Yamada. 2015. Formation of a Pacific bluefin tuna fishing ground on their spawning grounds around the Ryukyu Islands. In *Biology and Ecology of Bluefin Tuna*, edited by T. Kitagawa and S. Kimura, 123–36. Boca Raton, FL: CRC Press.

Okiyama, M. 1974. Occurrence of the postlarvae of bluefin tuna, *Thunnus thynnus,* in the Japan Sea. *Bulletin of the Japan Sea Regional Fisheries Research Laboratory* 25:89–97.

Okochi, Y., O. Abe, S. Tanaka, Y. Ishihara, and A. Shimizu. 2016. Reproductive biology of female Pacific bluefin tuna, *Thunnus orientalis,* in the Sea of Japan. *Fisheries Research* 174:30–39.

Pearcy, W. G., and R. R. Claeys. 1972. Zinc-65 and DDT residues in albacore tuna off Oregon in 1969. *California Marine Research Committee, CalCOFI Reports* 16:66–73.

Pearcy, W. G., J. P. Fisher, G. Anma, and T. Meguro. 1996. Species associations of epipelagic nekton of the North Pacific Ocean, 1978–1993. *Fisheries Oceanography* 5:1–20.

Phillips, D., and P. Eldridge. 2006. Estimating the timing of diet shifts using stable isotopes. *Oecologia* 147:195–203.

Proctor, C. H., R. E. Thresher, J. S. Gunn, D. J. Mills, I. R. Harrowfield, and S. H. Sie. 1995. Stock structure of the southern bluefin tuna *Thunnus maccoyii:* An investigation based on probe microanalysis of otolith composition. *Marine Biology* 122:511–26.

Ramos, R., and J. González-Solís. 2012. Trace me if you can: The use of intrinsic biogeochemical markers in marine top predators. *Frontiers in Ecology and the Environment* 10:258–66.

Revill, A., J. Young, and M. Lansdell. 2009. Stable isotopic evidence for trophic groupings and bio-regionalization of predators and their prey in oceanic waters off eastern Australia. *Marine Biology* 156:1241–53.

Rooker, J. R., D. H. Secor, V. S. Zdanowicz, and T. Itoh. 2001. Discrimination of northern bluefin tuna from nursery areas in the Pacific Ocean using otolith chemistry. *Marine Ecology Progress Series* 218:275–82.

Rooker, J. R., D. H. Secor, G. De Metrio, R. Schloesser, B. A. Block, and J. D. Neilson. 2008. Natal homing and connectivity in Atlantic bluefin tuna populations. *Science* 322:742–44.

Rooker, J. R., R. David Wells, D. G. Itano, S. R. Thorrold, and J. M. Lee. 2016. Natal origin and population connectivity of bigeye and yellowfin tuna in the Pacific Ocean. *Fisheries Oceanography* 25:277–91.

Secor, D. H., B. Gahagan, and J. R. Rooker. 2012. Atlantic bluefin tuna stock mixing within the US North Carolina recreational fishery, 2011–2012. International Commission for the Conservation of Atlantic Tunas (ICCAT) Standing Committee on Research and Statistics/2012/88, Madrid.

Seminoff, J. A., S. R. Benson, K. E. Arthur, T. Eguchi, P. H. Dutton, R. F. Tapilatu, and B. N. Popp. 2012. Stable isotope tracking of endangered sea turtles: Validation with satellite telemetry and $\delta^{15}$N analysis of amino acids. *PLOS ONE* 7:e37403.

Shiao, J.-C., T.-F. Yui, H. Høie, U. Ninnemann, and S.-K. Chang. 2009. Otolith O and C stable isotope compositions of southern bluefin tuna *Thunnus maccoyii* (Pisces: Scombridae) as possible environmental and physiological indicators. *Zoological Studies* 48 (1): 71–82.

Shiao, J. C., S. W. Wang, K. Yokawa, M. Ichinokawa, Y. Takeuchi, Y. G. Chen, and C. C. Shen. 2010. Natal origin of Pacific bluefin tuna *Thunnus orientalis* inferred from otolith oxygen isotope composition. *Marine Ecology Progress Series* 420:207–19.

Wang, C. H., Y. T. Lin, J. C. Shiao, C. F. You, and W. N. Tzeng. 2009. Spatio-temporal variation in the elemental compositions of otoliths of southern bluefin tuna *Thunnus maccoyii* in the Indian Ocean and its ecological implication. *Journal of Fish Biology* 75:1173–93.

Wells, R. D., J. R. Rooker, and D. G. Itano. 2012. Nursery origin of yellowfin tuna in the Hawaiian Islands. *Marine Ecology Progress Series* 461:187–96.

Wells, R. J. D., M. J. Kinney, S. Kohin, H. Dewar, J. R. Rooker, and O. E. Snodgrass. 2015. Natural tracers reveal population structure of albacore (*Thunnus alalunga*) in the eastern North Pacific. *ICES Journal of Marine Science* 72:2118–27.

Whitlock, R. E., M. K. McAllister, and B. A. Block. 2012. Estimating fishing and natural mortality rates for Pacific bluefin tuna (*Thunnus orientalis*) using electronic tagging data. *Fisheries Research* 119–120:115–27.

Whitlock, R. E., E. L. Hazen, A. Walli, C. Farwell, S. J. Bograd, D. G. Foley, M. Castleton, and B. A. Block. 2015. Direct quantification of energy intake in an apex marine predator suggests physiology is a key driver of migrations. *Science Advances* 1:e1400270.

CHAPTER 9

# Tagging to Reveal Foraging, Migrations, and Mortality of Pacific Bluefin Tuna

Rebecca E. Whitlock, Murdoch K. McAllister, and Barbara A. Block

## Introduction

Pacific bluefin tuna (*Thunnus orientalis*), hereafter PBT, is a highly migratory species, with the largest home range of any tuna in the genus *Thunnus*. PBT are distributed primarily throughout the temperate waters of the northern Pacific Ocean, but they also range into the western South Pacific (Collette and Nauen 1983, Bayliff 1994). Although genetic data indicate one PBT stock, two spawning areas are currently recognized. One is in the southwestern North Pacific off Taiwan and the other is in the Sea of Japan (Okiyama 1974, Chen et al. 2006, Tanaka et al. 2007). Unknown proportions of 1- and 2-year-old PBT migrate from the northern section of the western Pacific Ocean (WPO) to the northern section of the eastern Pacific Ocean (EPO) (Polovina et al. 1996). Tagging and stable isotope studies indicate that these transpacific migrants spend one to five years foraging in the California Current before returning to the WPO to spawn (Boustany et al. 2010, Block et al. 2011, Madigan et al. 2014, Tawa et al. 2017).

PBT are managed as a single Pacific-wide stock by the Western and Central Pacific Fisheries Commission and the Inter-American Tropical Tuna Commission. Commercial fisheries for PBT exist throughout their range. Annual reported oceanwide catches have varied between 8,000 and 35,000 tons since the early 1950s (IATTC 2005), averaging ~25,000 tons prior to 1980 and 15,000 tons thereafter. Fishing effort and catches in the EPO decreased between 1960 and 1990 but have rebounded since the late 1990s with the development of the PBT ranching industry off Baja Mexico (Aires-da-Silva et al. 2007; ISC 2016, table 3.2). Pacific bluefin tuna are also exploited by a variety of fisheries in the WPO, including purse seine, longline, and gillnet fisheries (Bayliff 2001). Although international management bodies are actively managing the harvest of this species, there have been questions regarding the effectiveness of proposed management measures, and concerns of collapse have emerged in the past decade. Significant reductions in international fishing effort have been called for to reduce mortality, but implementation of reductions and the monitoring of illegal, unreported, and unregulated (IUU) activity remains challenging in this international fishery. Depletion of the spawning stock biomass (spawning stock biomass in 2014/unfished spawning stock biomass) is estimated at 2.6%, while fishing mortality reference points evaluated in the most recent stock assessment indicate an overfished status (ISC 2016).

Electronic tagging of bluefin tunas across the globe has emerged as a powerful tool for gathering information about spatial and temporal dynamics that is relevant to fisheries management (Lutcavage et al. 2000; Block et al. 2001, 2005, 2011; Inagake et al. 2001; Royer et al. 2005; Kitagawa et al. 2007; Teo et al. 2007a,b; Walli et al. 2009; Boustany et al. 2010; Taylor et al. 2011; Galuardi and Lutcavage 2012; Whitlock et al. 2015). Biological data obtained from extensive electronic tagging have the capacity to improve our knowledge of key life history parameters, for example, age-specific migration, fishing and natural mortality rates (Kurota et al. 2009, Taylor et al. 2011, Whitlock et al. 2012, Wilson et al. 2015), and spatial habitat-use patterns related to foraging (Whitlock et al. 2015) or spawning (Hazen et al. 2016). Together with genetics and microconstituent and stable isotope data, tagging data can help distinguish stock of origin, which is key to making unbiased assessments of population-specific exploitation rates and depletion (Taylor et al. 2011; Kerr et al. 2012, 2016). In the Pacific, stable isotope analysis has provided novel insights on the age and time of year at which juvenile PBT enter the EPO and improved residency estimates (Madigan et al. 2012, 2014, 2018; Baumann et al. 2015; Tawa et al. 2017). Electronic tagging data can also provide detailed information about foraging behavior in endothermic bluefin tuna, using measurements of visceral heat generated during digestion from implanted archival tags (Gunn et al. 2001, Whitlock et al. 2013).

In this chapter, we illustrate the applications of electronic tagging with two studies on PBT: estimates of natural and fishing mortality and movement rates from a spatially structured mark-recapture model (Whitlock et al. 2012) and estimates of daily energetic intake in wild juvenile PBT using measurements of visceral heat generation made in the laboratory (Whitlock et al. 2013, 2015).

## Bayesian Mark-Recapture Model

In the following pages, we update the analysis of Whitlock et al. (2012) with archival tag recapture data, providing an example of how electronic tagging data can be used to estimate rates of fishing and natural mortality in a highly migratory fish species. For the purposes of analysis, we define three model areas: the northern EPO (1), the southern EPO (2), and the WPO (3) (Figure 9.1). The boundary between the northern and southern boxes in the EPO was assigned such that an approximately equal number of releases occurred on either side of the boundary, to facilitate estimation of movement rates between the EPO boxes. This division also coincides approximately with the southernmost limit of the USA's Exclusive Economic Zone (EEZ) in the Pacific Ocean, and as such, it has relevance with regard to fishing dynamics and fishery management.

### Data

We analyze recapture data from archival-tagged PBT released between 2002 and 2014 (last recapture in May 2016). Tagging experiments, inclusive of deployments utilizing archival tags and pop-up satellite archival tags (PSAT tags), were conducted under the Tagging of Pacific Pelagics (TOPP) Program

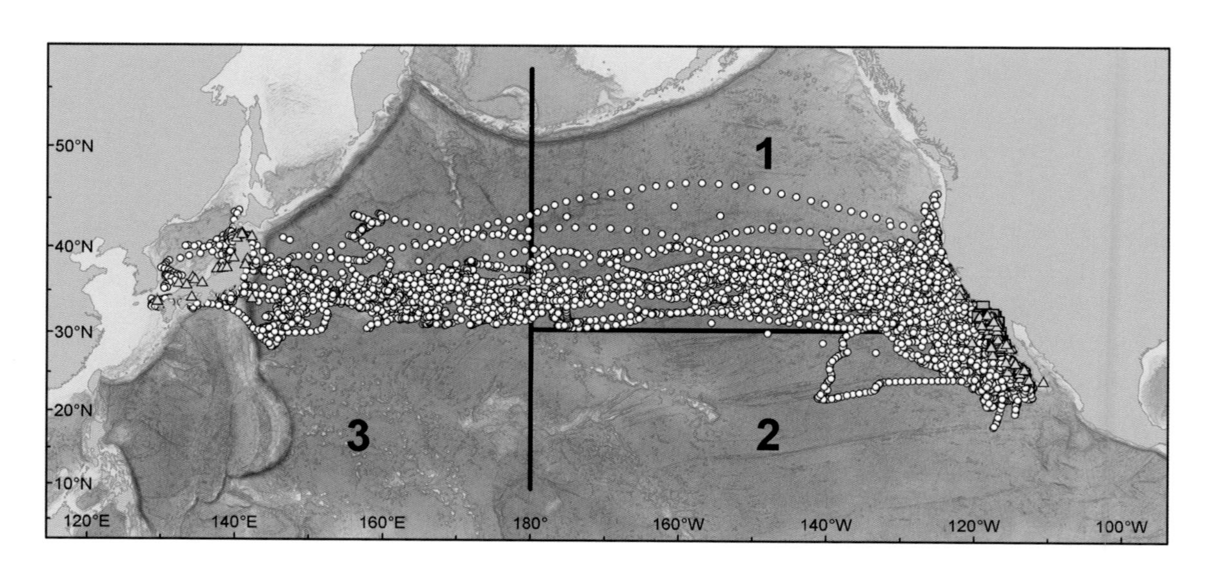


**Figure 9.1.** Spatial areas used in the Bayesian mark-recapture model for bluefin tuna overlaid with release (*orange squares*) and recapture (*yellow triangles*) locations and daily geolocation positions (*white circles*) for the archival tagging data set.

(see Kitagawa et al. 2007, Boustany et al. 2010, Block et al. 2011). Details of tagging protocols can be found in Whitlock et al. (2012) and Boustany et al. (2010). Data used in the mark-recapture model comprise release location (vessel GPS) data for 747 surgically implanted archival tags released off the Baja California coast. Recapture end-point GPS locations were available for 338 archival-tagged PBT of a total 371 recaptures (50%). Geolocation estimates of daily positions were obtained using light (longitude)–based algorithms and sea surface temperature (latitude)–based algorithms (Teo et al. 2004). A state-space modeling approach (Jonsen et al. 2005, Block et al. 2011) was used to refine daily position estimates and quantify the uncertainty associated with the positions. Quarterly locations (for quarters intermediate between release and recapture) were assigned as the model area in which maximal occupancy (number of days) occurred during each quarter. Curved fork lengths (CFLs) were measured in the tagging cradle (in cm) for all electronically tagged fish and converted to straight fork lengths (SFLs) using $SFL = -1.43 + 0.98CFL$. Ages at tagging were then assigned using an inverse von Bertalanffy equation with parameters taken from Shimose et al. (2009):

$$1.\ a = -0.254 - \left(\frac{1}{0.173}\right) log\left(1 - \frac{SFL}{249.6}\right).$$

Ages at release were converted to an age in quarters ($q$) for the model, which uses a quarterly time step:

$$2.\ q = trunc(4a),$$

where *trunc* denotes truncation to the nearest integer (Figure 9.2).

## Bayesian Mark-Recapture Model

We use an age-structured state-space Bayesian mark-recapture model, with a plus group for Pacific bluefin tuna of age 4 and older. The model describes the survival, movement, and capture of archival-tagged fish, with a quarterly time step (Q1, January–March; Q2, April–June; Q3, July–September; Q4, October–December). Quarterly positions from geolocation data (area of maximum occupancy during a quarter) were used to further inform estimates of movement parameters. This model was implemented by the addition of a state-space likelihood for the positions of individual archival-tagged PBT in each quarter (quarters intermediate between release and recapture; details provided below).

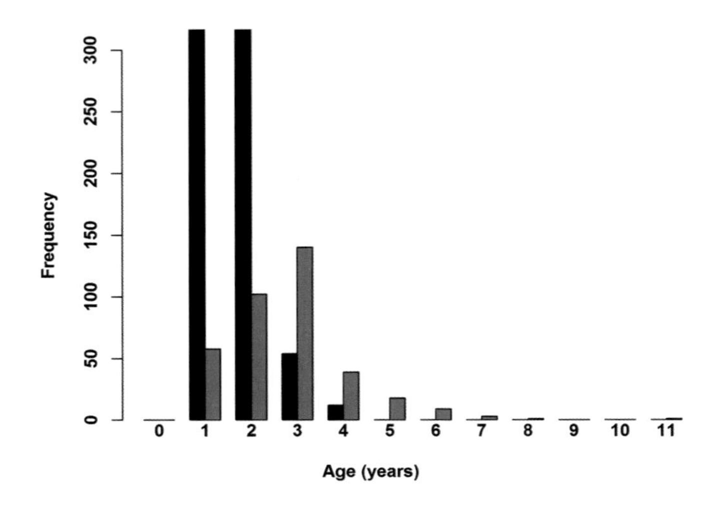


**Figure 9.2.** Frequency distributions for ages at release (*black*) and recapture (*red*) (based on age at release plus years at large).

Of recaptured archival tags, 91% were recovered in the EPO boxes, so this analysis is intended primarily to estimate fishing and natural mortality rates for the EPO. Rates of instantaneous fishing mortality (*F*) in the EPO areas were not disaggregated by fleet or gear type (approximately three-quarters of the tags were recovered by the Mexican purse seine fishery). *F*s were estimated by area and quarter, and an annual *F* multiplier was estimated and applied to the seasonal *F*s (i.e., the intra-annual seasonal pattern was assumed to be the same across years for the EPO [areas 1 and 2, Figure 9.1] and the WPO [area 3, Figure 9.1], respectively) (Equation 3). The *F* multiplier in the first year in the time series (2002) was set equal to 1, so that estimated *F* multipliers for subsequent years are relative to the 2002 value. We also tested three different configurations for the disaggregation of fishing mortality parameters by age—(i) age-invariant *F*, (ii) 2-parameter logistic selectivity function, and (iii) age-specific *F*—by computing the deviance information criterion (DIC) statistic (Spiegelhalter et al. 2002) for each configuration:

3. $\bar{F}_{y,s,j} = F_{s,j}\phi_{y,g(j)},$ (i)

$\bar{F}_{y,s,a,j} = F_{s,j}Fsel_a\phi_{y,g(j)},$ (ii)

where $Fsel_a = [1 + exp\ (-\sigma_{sel}\ (a - \mu_{sel}))]^{-1}$

$$\bar{F}_{y,s,a,j} = F_{s,a,j}\phi_{y,g(j)}, \qquad \text{(iii)}$$

where $F_{s,j}$ ($F_{s,a,j}$) is the quarterly fishing mortality rate for PBT in season $s$ and area $j$ (at age $a$ years) and $\phi_{y,g(j)}$ is the annual $F$ multiplier for year $y$ and area group $g(j)$. $g(j)$ takes the value 1 for areas 1 and 2 (EPO), and 2 for area 3 (WPO). To represent the general case, age ($a$) subscripts are kept for $F$ parameters in the following equations.

### Population Dynamics

In the equations that follow, $S$ denotes the number of seasons (4), whereas $J$ denotes the number of areas (3) in the archival tag model. $q$ denotes age in quarters, while $a(q)$ denotes age in years. The number of tagged fish of age $q$ quarters in area $i$ in year $y$ and quarter $s$, in the quarter of release, before fishing and natural mortality take place, is given by:

$$4.\ N_{y,s,q,k} = \sum_{j=1}^{J} R_{y,s,q,j}(1-\gamma)p_{s,j,k},$$

where $R_{y,s,q,j}$ is the number of tagged fish of age $q$ quarters released in area $j$ in year $y$ and quarter $s$, and $p_{s,j,k}$ is the probability of movement from area $j$ to area $k$ in quarter $s$. $\gamma$ is the fraction of fish dying from injuries induced by tagging (applied only to newly tagged and released fish). The number of tagged fish of age $q$ surviving at the beginning of season $s$ in year $y$ and area $k$ is given by:

$$5.\ N'_{y,s,q,k} = N_{y-1,S,q-1,k}\,e^{(-M_{a(q-1)}+F_{S,a(q-1),k}\phi_{y-1,g(k)})}\varepsilon_{y-1,a(q-1)} \qquad s=1$$

$$N'_{y,s,q,k} = N_{y,s-1,q-1,k}\,e^{(-M_{a(q-1)}+F_{s-1,a(q-1),k}\phi_{y,g(k)})}\varepsilon_{y,a(q-1)} \qquad s>1$$

where $M_{a(q)}$ is the rate of instantaneous natural mortality for fish of $a$ years old, scaled to the quarterly time step. $e^{(-M_{a(q)}+F_{s,a(q),k}\phi_{y,g(k)})}$ is the quarterly survival rate and $\varepsilon_{y,a(q)}$ is a term for process error in survival for fish of $a$ years old. Process error in age-specific quarterly survival rates is modeled using a symmetric uniform distribution centred around 1 with variance dependent on the length of the model time step and the total mortality rate, $Z_{y,s,a(q),k} = M_{a(q)} + \bar{F}_{y,s,a(q),k}$ (Michielsens et al. 2006). The number of tagged fish of age $q$ quarters in area $k$ in year $y$ and quarter $s$ before fishing and natural mortality take place is given by:

$$6.\ N_{y,s,q,k} = \sum_{j=1}^{J} N'_{y,s,q,j}\,p_{s,j,k},$$

and the predicted number of recaptured tagged fish of age $q$ in year $y$, quarter $s$ and area $k$, $C_{y,s,q,k}$ is given by:

$$7.\ C_{y,s,q,k} = N_{y,s,q,k} \frac{F_{s,a(q),k}\phi_{y,g(k)}}{Z_{y,s,a(q),k}} (1 - exp(-Z_{y,s,a(q),k}))\lambda_{g(k)},$$

where $\lambda_{g(k)}$ is the average fraction of archival tags that are reported in area group $g(k)$. The number of reported and recaptured tags in each quarter, area, and age group was assumed to follow a negative binomial distribution:

$$8.\ T_{y,s,q,k} \sim Negative - Binomial(C_{y,s,q,k}, \tau),$$

where $T_{y,s,q,k}$ is the observed number of recaptures of archival-tagged PBT of age $q$ quarters in year $y$, quarter $s$, and area $k$. $C_{y,s,q,k}$ is the mean of the negative binomial distribution, and $\tau$ is a parameter that controls the degree of overdispersion of the negative binomial distribution, where the distribution tends to a Poisson distribution as $\tau \rightarrow \infty$.

*Likelihood for Archival Track Positions*

The state-space likelihood for the movement of archival-tagged PBT is given by

state-equation

$$X_{i,t+1} \mid X_{i,t} \sim Multinomial\left(1,\ X_{i,t}\begin{bmatrix} p_{s[t],1,1} & p_{s[t],1,2} & p_{s[t],1,3} \\ p_{s[t],2,1} & p_{s[t],2,2} & p_{s[t],2,3} \\ p_{s[t],3,1} & p_{s[t],3,2} & p_{s[t],3,3} \end{bmatrix}\right),$$

where $X_{i,t}$ denotes the state of individual $i$ at time $t$.

Observation model

$$Y_{i,t} \mid X_{i,t} \sim Multinomial\left(1,\ X_{i,t}\begin{bmatrix} \pi & \frac{1-\pi}{2} & \frac{1-\pi}{2} \\ \frac{1-\pi}{2} & \pi & \frac{1-\pi}{2} \\ \frac{1-\pi}{2} & \frac{1-\pi}{2} & \pi \end{bmatrix}\right),$$

where $Y_{i,t}$ denotes the observation for individual $i$ at time $t$, and $\pi$ is the probability that the observed quarterly area is correctly assigned to the true

**Table 9.1.** State and observation vectors in the state-space likelihood for the movement of archival-tagged Pacific bluefin tuna

| Vector | Interpretation |
|---|---|
| Random state vector $X_{i,t}$ | |
| (1,0,0) | Alive in area 1 |
| (0,1,0) | Alive in area 2 |
| (0,0,1) | Alive in area 3 |
| Random observation vector $Y_{i,t}$ | |
| (1,0,0) | Majority of geolocations during quarter *s* in area 1 |
| (0,1,0) | Majority of geolocations during quarter *s* in area 2 |
| (0,0,1) | Majority of geolocations during quarter *s* in area 3 |

area (it is assumed that the probability of misclassification to either of the two other areas is equal) (Table 9.1).

### Priors

The prior applied to interannual *F*s was uninformative, with an a priori assumption of equal *F* across seasons in both the EPO and the WPO ($F_{s,k}$ and $F_{s,a,k}$ parameters [Table 9.2]). Priors for *F* multipliers in the EPO between 2002 and 2006 were based on the ratios of estimates of annual fishing effort in this region to estimated fishing effort in 2002 (Aires-da-Silva et al. 2007). EPO *F* multipliers from 2007 onward and WPO *F* multipliers were given a prior with a median of 1. Prior medians for age-specific rates of annual instantaneous natural mortality (*M*) in the archival tag model follow values used in the International Scientific Committee for Tuna and Tuna-like Species in the North Pacific Ocean (ISC) stock assessment (ISC 2016): 0.40 $yr^{-1}$ (95% prior probability interval 0.12–1.38) for age-1 PBT, and 0.25 $yr^{-1}$ (95% prior probability interval 0.07–0.85) for ages 2, 3, and 4+. A prior standard deviation of 0.63 (standard deviation of log [*x*]) was used for all *M* priors. Separate tag reporting rates were estimated for fisheries operating in the EPO and the WPO. Because we lacked independent information about WPO and EPO reporting rates, we chose to apply the same prior probability density function (PDF) in both areas. An uninformative Beta prior was used for archival tag reporting rates in the EPO and the WPO (Table 9.2). An informative prior was placed on the probability of mortality resulting from archival tag implantation, $\gamma$ (Table 9.2). This prior had a

**Table 9.2.** Priors in the mark-recapture model for Pacific bluefin tuna. Standard deviations of lognormal distributions are given as the standard deviation of log ($x$)

| Parameter | Prior | Median | Standard deviation | *F* Model |
|---|---|---|---|---|
| $F_{s,k}$ | *Lognormal*(−2.00,0.50) | 0.14 | 0.71 | i and ii |
| $F_{s,a,k}$ | *Lognormal*(−2.00,0.50) | 0.14 | 0.71 | iii |
| $\mu_{sel}$ | *Normal*(2.00,0.20) | 2.00 | 0.45 | ii |
| $\sigma_{sel}$ | *Lognormal*(1.39,0.20) | 4.01 | 0.45 | ii |
| $M_1$ | *Lognormal*(−0.92,0.40) | 0.40 | 0.63 | All |
| $M_2$ | *Lognormal*(−1.39,0.40) | 0.25 | 0.63 | All |
| $M_3$ | *Lognormal*(−1.39,0.40) | 0.25 | 0.63 | All |
| $M_4$ | *Lognormal*(−1.39,0.40) | 0.25 | 0.63 | All |
| $p_{s,j,k}$ | *Dirichlet*($a_p$) | 0.25 | 0.30 | All |
| $\gamma$ | *Beta*(1,19) | 0.036 | 0.047 | All |
| $\lambda_1$ | *Beta*(2,2) | 0.50 | 0.22 | All |
| $\lambda_2$ | *Beta*(2,2) | 0.50 | 0.22 | All |
| $\varphi_{2,1}$ | *Lognormal*(0.05,0.36) | 1.05 | 0.60 | All |
| $\varphi_{3,1}$ | *Lognormal*(0.58,0.36) | 1.79 | 0.60 | All |
| $\varphi_{4,1}$ | *Lognormal*(0.33,0.36) | 1.39 | 0.60 | All |
| $\varphi_{5,1}$ | *Lognormal*(0.81,0.36) | 2.25 | 0.60 | All |
| $\varphi_{6:15,1}$ | *Lognormal*(0.00,0.36) | 1.00 | 0.60 | All |
| $\varphi_{2:15,1}$ | *Lognormal*(0.00,0.36) | 1.00 | 0.60 | All |
| $\pi$ | *Beta*(95,5) | 0.95 | 0.02 | All |
| $\tau$ | *Uniform*(0.1,100) | 50.0 | 28.9 | All |

median of 0.05, based on an estimate of catch-and-release mortality for ABT of 0.05 from Stokesbury et al. (2011).

Inference was performed using JAGS (Just Another Gibbs Sampler) version 4.2.0 (Plummer 2015). The first 40,000 iterations were discarded as burn-in, after which 20,000 more iterations were kept and thinned at an interval of 10 to yield a final sample of 2,000 iterations. Two chains were run in parallel; convergence was checked using the Gelman-Rubin diagnostic and by visual inspection of trace plots.

## Results

DIC statistics for the three fishing mortality configurations were as follows: (i) age-invariant $F$, DIC = 1185, (ii) logistic selectivity, DIC = 1232, (iii) age-specific $F$, DIC = 1301. Results from the mark-recapture model presented in this section are from the age-invariant $F$ model, which had the lowest DIC statistic (Figure 9.3).

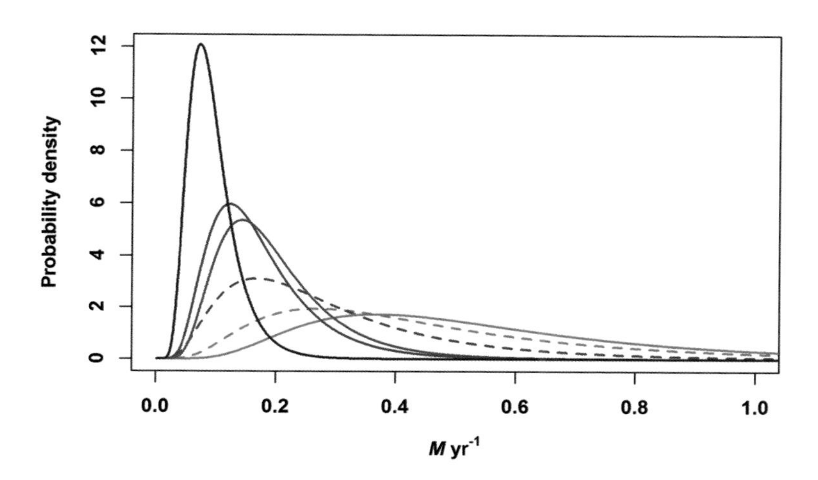


**Figure 9.3.** Probability density functions for the rate of annual instantaneous natural mortality (*M*). (*Dashed red line*, prior age 1; *dashed blue line*, prior ages 2, 3 and 4+. *Solid red line*, posterior age 1; *solid purple line*, posterior age 2; *solid blue line*, posterior age 3; *solid dark blue line*, posterior age 4.)

Priors for age-specific *M*s were updated by the tagging data, with a greater degree of updating (lower posterior coefficients of variation [CVs]) for PBT aged 2 and older (Figure 9.3). Estimated posterior median *M*s (CVs in parentheses) by age were as follows: age 1 0.50 $yr^{-1}$ (0.58), age 2 0.18 $yr^{-1}$ (0.48), age 3 0.16 $yr^{-1}$ (0.50), and age 4 0.09 $yr^{-1}$ (0.43). The posterior for age-1 PBT had a higher median than is used in ISC's stock assessment (0.50 $yr^{-1}$ vs. 0.40 $yr^{-1}$) but was fairly similar to the prior, possibly because relatively few fish were recaptured as 1-year-olds. Posteriors for age-2 and older PBT indicated a lower rate of natural mortality than is currently assumed in the stock assessment, with 77%, 84%, and 99% of the posterior probability associated with lower values than that used in the assessment (0.25 $yr^{-1}$) for age 2, 3, and 4+ PBT, respectively.

Estimates of intra-annual quarterly *F*s revealed a strong season-area pattern to the EPO fishery, with higher estimated fishing mortality rates in quarters 1 and 4 in the northern EPO and in quarters 2 and 3 in the southern EPO (Figure 9.4). Posterior median quarterly *F*s (the product of the estimated intra-annual fishing mortality rate and *F* multiplier) ranged between 0.02 and 0.61 for the northern EPO, between 0.01 and 3.07 for the southern EPO, and between 0.02 and 0.23 for the WPO.

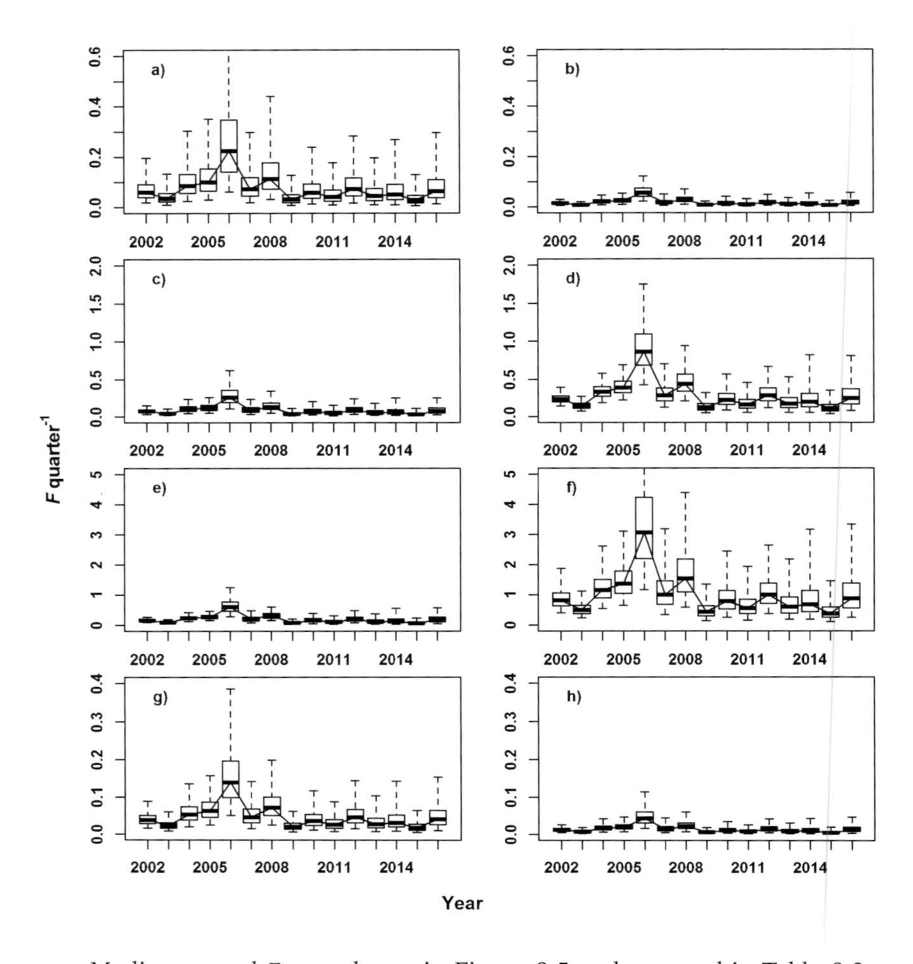


**Figure 9.4.** Posterior distributions for rates of quarterly instantaneous fishing mortality in the EPO, separated into the northern EPO, area 1 (*left column*); and the southern EPO, area 2 (*right column*). Panels (*a*) and (*b*) represent quarter 1, (*c*) and (*d*) represent quarter 2, (*e*) and (*f*) represent quarter 3, and (*g*) and (*h*) represent quarter 4.

Median annual *F*s are shown in Figure 9.5 and reported in Table 9.3. Annual *F*s ranged between 0.16 $yr^{-1}$ and 1.30 $yr^{-1}$ in the northern EPO (Figure 9.5a; Table 9.3). In the southern EPO, annual *F*s ranged between 0.51 $yr^{-1}$ and 4.07 $yr^{-1}$ (Figure 9.5b, Table 9.3). Areas 1 and 2 showed a similar temporal pattern of estimated annual *F*s, with the highest values occurring in 2006 and 2008 (Figure 9.5; Table 9.3). Median annual *F*s ranged between 0.24 $yr^{-1}$ and 0.61 $yr^{-1}$ in the WPO (Figure 9.5c). The annual pattern in WPO *F*s was fairly flat, reflecting the relatively sparse recapture information in this area (Figure 9.5c). Posterior correlations between estimates of annual *F* and *M* were low, with a maximum absolute pairwise correlation of 0.07. The maximum absolute pairwise correlation between estimated annual *F*s and tag reporting rates was 0.22.

Estimated quarterly movement rates within the EPO showed a strong seasonal pattern (Figure 9.6). During winter (quarter 4) and spring (quarter 1), PBT moved south from area 1 to area 2 (Figure 9.6, row 2), while

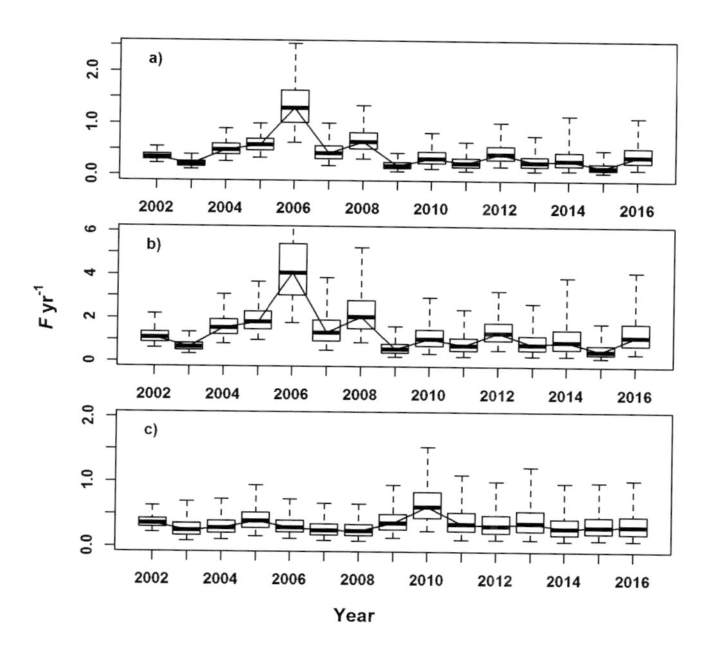


**Figure 9.5.** Posterior distributions for rates of annual instantaneous fishing mortality in the (*a*) northern EPO, (*b*) southern EPO, and (*c*) WPO.

**Table 9.3.** Posterior median annual instantaneous fishing mortality rates, *F*, $yr^{-1}$ from the mark-recapture model for Pacific bluefin tuna

| Year | Northern EPO | Southern EPO | WPO |
|---|---|---|---|
| 2002 | 0.35 (0.08) | 1.09 (0.42) | 0.36 (0.11) |
| 2003 | 0.21 (0.07) | 0.67 (0.27) | 0.25 (0.16) |
| 2004 | 0.49 (0.16) | 1.55 (0.59) | 0.29 (0.16) |
| 2005 | 0.58 (0.17) | 1.82 (0.72) | 0.39 (0.20) |
| 2006 | 1.30 (0.51) | 4.07 (2.05) | 0.29 (0.17) |
| 2007 | 0.42 (0.21) | 1.33 (0.89) | 0.25 (0.15) |
| 2008 | 0.65 (0.26) | 2.07 (1.15) | 0.24 (0.15) |
| 2009 | 0.18 (0.10) | 0.58 (0.39) | 0.36 (0.21) |
| 2010 | 0.33 (0.19) | 1.05 (0.70) | 0.61 (0.34) |
| 2011 | 0.23 (0.15) | 0.74 (0.58) | 0.35 (0.27) |
| 2012 | 0.42 (0.22) | 1.33 (0.71) | 0.32 (0.25) |
| 2013 | 0.26 (0.18) | 0.80 (0.66) | 0.36 (0.31) |
| 2014 | 0.29 (0.29) | 0.91 (1.03) | 0.28 (0.25) |
| 2015 | 0.16 (0.12) | 0.51 (0.47) | 0.31 (0.25) |
| 2016 | 0.37 (0.28) | 1.17 (1.04) | 0.31 (0.27) |

*Notes*: Standard deviations in parentheses. EPO = eastern Pacific Ocean; WPO = western Pacific Ocean.

**Figure 9.6.** Posterior distributions for quarterly movement rates. Rows correspond to the area from which movements originate, while columns correspond to the quarter of the year (Q1, January–March; Q2, April–June; Q3, July–September; Q4, October–December).

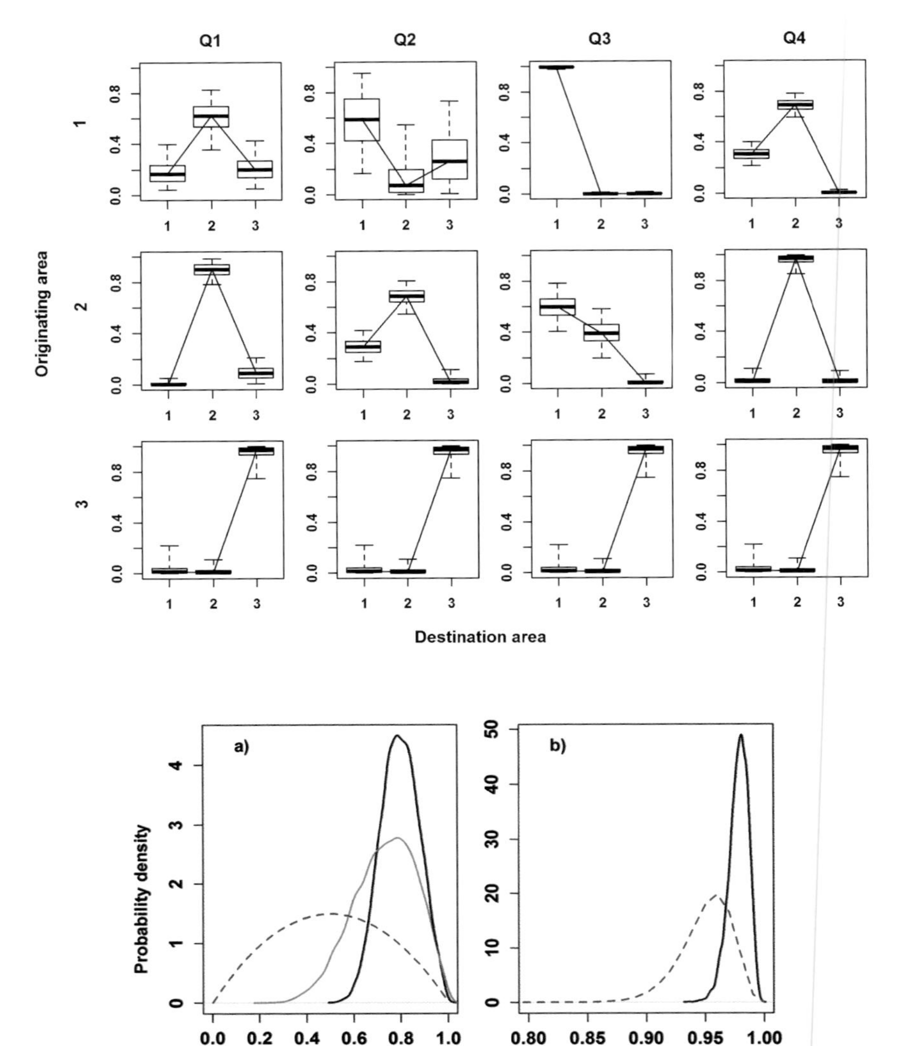


**Figure 9.7.** (*a*) Probability density functions (PDFs) for tag reporting rates. Prior, *dashed blue line*; posterior western Pacific Ocean, *solid red line*; posterior eastern Pacific Ocean, *solid black line*. (*b*) PDFs for the probability that an archival tag is correctly detected in area k (prior, *dashed blue line*; posterior, *solid black line*).

during quarters 2 and 3 (April to September), residency in area 1 (northern EPO) was associated with the highest rate (Figure 9.6, row 1). Residency in area 2 was predominant in quarters 1, 2, and 4 (Figure 9.6, row 2), with a higher rate of movement northward into area 1 only in quarter 3 (July to September). A high rate of residency in area 3 (WPO) was estimated for all seasons (Figure 9.6, row 3).

Priors for archival tag reporting rates were also updated, particularly for the EPO (Figure. 9.7a), to yield posterior PDFs with medians of 0.84 (EPO) (standard deviation of 0.07) and 0.79 (WPO) (standard deviation of

0.12). The prior for the fraction of tagging-induced mortalities was updated by the archival tag data to a more precise beta distribution with a median of 0.023 and standard deviation of 0.029, relative to the prior mean of 0.06 and standard deviation of 0.05.

## Quantifying Foraging and Energy Intake in Wild Juvenile Pacific Bluefin Tuna

In bluefin tunas, the metabolic heat generated by digestion, called the heat increment of feeding (HIF), is measurable using viscerally implanted archival tags (Carey et al. 1984) and is strongly correlated with the energetic value of a meal (Gunn et al. 2001, Whitlock et al. 2013). Visceral warming has been used to measure feeding frequency and relative intake size (kg) of feeds in southern bluefin tuna (*Thunnus maccoyii*) in the Indian Ocean (Bestley et al. 2008). Using only estimates of feeding frequency or presence/absence of feeding events may fail to reveal the true variation in energy intake if the magnitude of feeding events is variable. Whitlock et al. (2015) applied experimental estimates of HIF generated per kcal ingested to quantify energy intake in 144 wild PBT in the California Current Large Marine Ecosystem (CCLME). Estimates of HIF $kcal^{-1}$ from a hierarchical Bayesian analysis of laboratory measurements made on similar-sized bluefin tuna in a controlled tank setting at a range of ambient temperatures (Whitlock et al. 2013) were used to predict energy intake for more than 39,000 daily HIF measurements from archival tags implanted in wild PBT. By accounting for the magnitude of feeding events and the effect of ambient temperature on HIF (whereby greater HIF $kcal^{-1}$ is observed at cooler ambient temperatures, consistent with greater heat conservation or reduced dissipation at low temperatures), a more accurate and robust measure of energy acquisition and foraging success in wild bluefin tuna was possible.

Measurements of energy intake in PBT revealed spatial and temporal heterogeneity in foraging success within the CCLME (Figures 9.8 and 9.9). Seasonally, the highest mean daily energy intake occurred in summer (June and July), when fish are distributed between 23°N and 34°N off the coast of the Baja peninsula up to waters of the Southern California Bight (Figures 9.8 and 9.9). Lower energy intake was observed during late summer (August and September), when PBT are moving up through the Southern California Bight (28°N to 32°N). A second seasonal peak in energy intake

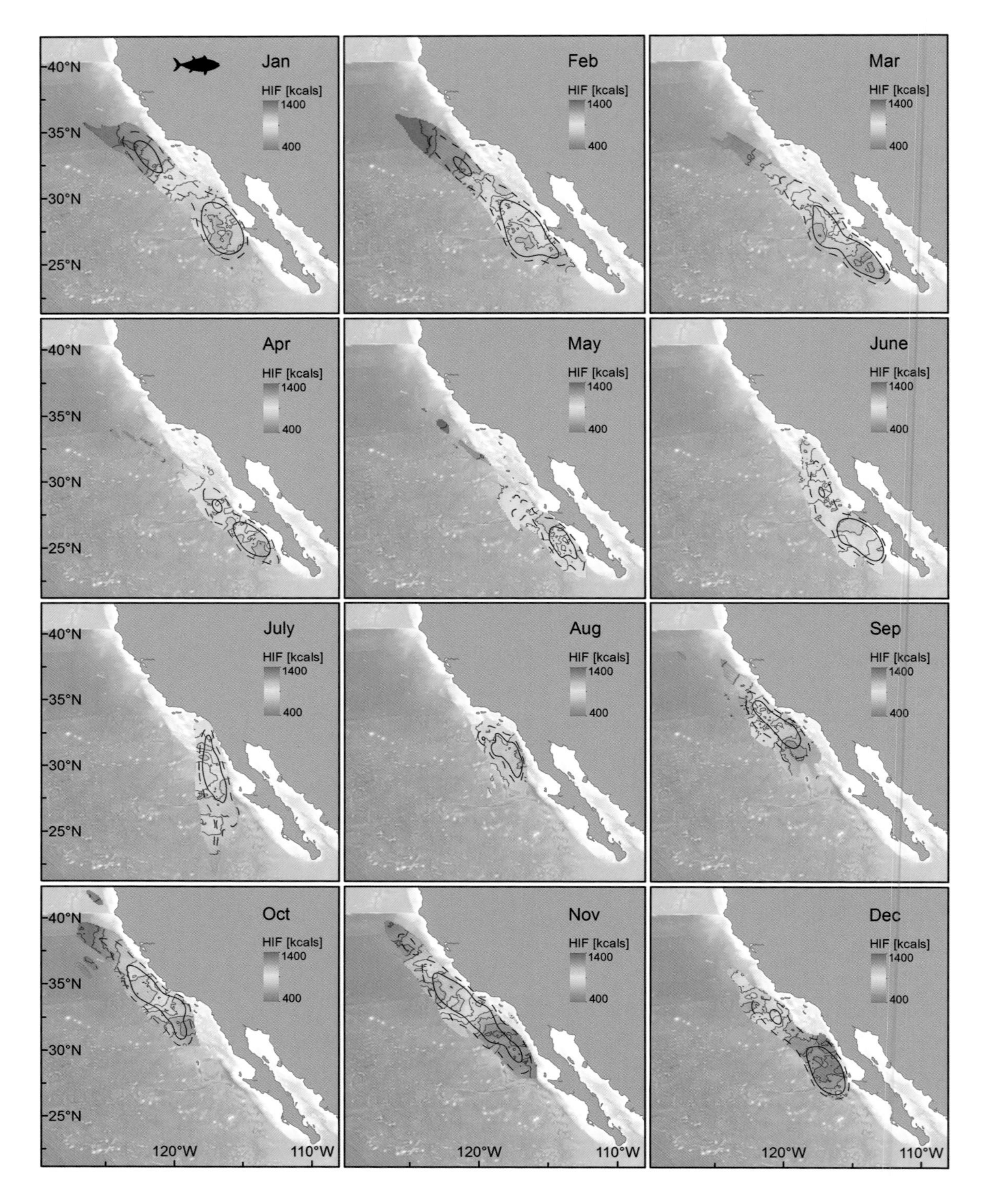


**Figure 9.8.** Predicted values for average daily energy intake (kcal) by month from the final generalized additive mixed model fitted to heat increment of feeding (HIF) data between 2002 and 2009 (data from 144 archival-tagged Pacific bluefin tuna). Predicted values are plotted within the 90% utilization density (UD) contour; the 50% (*solid line*) and 75% (*dashed line*) UD contours are indicated. *Figure reproduced from Whitlock et al. 2015.*

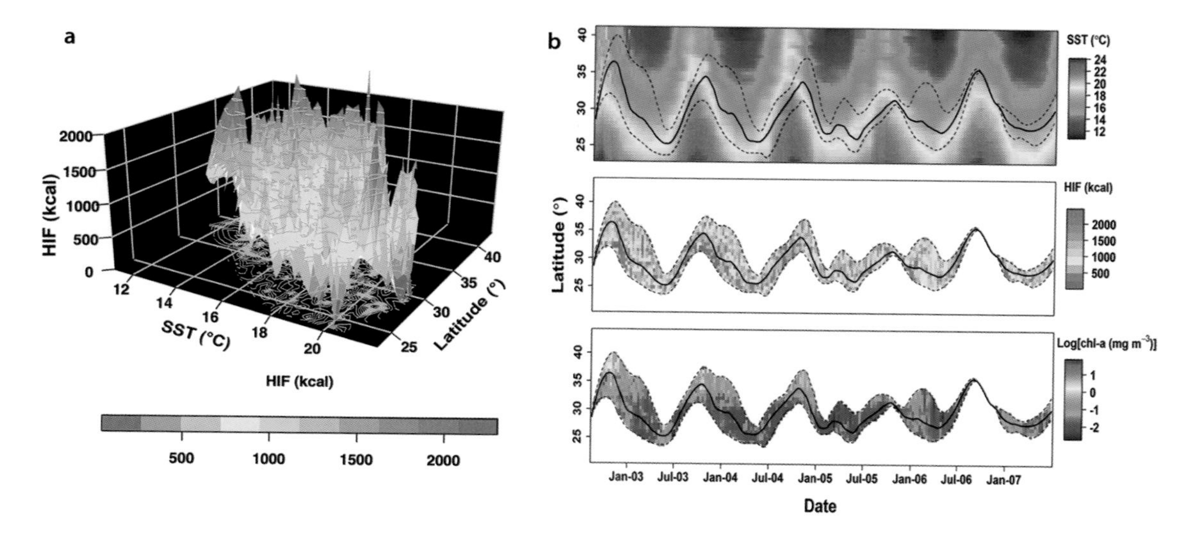


**Figure 9.9.** Heat increment of feeding (HIF) in relation to sea surface temperature (SST), latitude, and chlorophyll-*a*. (*a*) Three-dimensional contour plot for observed daily energy intake in 2003 (interpolated and smoothed for visualization) against mean daily SST and latitude. (*b*) Latitudinal distribution of 144 tagged bluefin tuna in the California Current, 2002 to 2007. Date versus latitude with remotely sensed mean daily SST (°C) indicated by the color scale (*top*). Date versus latitude with median daily energy intake (kcal) indicated by the color scale (*center*). Date versus latitude with the logarithm of median daily chlorophyll-*a* concentration (mg $m^{-3}$) indicated by the color scale (*bottom*). The solid black line in each panel denotes the median latitude of archival-tagged tuna, whereas the dashed lines show the 2.5th and 97.5th percentiles for the latitudinal distribution. *Figure reproduced from Whitlock et al. 2015.*

was observed in autumn (October and November), as the PBT reach their northernmost distribution at latitudes between 34°N and 40°N (Figures 9.8 and 9.9), coinciding with high primary productivity and cool water temperatures in this region (Figure 9.9). The lowest mean energy intake occurred in December, when most PBT have returned south of 34°N (Figures 9.8 and 9.9). High energy intake was observed in January and February, when the bluefin's distribution extended northward again. In autumn and winter (October to February), the highest energy intake values were observed in the northernmost part of the tuna's range (Figure 9.9).

Generalized additive mixed models (Wood 2006) fitted to a suite of environmental covariates showed that energy intake in wild PBT is correlated with sea surface temperature (SST), isothermal layer depth (ILD), relative light level, primary productivity, tuna body length, eddy kinetic energy, day of the year, and location (longitude and latitude) (Whitlock et al. 2015).

Energy intake was predicted to be highest for lower values of SST and ILD, and at intermediate levels of primary productivity, as measured by chlorophyll-*a* (Whitlock et al. 2015). Age-specific patterns of habitat use indicate thermal niche expansion in older and larger PBT, resulting in potentially higher energetic reward for large fish able to exploit seasonally available energy-rich cold waters off central and northern California in October and November (Whitlock et al. 2015).

## Discussion

Electronic tagging data provide new opportunities for improving our understanding of the spatial and temporal components of fish population dynamics. They offer information on the movements, stock structure, life history, and feeding ecology at finer spatial scales than were previously possible in tunas (e.g., Bestley et al. 2010, Kurota et al. 2009, Boustany et al. 2010, Whitlock et al. 2015). Spatially explicit assessment models (e.g., Kurota et al. 2009, Goethel et al. 2011, Taylor et al. 2011) can accommodate information from electronic tagging to account for differential habitat use by fish at different life history stages or by different populations, can describe the spatial overlap between populations and fishing fleets to improve exploitation rate estimates, and can evaluate the implications of habitat loss or modification. We used a spatial model for estimation of fishing ($F$) and natural ($M$) mortality rates for PBT, since this is expected to result in less biased estimates of $F$ and other parameters (e.g., Carruthers et al. 2011); for example, the spatial model accounts for the differing abundances and recapture rates of tagged PBT in different quarters and areas that will affect estimates of $F$.

Seasonal movement patterns estimated in the Bayesian mark-recapture model were consistent with patterns of movement described elsewhere (e.g., Kitagawa et al. 2007, Boustany et al. 2010). The spatial box model utilized here captured the movements of PBT that were tagged primarily in the third quarter off the coast of Baja, Mexico, and subsequently moved northward. The northward movement is thought to be associated with increasing surface water temperatures and a decrease in upwelling and associated production (Block et al. 2011). PBT achieve their most northerly locations along the western coast of North America late in the third and fourth quarters of the year and move southward or offshore in winter months (Boustany et al. 2010, Whitlock et al. 2015). The movement rate estimates from archival tag data demonstrate one of the key advantages of electronic tags over conventional tags, namely that if geolocation is feasible, a data set that is highly informative about movement rates and habitat use can be generated, resulting in improved precision. In addition, locations obtained by geolocation are independent of fisheries and should therefore constitute a less biased source of information about the spatial distribution of fish populations than recapture locations, which will to some extent be affected by the spatial distribution of fishing effort.

Electronic tagging data can be highly informative about rates of natural ($M$) and fishing ($F$) mortality when the tag-reporting rate can be reliably estimated. In the case of archival-tagged PBT, a high tag recovery rate combined with long times at large yielded an informative data set with respect to $M$. The archival tagging data updated prior PDFs for natural mortality for PBT of ages 2, 3, and 4+ but were less informative about $M$ for age-1 PBT, likely owing to the higher proportion of tags recaptured from releases of older fish; thus there is more information available about $M$ from the distribution of times at large for older PBT. $M$ is a key parameter in stock-assessment models (Brodziak et al. 2011), underpinning potential maximum sustainable yield (Beddington and Cooke 1983) and a population's response to harvest. $M$ has commonly been identified as a key source of uncertainty in fish stock assessments (Lapointe et al. 1989, Hilborn and Walters 1992), among them the stock assessment for PBT. Despite this uncertainty, age-specific $M$s are treated as known, fixed parameters in the assessment for PBT, potentially increasing the risk of overexploitation. Currently, only the value of $M$ for age-0 PBT has been empirically derived using tagging data (Takeuchi and Takahashi 2006). Values for fish age 1 and older have been derived using methods based on life history and estimates for southern bluefin tuna (*Thunnus maccoyii*) (Polacheck et al. 1997). Posterior median estimates of $M$ for ages 2 and above from this study were consistently lower than the corresponding value (0.25 $yr^{-1}$) used in recent stock assessments (e.g., ISC 2016). An $M$ of 0.25 $yr^{-1}$ for age-2 and older PBT also appears to be incongruous with the reported maximum age of 26 years for this species (Shimose et al. 2009) and size-frequency distribution observed in the Taiwanese longline fishery (PBT of 200–250+ cm, aged 10 and older) (ISC 2016). A sensitivity analysis for $M$ was conducted in the most recent assessment, although the value corresponding to the low $M$ scenario was 0.225 $yr^{-1}$ (ISC 2016, figure 5-14), which is still considerably higher than the estimates from this analysis. Using empirical estimates of age-specific

*M*s from electronic tagging would help to reduce potential bias in stock reconstructions from the PBT stock assessment, while accounting for the uncertainty associated with estimates of *M* would be more consistent with a precautionary approach. Moreover, integrating tagging data into the stock-assessment model for PBT should improve the accuracy of estimated fishing mortality rates and abundance. This would be desirable in the context of effective regulation of fishing effort for PBT fisheries.

Direct estimates of energy intake in a highly migratory pelagic predator present a number of opportunities for future research, including the bioenergetics of migration and growth, integration of the preyscape (for example, sardine density) with oceanography and feeding success, and predicting how environmental change will affect the CCLME ecosystem via foraging and migration in an apex predator.

Identification of hot spots (for example foraging or spawning grounds) has important implications for the management of highly migratory marine species such as *T. orientalis* because of the potential for increased interaction with fisheries (and higher catchability; Arreguín-Sánchez 1996) when aggregations are targeted because of predictably high fish densities. Interestingly, the highest estimated fishing mortality rates for PBT, in the southern EPO during quarter 3 (July to September), followed by quarter 2 (April to June), coincide with the months when energy intake is estimated to be highest during the year (June and July; Whitlock et al. 2015), and the spatial distribution of PBT is relatively constrained relative to other times of year (Figure 9.8). Further work is needed to investigate whether intensive foraging behavior could be associated with higher catchability in PBT. Identifying hot spots can also aid in the development of spatially explicit fisheries management strategies or in prioritizing habitats for marine zoning (Hooker and Gerber 2004, Hazen et al. 2013). The ability to identify patterns of habitat use and areas of potential aggregation is particularly important for the management of severely depleted migratory stocks such as Pacific bluefin tuna, for which development of spatial management strategies is of high priority. The information on fisheries exploitation rates together with patterns of spatial habitat use, residency, and migration that is provided by electronic tags makes them an indispensable tool in the development of sound management measures for rebuilding PBT stocks.

## Acknowledgments

The authors thank the captains, T. Dunn, N. Kagawa, and B. Smith, and crew of the fishing vessel Shogun for their help with the capture, archival tagging, and release of wild Pacific bluefin tuna. Thanks go to the technical staff of the Tuna Research and Conservation Centre (R. Schallert, J. Noguiera, L. Rodriguez, and students), A. Norton, and the many teams that helped in the tagging efforts through the years. The authors are grateful to the Mexican government for permitting access to bluefin tunas in Mexican waters for tagging and release and to T. Baumgartner McBride, Department of Biological Oceanography, Centro de Investigación Científica y de Educación Superior de Ensenada University, for his support. The authors would also like to thank G. Lawson for making available his code to extract isothermal layer depth and for helpful discussions in the early stages of this work, and A. Swithenbank for assistance in automating codes to run through the Pacific bluefin tuna archival tag data set.

### REFERENCES

Aires-da-Silva, A., M.G. Hinton, and M. Dreyfus. 2007. *Standardized catch rates for Pacific bluefin tuna caught by United States- and Mexican-flag fisheries in the Eastern Pacific Ocean (1960–2006)*. International Scientific Committee for tuna and tuna-like species in the North Pacific Ocean (ISC) working group paper. ISC/07/PBF-3/01.

Arreguín-Sánchez, F. 1996. Catchability: A key parameter for fish stock assessment. *Reviews in Fish Biology and Fisheries* 6 (2): 221–42.

Baumann, H., R.J.D. Wells, J.R. Rooker, S. Zhang, Z. Baumann, D.J. Madigan, H. Dewar, et al. 2015. Combining otolith microstructure and trace elemental analyses to infer the arrival of juvenile Pacific bluefin tuna in the California current ecosystem. *ICES Journal of Marine Science* 72 (7): 2128–38.

Bayliff, W.H. 1994. A review of the biology and fisheries for northern bluefin tuna, *Thunnus thynnus*, in the Pacific Ocean. *FAO Fisheries Technical Paper* 336:244–95.

Bayliff, W.H. 2001. Status of bluefin tuna in the Pacific Ocean. IATTC Stock Assessment Report 1.

Beddington, J.R., and J.G. Cooke. 1983. On the potential yield of fish stocks. *FAO Fisheries Technical Paper* 242:1–47.

Bestley, S., T.A. Patterson, M.A. Hindell, and J.S. Gunn. 2008. Feeding ecology of wild migratory tunas revealed by archival tag records of visceral warming. *Journal of Animal Ecology* 77 (6): 1223–33.

Bestley, S., T.A. Patterson, M.A. Hindell, and J.S. Gunn. 2010. Predicting feeding success in a migratory predator: Integrating telemetry, environment, and modeling techniques. *Ecology* 91:2373–84.

Block, B.A., H. Dewar, S.B. Blackwell, T.D. Williams, E.D. Prince, C.J. Farwell, A. Boustany, et al. 2001. Migratory movements, depth preferences, and thermal biology of Atlantic bluefin tuna. *Science* 293:1310–14.

Block, B.A., S.L.H. Teo, A. Walli, A. Boustany, M.J.W. Stokesbury, C.J. Farwell, K.C. Weng, H. Dewar, and T.D. Williams. 2005. Electronic tagging and population structure of Atlantic bluefin tuna. *Nature* 434:1121–27.

Block, B.A., I.D. Jonsen, S.J. Jorgensen, A.J. Winship, S.A. Shaffer, S.J. Bograd, E.L. Hazen, et al. 2011. Tracking apex marine predator movements in a dynamic ocean. *Nature* 475:86–90. doi:10.1038/nature10082.

Boustany, A.M., R. Matteson, M. Castleton, C. Farwell, and B.A. Block. 2010. Movements of pacific bluefin tuna (*Thunnus orientalis*) in the Eastern North Pacific revealed with archival tags. *Progress in Oceanography* 86:94–104.

Brodziak, J., J. Ianelli, K. Lorenzen, and R. Methot Jr. 2011. Estimating natural mortality in stock assessment applications. Technical report, NOAA Technical Memorandum NMFSF/SPO-119.

Carey, F.G., J.W. Kanwisher, and E.D. Stevens. 1984. Bluefin tuna warm their viscera during digestion. *Journal of Experimental Biology* 109 (1): 1–20.

Carruthers, T.R., M.K. McAllister, and N.G. Taylor. 2011. Spatial surplus production modeling of Atlantic tunas and billfish. *Ecological Applications* 21:2734–55.

Chen, K.S., P. Crone, and C.C. Hsu. 2006. Reproductive biology of female Pacific bluefin tuna *Thunnus orientalis* from the south-western North Pacific Ocean. *Fisheries Science* 72:985–94.

Collette, B.B., and C.E. Nauen. 1983. FAO Species Catalogue, vol. 2, Scombrids of the world. *FAO Fisheries Synopsis* 125 (2): 1–137. www.fao.org/docrep/009/ac478e/ac478e00.htm.

Galuardi, B., and M. Lutcavage. 2012. Dispersal routes and habitat utilization of juvenile Atlantic bluefin tuna, *Thunnus thynnus*, tracked with mini PSAT and archival tags. *PLOS ONE* 7 (5): e37829.

Goethel, D.R., T.J. Quinn, and S.X. Cadrin. 2011. Incorporating spatial structure in stock assessment: Movement modeling in marine fish population dynamics. *Reviews in Fisheries Science* 19 (2): 119–36.

Gunn, J., J. Hartog, and K. Rough. 2001. The relationship between food intake and visceral warming in southern bluefin tuna (*Thunnus maccoyii*). In *Electronic Tagging and Tracking in Marine Fisheries*, edited by J.R. Sibert and J.L. Nielsen, 109–30. Dordrecht, Netherlands: Kluwer Academic.

Hazen, E.L., R.M. Suryan, J.A. Santora, S.J. Bograd, Y. Watanuki, and R.P. Wilson. 2013. Scales and mechanisms of marine hotspot formation. *Marine Ecology Progress Series* 487:177–83.

Hazen, E.L., A.B. Carlisle, S.G. Wilson, J.E. Ganong, M.R. Castleton, R.J. Schallert, and B.A. Block. 2016. Quantifying overlap between the Deepwater Horizon oil spill and predicted bluefin tuna spawning habitat in the Gulf of Mexico. *Scientific Reports* 6:33824.

Hilborn, R., and C.J. Walters. 1992. *Quantitative Fisheries Stock Assessment: Choice, Dynamics, and Uncertainty.* New York: Chapman and Hall.

Hooker, S.K., and L.R. Gerber. 2004. Marine reserves as a tool for ecosystem-based management: The potential importance of megafauna. *Bioscience* 54 (1): 27–39.

IATTC (Inter-American Tropical Tuna Commission). 2005. Tunas and billfishes in the Eastern Pacific Ocean in 2004. Fishery Status Report No. 3, La Jolla, California.

Inagake, D., H. Yanada, K. Segawa, M. Okazaki, A. Nitta, and T. Itoh. 2001. Migration of young bluefin tuna, *Thunnus orientalis* Temminck et Schlegal, through archival tagging experiments and its relation with oceanographic conditions in the western North Pacific. *Bulletin of the National Research Institute of Far Seas Fisheries* 38:53–81.

ISC (International Scientific Committee for Tuna and Tuna-Like Species in the North Pacific Ocean). 2016. Annex 9: 2016 Pacific Bluefin Tuna Stock Assessment, July 13–18, 2016, Hokkaido, Japan.

Jonsen, I.D., J.M. Flemming, and R.A. Myers. 2005. Robust state-space modeling of animal movement data. *Ecology* 86:2874–80.

Kerr, L.A., S.X. Cadrin, and D.H. Secor. 2012. Evaluating population effects and management implications of mixing between Eastern and Western Atlantic bluefin tuna stocks. *ICES CM* 13. www.ices.dk/sites/pub/CM%20Doccuments/CM-2012/N/N13 12.pdf.

Kerr, L.A., S.X. Cadrin, D.H. Secor, and N.G. Taylor. 2016. Modeling the implications of stock mixing and life history uncertainty of Atlantic bluefin tuna. *Canadian Journal of Fisheries and Aquatic Sciences* 999:1–15. doi.org/10.1139/cjfas-2016-0067.

Kitagawa, T., A.M. Boustany, C.J. Farwell, T.D. Williams, M.R. Castleton, and B.A. Block. 2007. Horizontal and vertical movements of juvenile Pacific bluefin tuna (*Thunnus orientalis*) in relation to seasons and oceanographic conditions. *Fisheries Oceanography* 16:409–21.

Kurota, H., M.K. McAllister, G.L. Lawson, J.I. Nogueira, S.L.H. Teo, and B.A. Block. 2009. A sequential Bayesian methodology to estimate movement and exploitation rates using electronic and conventional tag data: Application to Atlantic bluefin tuna (*Thunnus thynnus*). *Canadian Journal of Fisheries and Aquatic Sciences* 66:321–42.

Lapointe, M.F., R.M. Peterman, and A.D. MacCall. 1989. Trends in fishing mortality rate along with errors in natural mortality can cause spurious time trends in fish stock abundances estimated by virtual population analysis (VPA). *Canadian Journal of Fisheries and Aquatic Sciences* 46:2129–39.

Lutcavage, M.E., R.W. Brill, G.B. Skomal, B.C. Chase, J.L. Goldstein, and J. Tutein. 2000. Tracking adult North Atlantic bluefin tuna (*Thunnus thynnus*) in the northwestern Atlantic using ultrasonic telemetry. *Marine Biology* 137 (2): 347–58.

Madigan, D.J., S.Y. Litvin, B.N. Popp, A.B. Carlisle, C.J. Farwell, and B.A. Block. 2012. Tissue turnover rates and isotopic trophic discrimination factors in the endothermic teleost, Pacific bluefin tuna (*Thunnus orientalis*). *PLOS ONE* 7 (11): e49220.

Madigan, D.J., Z. Baumann, A.B. Carlisle, D.K. Hoen, B.N. Popp, H. Dewar, O.E. Snodgrass, B.A. Block, and N.S. Fisher. 2014. Reconstructing trans-oceanic migration patterns of 475 Pacific bluefin tuna using a chemical tracer toolbox. *Ecology* 95: 1674–83.

Madigan, D.J., Z. Baumann, A.B. Carlisle, O.E. Snodgrass, H. Dewar, and N.S. Fisher. 2018. Isotopic insights into migration patterns of Pacific bluefin tuna in the eastern Pacific

Ocean. *Canadian Journal of Fisheries and Aquatic Sciences* 75 (2): 260–70. doi.org/10.1139/cjfas-2016-0504.

Michielsens, C.G., M.K. McAllister, S. Kuikki, T. Pakarinen, L. Karlsson, A. Romakkaniemi, I. Pera, and S. Mäntyniemi. 2006. A Bayesian state-space mark-recapture model to estimate exploitation rates in mixed stock fisheries. *Canadian Journal of Fisheries and Aquatic Sciences* 63:321–34.

Okiyama, M. 1974. Occurrence of the postlarvae of bluefin tuna, *Thunnus thynnus*, in the Japan Sea. *Bulletin of the Japan Sea Regional Fisheries Research Laboratory* 25:89–97.

Plummer, M. 2015. *JAGS version 4.0.0 user manual*. Sourceforge.

Polacheck, T., W.S. Hearn, C. Miller, W. Whitelaw, and C. Stanley. 1997. Updated estimates of mortality rates for juvenile SBT from multi-year tagging of cohorts. CCSBT-SC/9707/26.

Polovina, J.J. 1996. Decadal variation in the trans-Pacific migration of northern bluefin tuna (*Thunnus thynnus*) coherent with climate-induced change in prey abundance. *Fisheries Oceanography* 5:114–19.

Royer, F., J.M. Fromentin, and P. Gaspar. 2005. A state-space model to derive bluefin tuna movement and habitat from archival tags. *Oikos* 109 (3): 473–84.

Shimose, T., T. Tanabe, K. Chen, and C. Hsu. 2009. Age determination and growth of Pacific bluefin tuna, *Thunnus orientalis*, off Japan and Taiwan. *Fisheries Research* 100: 134–39.

Spiegelhalter, D.J., N.G. Best, B.P. Carlin, and A. Van Der Linde. 2002. Bayesian measures of model complexity and fit. *Journal of the Royal Statistical Society: Series B (Statistical Methodology)* 64 (4): 583–639.

Stokesbury, M.J.W., J.D. Neilson, E. Susko, and S.J. Cooke. 2011. Estimating mortality of Atlantic bluefin tuna (*Thunnus thynnus*) in an experimental recreational catch-and-release fishery. *Biological Conservation* 144:2684–91.

Takeuchi, Y., and M. Takahashi. 2006. Estimate of natural mortality of age 0 Pacific bluefin tuna from conventional tagging data. ISC/06/PBF-WG/07.

Tanaka, Y., M. Mohri, and H. Yamada. 2007. Distribution, growth, and hatch date of juvenile Pacific bluefin tuna, *Thunnus orientalis* in the coastal area of the Sea of Japan. *Fisheries Science* 73:534–42.

Tawa, A., T. Ishihara, Y. Uematsu, T. Ono, and S. Ohshimo. 2017. Evidence of westward transoceanic migration of Pacific bluefin tuna in the Sea of Japan based on stable isotope analysis. *Marine Biology* 164 (4): 94.

Taylor, N.G., M.K. McAllister, G.L. Lawson, T. Carruthers, and B.A. Block. 2011. Atlantic bluefin tuna: A novel multi-stock spatial model for assessing population biomass. *PLOS ONE* 6 (12): 1–10.

Teo, S.L.H., A. Boustany, S. Blackwell, A. Walli, K.C. Weng, and B.A. Block. 2004. Validation of geolocation estimates based on light level and sea surface temperature from electronic tags. *Marine Ecology Progress Series* 283:81–98.

Teo, S.L., A. Boustany, H. Dewar, M.J. Stokesbury, K.C. Weng, S. Beemer, A.C. Seitz, et al. 2007a. Annual migrations, diving behavior, and thermal biology of Atlantic bluefin tuna, *Thunnus thynnus*, on their Gulf of Mexico breeding grounds. *Marine Biology* 151 (1): 1–18.

Teo, S. L., A.M. Boustany, and B.A. Block. 2007b. Oceanographic preferences of Atlantic bluefin tuna, *Thunnus thynnus,* on their Gulf of Mexico breeding grounds. *Marine Biology* 152 (5): 1105–19.

Walli, A., S.L. Teo, A. Boustany, C.J. Farwell, T. Williams, H. Dewar, E. Prince, and B.A. Block. 2009. Seasonal movements, aggregations and diving behavior of Atlantic bluefin tuna (*Thunnus thynnus*) revealed with archival tags. *PLOS ONE* 4 (7): e6151.

Whitlock, R.E., M.K. McAllister, and B.A. Block. 2012. Estimating fishing and natural mortality rates for Pacific bluefin tuna (*Thunnus orientalis*) using electronic tagging data. *Fisheries Research* 119–120:115–27. Corrigendum in *Fisheries Research* 184 (September 2016): 248–53.

Whitlock, R.E., A. Walli, P. Cermeño, L.E. Rodriguez, C. Farwell, and B. A. Block. 2013. Quantifying energy intake in Pacific bluefin tuna (*Thunnus orientalis*) using the temperature increment of feeding. *Journal of Experimental Biology* 216:4109–23.

Whitlock, R.E., E. Hazen, A. Walli, C. Farwell, S.J. Bograd, D.G. Foley, M. Castleton, and B.A. Block. 2015. Direct quantification of energy intake in an apex marine predator suggests physiology is a key driver of migrations. *Science Advances* 1 (8): e1400270.

Wilson, S.G., I.D. Jonsen, R.J. Schallert, J.E. Ganong, M.R. Castleton, A.D. Spares, A.M. Boustany, M.J.W. Stokesbury, and B.A. Block. 2015. Tracking the fidelity of Atlantic Bluefin Tuna released in Canadian waters to the Gulf of Mexico spawning grounds. *Canadian Journal of Fisheries and Aquatic Sciences* 72 (11): 1700–1717.

Wood, S. 2006. *Generalized Additive Models: An Introduction with R.* Boca Raton, FL: CRC Press.

# SOUTHERN

CHAPTER 10

# Keys to Advancing the Management of Southern Bluefin Tuna, *Thunnus maccoyii*

Jessica H. Farley, Ann L. Preece, Mark V. Bravington, J. Paige Eveson, Campbell R. Davies, Karen Evans, Toby A. Patterson, Naomi P. Clear, Peter M. Grewe, Jason R. Hartog, Richard M. Hillary, Alistair J. Hobday, Matthew J. Lansdell, and Craig H. Proctor

## Introduction

Strategic research and long-term monitoring programs that provide targeted, high-quality data are the foundations for science-based fishery management. In the case of southern bluefin tuna (SBT), *Thunnus maccoyii*, there has been a long and sustained commitment to research and monitoring and the development of new techniques that improve the quality and cost-effectiveness of data collection. This strategic approach has focused the science on the collection of reliable fisheries-dependent data, on the development of methods for collecting fishery-independent data, on improving fundamental biological knowledge, and on developing new statistical methods, with the ultimate aim of directly improving advice for management. With these foundations in place, the Commission for the Conservation of Southern Bluefin Tuna Extended Scientific Committee (CCSBT ESC) has been able to develop and implement (in 2011) a management procedure that provides science-based global catch recommendations as a central part of the CCSBT's rebuilding plan (Hillary et al. 2016).

This chapter provides an overview of the range of research that has been completed as part of this strategic program and how it directly informs the management of the fishery for SBT. We first summarize the history of the commercial fishery and then focus on the science delivered in four key areas: biology, movement and migration, recruitment monitoring, and close-kin abundance estimation. We explain the motivation for each research area, how the data are used for advice on stock status and recommending catch levels through the management procedure, and future research needs.

## History of the Southern Bluefin Tuna Fishery and Management

Industrial-scale fishing of SBT commenced in the 1950s, with the Japanese longline fishery operating in the Indian Ocean and extending east into the Pacific, and the Australian surface fishery operating in Australian coastal waters, predominantly on the southeast coast but also in the Great Australian Bight (GAB). Catches of SBT by the two nations increased rapidly to approximately 80,000 metric tons in 1960, with the Japanese longline fishery dominating catches in the 1950s and then declining after 1961. While Japanese catches declined, Australia's rose as the fishery increased catches in the GAB, peaking in 1982. New Zealand entered the fishery in the 1980s with annual catches generally < 500 metric tons. By the mid-1980s, a major component of the Australian surface fishery on its southeast coast had collapsed (Anonymous 1986), and Japan's catches had declined to approximately 20,000 metric tons. Stock assessments and analysis of tag-return data from the early 1980s indicated that harvest rates for SBT had been very high, and there were concerns that this heavy fishing pressure had resulted in a substantial decline in both adult and juvenile numbers (Hampton et al. 1984, Majkowski and Eckert 1986). The spatial extent of the high-seas longline fishery decreased through the following decades (Figure 10.1), although it is unclear whether this was the result of range contraction or changes in fishing practices by Japan (Farley et al. 2007).

In 1982 informal trilateral fishing agreements were established between Australia, Japan, and New Zealand. In 1986 they agreed on catch limits, although the limits did not restrict catches until 1989. In 1994 this informal arrangement was formalized as the Convention for the Conservation of Southern Bluefin Tuna (CCSBT). Since then, it has grown to include eight members (Australia, the European Union, the Fishing Entity of Taiwan, Indonesia, Japan, Republic of Korea, New Zealand, and South Africa) and one cooperating nonmember (the Philippines).

Successive stock assessments showed spawning stock biomass had declined below commonly accepted reference-point levels (Hillary et al. 2016). The lowest recruitment estimated by the stock assessment occurred in 1999–2000 (Figure 10.2); since then, recruitment estimates have improved as a result of reduction in total removals owing to cessation of substantial overquota catches in the longline fleets and to total allowable catch (TAC) cuts that occurred in response to concerns of high depletion and unreported

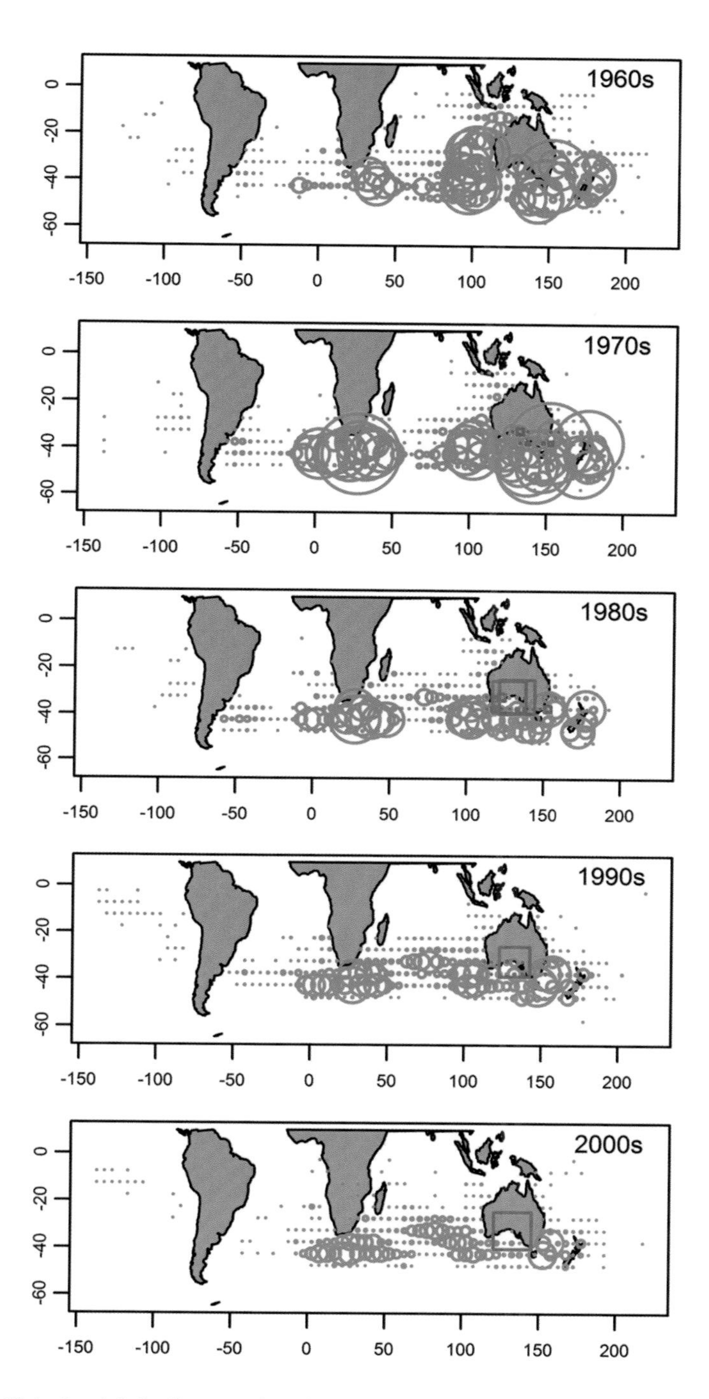


**Figure 10.1.** Spatial distribution of catches over time, aggregated by 5×5-degree latitude-longitude grid cells (including only cells where > 10 fish were caught). Blue circles represent longline catches, and red squares represent surface fishery catches. The area of the circle (*square*) is proportional to the maximum longline (surface) catch in a grid cell across all decades.

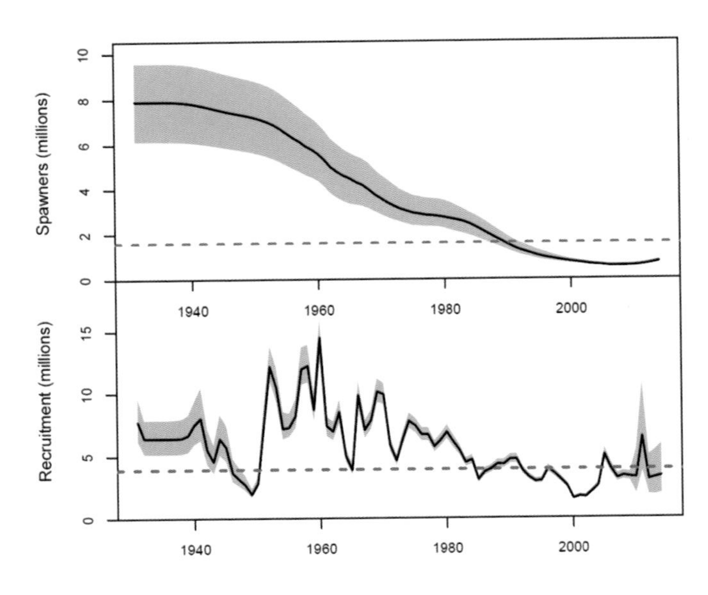


**Figure 10.2.** Spawner abundance (*top*) and recruitment (*bottom*) and 90% confidence interval (*grey shading*) for the reference set of 320 operating models for the 2014 stock assessment. For spawner abundance, the dotted magenta line represents 20% of the median of initial abundance estimates. For recruitment, it represents 50% of the median of initial recruitment estimates. *Hillary et al. 2014.*

catches (Anonymous 2006, 2016a). The detection of large unreported catches over two decades, from approximately 1986 (Anonymous 2006, Polacheck and Davies 2008), added to uncertainties in catch and effort data and interpretation of catch-per-unit-effort (CPUE) indices. The stock assessments use a set of scenarios for CPUE that cover hypotheses for impacts of the overcatch on CPUE indices. The most recent assessment of stock status, in 2014, based on a reference set of models, estimated the current spawning biomass to be 8%–12% of initial spawning biomass (Figure 10.2).

A management procedure was adopted by the CCSBT in 2011 as part of the rebuilding plan for the stock, and it has been used to set global catches since 2012 (Anonymous 2011; Hillary et al. 2016; chapter 11, this volume). The CCSBT management procedure uses the fishery-independent recruitment index from the aerial survey of juveniles (Eveson and Farley 2016) and an abundance index from standardized Japanese longline CPUE data (Itoh and Takahashi 2016) as the monitoring series. The operating models (320 in the reference set) used to test the management procedure integrate a number of sources of data and a grid of important uncertainties in key parameters of the population dynamics of the system (Anonymous 2014a). The operating models are updated every three years to provide advice on stock status, and the management procedure review process is scheduled to occur after six years (Anonymous 2011, 2014a). The management procedure model is fixed, since its adoption, but the input data are updated every three years to provide TAC advice (Anonymous 2011, 2013a, 2016a). High-quality, informative data, sustained through a commitment to strategic and cost-effective monitoring methods, have provided the foundations for an internationally agreed-upon rebuilding plan for the SBT stock based on the scientifically tested management procedure.

## Advances in Biology and Life History Research

Understanding of the biology and life history of SBT has increased substantially over the past two decades through the collection of high-quality data on age, growth, and reproduction. Strong international collaboration has been a key factor in the success of this research, and a primary benefit has been the establishment and coordination of sample collection programs. These programs led to the routine collection of length data and otoliths by CCSBT member countries, as well as other biological samples, such as gonads, for specific projects.

### Long-Term Monitoring Programs

Large-scale targeted data and sample collection programs have been a key component of the overall strategy to obtain accurate and timely fishery and biological information for SBT. Central to this effort was the collaboration of scientists, industry, and managers in Australia, Japan, and New Zealand in the early 1990s. This collaboration included the deployment of trained independent observers on Japanese high-seas longline vessels to collect fisheries data and biological samples as part of the Real Time Monitoring Program (RTMP). The collection of Japanese longline catch and effort data in the RTMP was particularly beneficial as it facilitated faster collation of catch data from the fishery; this allowed for CPUE analysis of the most recent year of data to be included in stock-assessment models. At the same time, observers were deployed on joint-venture Japanese longline vessels in the Australian Fishing Zone and were tasked with collecting similar data, and a port-based catch-monitoring program aimed at collecting fisheries data and biological samples from SBT caught on the spawning ground was established in Indonesia. In the mid-1990s, port-based monitoring of the Taiwanese SBT catch in the Indian Ocean began, and a scientific observer pilot program was established in 2001 (Farley et al. 2001, Anonymous 2005). These programs facilitated the routine collection of catch and effort data, length data, biological samples for demographic (see below) and diet studies, deployment of conventional and electronic tags for establishing the spatial dynamics of juvenile and adult SBT (see the sections "Spatial Dynamics—An Overview and Implications for Management" and "Long-Term Recruitment-Monitoring Data for the Management-Procedure and Stock-Assessment Models"), and samples for close-kin mark-recapture (CKMR) estimates of spawning biomass (see the section "Close-Kin Mark-Recapture").

## Age and Growth

Early information on SBT age and growth was obtained from a variety of sources including analyses of length frequency and tag-return data (see review by Shimose and Farley 2015). The conventional tagging experiments (see also "Long-Term Recruitment-Monitoring Data for the Management-Procedure and Stock-Assessment Models") showed that SBT exhibit complex growth patterns and that juvenile growth rates increased significantly between the 1960s and 1980s (Leigh and Hearn 2000, Hearn and Polacheck 2003). At the time, direct age estimates from hard parts were not available, and as a result, the estimated growth curve and the catch-at-age matrix were highly uncertain.

In the 1990s, techniques were developed to estimate the daily and annual age of SBT from otoliths (Jenkins and Davis 1990, Itoh and Tsuji 1996, Rees et al. 1996, Gunn et al. 2008). Annual age estimates were validated using marginal increment analysis (up to age 4), a mark-recapture experiment (up to age 6), and bomb radiocarbon analysis (up to age 34) (Thorogood 1987, Kalish et al. 1996, Clear et al. 2000). In 2002 an age estimation workshop was held to standardize annual aging protocols of otoliths among CCSBT members, and an age-determination manual was produced (Anonymous 2002), which allowed for the routine collection of age data by individual members.

Analysis of otoliths has shown that SBT are a long-lived species (up to 40 years) and that life expectancy is similar for males and females (Farley et al. 2007, Gunn et al. 2008). Growth is most rapid in the first few years of life, and asymptotic length is reached at ~15 years. SBT exhibit sexual dimorphism in growth; males grow faster, on average, than females in age classes <7–9 years (Farley et al. 2007, Lin and Tzeng 2010). This dimorphism contributes to a skewed sex ratio, favoring males in length classes >170 cm fork length (FL) (Farley et al. 2007).

The age distribution of SBT within catches varies significantly among areas and fleets, but the demographics are consistent with seasonal and age-related changes in migration (Farley et al. 2007, Shiao et al. 2008). Juveniles aged 2–4 years dominate Australian catches in the GAB during the austral summer and Taiwanese catches across the central Indian Ocean in winter (30°S–35°S). Relatively fewer juveniles are caught around New Zealand, suggesting that most do not migrate east of the Tasman Sea (Farley et al. 2007; see also the section "Spatial Dynamics—An Overview and Implications for Management"). SBT aged ~3–40 years are caught on the feeding grounds across the southern Indian and Pacific oceans and into the eastern Atlantic Ocean, although the majority that are caught are < 15 years. Only adults aged ~8–30 years are caught on the spawning ground in the northeastern Indian Ocean. The size and age distribution of the spawning population has changed substantially over time, however (Farley et al. 2014). In the early 2000s, the mean size of SBT caught by the Indonesian longline fishery on the spawning ground decreased substantially, probably owing to a pulse of recruitment of small/young fish to the spawning-ground fishery. A second pulse of young fish appeared in the mid-2000s.

An integrated growth model developed during the 2000s combined growth information from tagging with length frequency and direct aging data. It also incorporated the transition in growth from juvenile to subadults and seasonal variation in growth (Leigh and Hearn 2000, Hearn and Polacheck 2003, Eveson et al. 2004, Laslett et al. 2004, Polacheck et al. 2004). Each data source provides growth information for different life history stages, and in combination they provided a more robust estimate of length-at-age. Using this integrated growth model, Polacheck et al. (2004) confirmed there had been a substantial increase in growth rates of juveniles (up to age 4) between the 1960s and the 1980s, as previously suggested by Leigh and Hearn (2000), and showed that growth continued to increase in the 1990s. This change in growth was most likely a density-dependent response to the decline in population size over the period; however, changes in growth owing to environmental factors could not be excluded. Further work by Eveson et al. (2005) suggested that growth of juveniles in the early 2000s was similar to that in the early 1990s, perhaps continuing to increase.

The integrated growth model of Eveson et al. (2004) is currently used to estimate a catch-at-age matrix using cohort slicing, which is then incorporated into the operating models used in the periodic assessments of stock status and the testing and selection of management procedures. Direct aging data are currently used to derive the catch-at-age for the Indonesian fishery only, with cohort slicing used for the Australian surface fishery and catch-at-length used for the other fisheries. Sex-specific growth curves are not currently included in the stock-assessment models used by the CCSBT because the required data (size, age, and sex) are not available, although the data are available for the Indonesian fishery.

## Reproduction and Maturity

Larval surveys from the 1950s to the 1980s and the examination of ovaries since the 1960s indicate that SBT spawn in an area of the tropical northeast Indian Ocean between the continental shelf of Australia and the islands of the Indonesian archipelago between Java and Sumba (Shingu 1978, Nishikawa et al. 1985; Figure 10.3). Spawning occurs between September and April, with two peaks in activity, in September–October and February–March (Davis and Farley 1995, Farley and Davis 1998). Individual fish, however, do not spawn for the entire season, with postspawning females occurring as early as October (Farley and Davis 1998). Females leave the spawning ground immediately after spawning is complete, and they are replaced by new arrivals. Examination of batch fecundity and spawning frequency suggests that females, while they are on the spawning ground, will cycle between spawning and resting episodes. When spawning, females are capable of spawning daily and release an average of 6.5 million oocytes per spawning event. Relative batch fecundity and the duration of spawning and nonspawning episodes increase with fish size. Relative batch fecundity of a 160 cm fish was estimated at 66% of a 190 cm fish. The average number of sequential spawning and nonspawning events for a 150 cm female was estimated at 3.6 and 1.3 days, respectively, compared with 6.9 and 2.2 days for a 190 cm female (Farley et al. 2015).

There remains uncertainty about the size and age at which SBT mature and the functional form of the maturity ogive. Previous estimates of the length/age at 50% maturity for female SBT converged between 152 and 162 cm and between 10 and 12 years old. Up until 2013, the SBT operating models used a "knife-edge" maturity relationship, which specified that 0–9-year-olds made no contribution to the spawning biomass or reproduc-

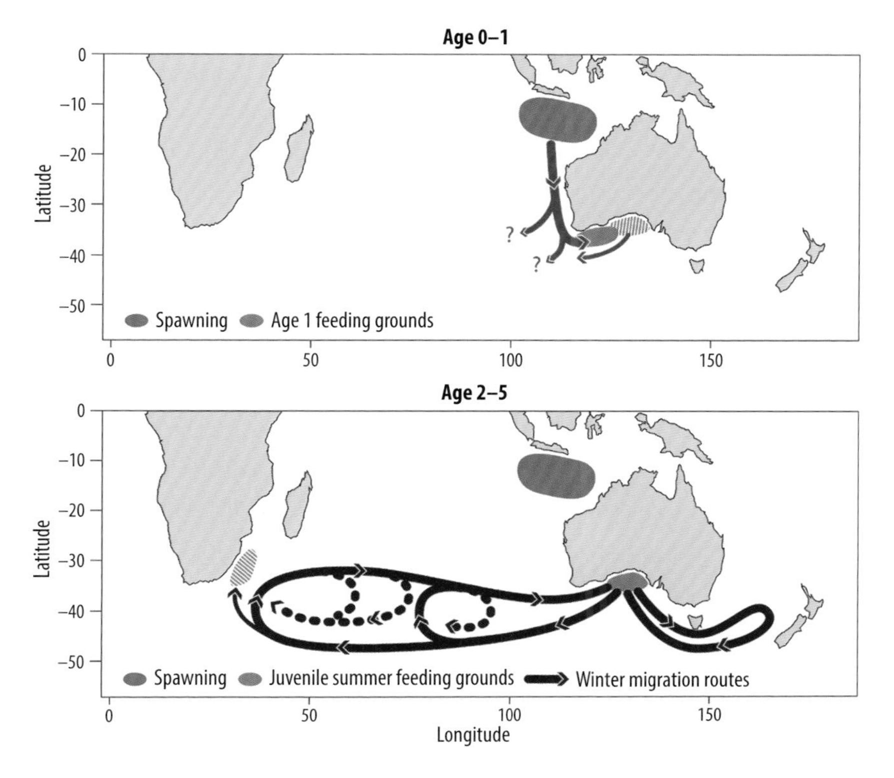


**Figure 10.3.** General migration patterns and age distribution of juvenile southern bluefin tuna with age-1 movements from the spawning ground (*upper panel*) and age-2–5 movements from the Great Australian Bight (*lower panel*). *Hobday et al. 2015.*

tive output of the population, and 10+-year-olds all contributed in proportion to their weight. In 2013 the CCSBT adopted a new maturity ogive using available biological information, to give a spawning potential by age for use in the 2014 stock assessment (Anonymous 2013b). The new maturity ogive was required in the stock assessment to match the assumptions used in analysis of CKMR data (see the section "Consideration of Close-Kin Mark-Recapture for Pacific and Western Atlantic Bluefin Tuna") integrated into the assessment models in 2012 (Hillary et al. 2012). The ogive was based on the proportion of the population that had recruited onto the spawning ground, which was calculated by comparing the length frequency data of SBT caught on the spawning ground in the Indonesian longline fishery with the data from the Japanese longline catch off the spawning ground (Davis et al. 2001). Length at 50% maturity was estimated for the years 1995 to 2000 to be between 158.4 and 163.1 cm FL (~11–12 years).

### Key Uncertainties and Ongoing Needs

The CCSBT ESC recognizes the lack of an independent estimate of maturity as well as the importance of obtaining an updated and unbiased estimate of the proportion of the population that is sexually mature by age and length (Anonymous 2013a). Recently, as part of the CCSBT Scientific Research Program (SRP), the CCSBT ESC supported a proposal for member countries to collect SBT ovaries and otoliths from females caught on feeding grounds during the nonspawning months to estimate an unbiased maturity schedule. The presence of "maturity markers" in histological sections of the ovaries will be used to distinguish mature resting SBT from immature SBT females, following protocols developed by Farley et al. (2013). If successful, this approach could be applied to other tuna species.

The development of age-length keys and an estimation of the age composition of juvenile and adult catches are high priorities for the current CCSBT SRP. CCSBT members are required to collect otoliths from SBT in their catch and to provide direct age data to the CCSBT. Continued monitoring of juvenile growth rates, the size and age distribution of the spawning stock, and adult reproductive parameters is important for assessing the productivity of the SBT stock. Estimating total annual fecundity as a function of length (or age) remains difficult for SBT because average time on the spawning ground by individuals is unknown. Archival tagging of adults may provide a useful method to determine residency and spawning behavior. Interannual variation in relative annual fecundity could then be estimated for SBT using relative annual fecundity-at-size and the size distribution of the spawning stock (Farley et al. 2015).

## Spatial Dynamics—an Overview and Implications for Management

Broad-scale information on the distribution of SBT prior to the 1990s was based on catch data, fishery-dependent tagging programs, and a limited number of larval surveys (Nishikawa et al. 1985; Davis et al. 1990a,b; Caton 1991). Data from these sources suggested that although the population was considered a single stock, there was likely some spatial and temporal structuring of age cohorts of SBT, which, depending on the extent of the structuring, could have implications for assessments of the stock and its rebuilding. The development of electronic tags in the 1990s provided a novel means by which to directly investigate questions about movement and spatial mixing and also provided information on habitat use and behavior. Advances in light-based geolocation methods for archival tags to estimate not only a most probable track but also its uncertainty (e.g., Nielsen and Sibert

2007, Basson et al. 2016) have made tag-based studies of movement and habitat more robust.

### Demographic Variability in Distribution and Movements

Although catch data helped to clarify that there is a single spawning ground for SBT, they provide less information about individual lengths of time on the spawning ground and the spatial and behavioral dynamics of adult fish while they are present. Further, although available electronic tagging data support the idea that once the seasonal spawn is complete, adult SBT move rapidly from the spawning ground into southern temperate waters (Itoh et al. 2002), the specific paths they take are largely unknown.

Similarly, little is known of the spatial dynamics of SBT larvae on the spawning ground. Studies have shown that larvae (4–7 mm) are largely restricted to the mixed layer and drift passively with currents until the age of approximately 20 days, when they become capable of independent movement (Davis et al. 1990a,b; Jenkins and Davis 1990). It is generally thought that they then move south from the spawning ground, facilitated by the Leeuwin Current (Ridgway and Condie 2004), and after several months, they reach the waters off southwestern Australia (Hobday et al. 2015). An unknown fraction, but likely the majority, then move into the continental shelf waters south of Western Australia at around 1 year of age, and the remainder are thought to migrate west.

As age-1 juveniles grow, they gradually move eastward into the western sections of the GAB, where some remain throughout the austral winter (Hobday et al. 2009b, Fujioka et al. 2010a) and through the following summer months. For the first 4–5 years of life, the majority of juvenile SBT are considered to frequent the GAB during the austral summer, generally arriving from late spring to early summer (November through January; Basson et al. 2012). Here, they feed predominantly on seasonal high densities of small pelagic fish such as the Australian sardine (*Sardinops sagax*) (Kemps et al. 1998, Ward et al. 2006, Itoh et al. 2011). While in the GAB, juvenile SBT aggregate in large schools, which spend substantial time in the upper 100 m of the water column, mostly during the day (Bestley et al. 2009). These schools generally comprise similarly sized conspecifics, although, as the overall population size declined from the 1960s to the 1990s, schools containing mixed sizes have become more prevalent (Dell and Hobday 2008).

Commercial purse seine operations take advantage of the surfacing behavior of juvenile SBT in the GAB by using spotter planes to locate and target surface schools (Basson and Farley 2014, Ellis and Kiessling 2016). Similarly, the scientific aerial survey uses this behavior to derive a relative abundance index of juveniles (Cowling et al. 2003, Eveson and Farley 2016; see also the section "Recruitment Monitoring: Aerial Survey and Gene Tagging Methods"). Both the scientific aerial survey and commercial spotting indices suggest that the distribution of juveniles has shifted into eastern parts of the GAB in recent years (Eveson and Farley 2016). It should be noted that the aerial survey is conducted over the extent of the continental shelf, and therefore it does not provide information on the potential abundance and distribution of juvenile SBT in waters beyond this.

During the austral autumn and winter months, most juvenile SBT leave the GAB, moving either west into the Indian Ocean or east into the Tasman Sea, where they spend the winter months before returning to the GAB for the following summer (Figure 10.5; Bestley et al. 2009, Basson et al. 2012). This departure from the GAB coincides with a decline in seasonal upwelling events in the eastern GAB, particularly those that occur off the southern tip of Kangaroo Island, the eastern coastline of South Australia, and the western coastline of Victoria (Schahinger 1987, Middleton and Bye 2007). Many individuals will undertake occasional forays out of the GAB into neighboring areas before their final departure to winter grounds (Hobday et al. 2015). Departure schedules are highly variable, with individuals departing the GAB as early as March and as late as August (Bestley et al. 2009, Basson et al. 2012). Migration paths into the Indian Ocean and the Tasman Sea do not appear to be consistent between individuals, and individuals do not necessarily migrate to the same location each year (Basson et al. 2012).

During this cyclical movement between the GAB and the Indian Ocean, juvenile SBT disperse across subantarctic to subtropical regions, occupying waters with sea surface temperatures of 8–22°C (Figure 10.4). Individuals demonstrate a pattern of relatively high foraging success throughout their migratory range, a pattern suggesting that juveniles search for prey extensively while migrating; this is a strategy that would be highly efficient for finding dispersed or isolated patches of prey species (Bestley et al. 2010). As has been observed in many other tunas (e.g., bigeye tuna, *Thunnus obesus;* Evans et al. 2008), juvenile SBT dive deeper during a full moon than during other periods in the lunar cycle (Bestley et al. 2009); this is thought to be in response to variability in the diel migrations of their prey in the deep scattering layer (Josse et al. 1998, Luo et al. 2000).

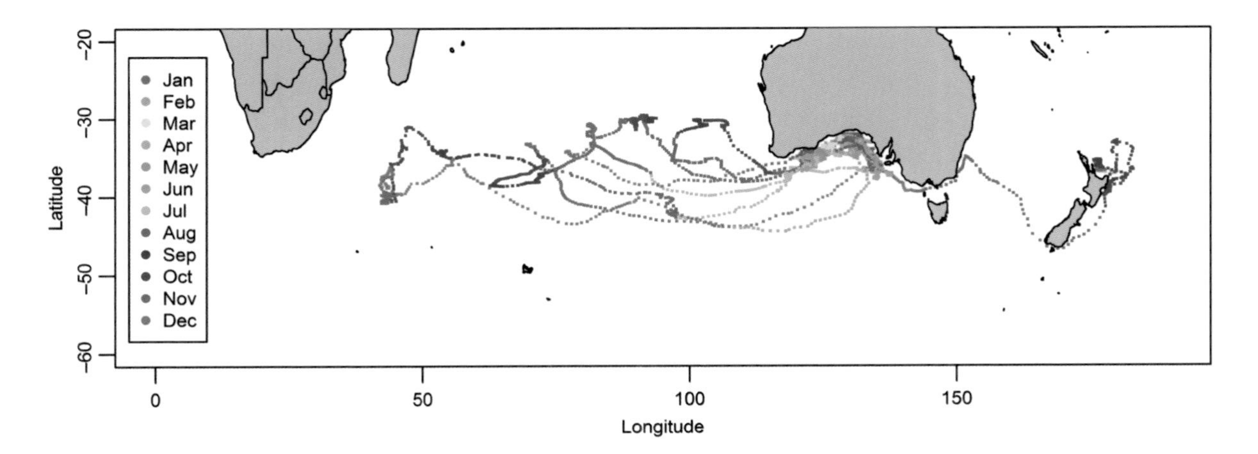


**Figure 10.4.** Track locations of juvenile southern bluefin tuna recorded by archival tags. Locations are colored by month (and jittered for plotting). *Basson et al. 2012.*

Conventional tagging data from the 2000s identified that a smaller proportion of juvenile SBT undertook movements eastward into the Tasman Sea or into the far western Indian Ocean off the east coast of South Africa than were observed in the 1990s (Basson et al. 2012). Although the reasons for these changes are still unclear, they may be associated with a declining population and associated range contraction, changing environmental conditions, or a change in preferred habitat (Basson et al. 2012, Briscoe et al. 2017).

Few fish older than 5 years of age are caught commercially within the GAB or have been encountered by conventional tagging programs operating in the GAB, suggesting an ontogenetic shift in the movement patterns of subadult fish, with individuals ceasing to return to the shelf waters of the GAB during the austral summer months. Little is known of the movement patterns of these subadult SBT, but commercial catch data suggest these animals disperse throughout southern temperate waters. Preliminary tagging of 130–140 cm SBT in the southwest Indian Ocean has recorded individuals moving westward around South Africa and into the eastern Atlantic Ocean (CSIRO, unpublished data).

Both subadult and adult SBT (> 10 years) are found seasonally during the winter in the Tasman Sea, where they feed on a mixture of fish and squid species (Young et al. 1997). There is evidence that subadults and young, mature adults remain in the Tasman Sea year round (Evans et al. 2012). The presence of young and potentially mature fish on foraging grounds during the austral summer suggests that some SBT may not migrate to the spawning ground each year (i.e., individual skip spawners).

Similar to juveniles' pattern, the spawning migration schedule of adults departing the Tasman Sea after foraging over the winter months is extended, with individuals departing the Tasman Sea from September through December (Figure 10.5; Patterson et al. 2008, Evans et al. 2012). However, once movement to the spawning ground is initiated, migration is direct and relatively quick (one fish traveled ~9,000 km over 113 days; Patterson et al. 2008). Individuals tagged in the Tasman Sea have been estimated to arrive at the spawning ground in the second half of the season, potentially contributing to the second peak in spawning-ground abundance (Evans et al. 2012; see also the section "Reproduction and Maturity"). The cues triggering the timing of migration of adults to the spawning ground are unclear but may be associated with seasonal and regional changes in oceanography, resulting in the spatial movement of suitable foraging areas away from the Tasman Sea and into areas south and west (Patterson et al. 2008).

Adult SBT also move through a wide range of water temperatures as they migrate from their winter foraging ground to the spawning ground, with tags recording temperatures of 2.6°C–30.4°C (e.g., Figure 10.6). Overall, adult SBT have been observed to occupy water below 18.5°C 30% of the time, water at or below 20°C 50% of the time, and water at or below 21°C 90% of the time (Patterson et al. 2008). Although observations of the behavior of adults on the spawning ground are limited, when in the region, time spent in surface waters less than 150 m has been observed to increase. This transition to surfacing behavior occurs in association with the distribution of sea surface temperatures greater than 24°C, which is the temperature required for successful spawning activity (Schaefer 2001, Itoh et al. 2002, Patterson et al. 2008).

### Application of Spatial Dynamics Information in Management

Despite rapid advances in both electronic tagging technologies and the methods used to characterize the movements of pelagic species, direct application of data from electronic tags in fishery assessment and management remains rare (Sippel et al. 2014). In this regard, SBT is one of the few species for which information on the spatial dynamics of individuals has been applied to the review of research projects and to domestic fishery management.

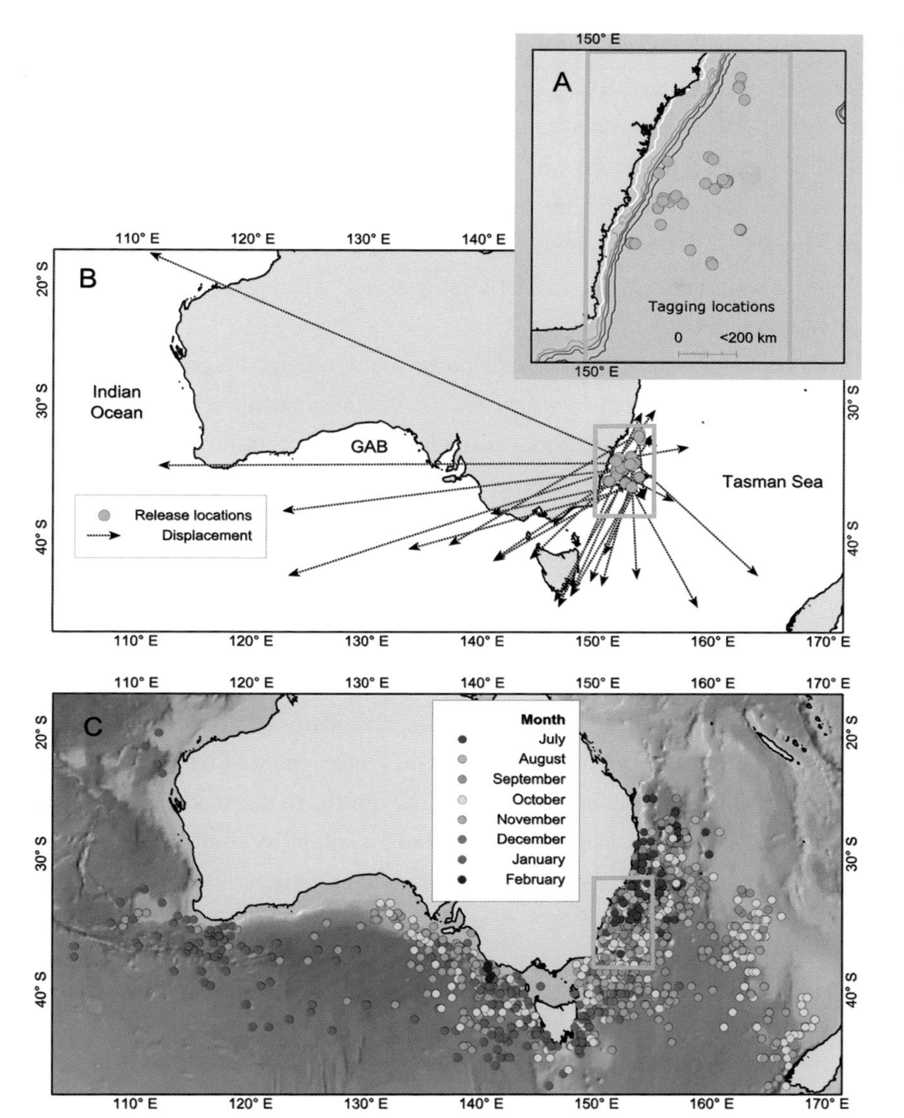


Figure 10.5. Movements of adult southern bluefin tuna recorded using pop-up satellite archival tags. *Patterson et al. 2008.*

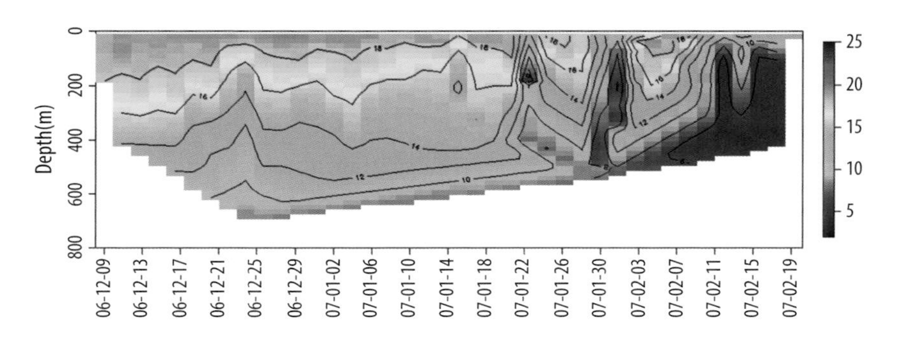


Figure 10.6. Interpolated temperature-at-depth data (December 2007–February 2007) derived from temperature-depth profiles recorded by a pop-up satellite tag deployed on a 142 cm (length to caudal fork) southern bluefin tuna in the Indian Ocean.

An acoustic tagging study was used to investigate a decline in abundance indices derived from a ship-based acoustic survey operating in the western GAB during the second half of the 1990s and through the 2000s (Hobday et al. 2009b, Fujioka et al. 2010b). The study documented high individual variability in the movement patterns and residence of juvenile SBT in the area of the survey; this would result in high uncertainty in any abundance index derived from the ship-based survey. Consequently, the survey was ceased.

In eastern Australia, information on the temperature preference (surface and subsurface water temperatures) in combination with information on the movements of subadults and adults has been used to generate spatial maps of the likely occurrence of SBT (Hobday and Hartmann 2006). The Australian Fisheries Management Authority uses this information to inform the spatial management of a longline fishery operating in the Eastern Tuna and Billfish Fishery. Several zones of varying access by the longline fishery are set, with the aim of reducing unwanted capture of SBT by fishers targeting yellowfin tuna (*Thunnus albacares*) who do not have quota to catch SBT (Hobday et al. 2009a, 2010). This form of spatial management is implemented each year over the austral winter months when SBT occur in the region, although in recent years the advent of electronic monitoring of catches within the fishery, and a greater availability of SBT quota, have reduced the importance of the habitat prediction system in setting the management zones.

A similar habitat prediction approach has been developed for juvenile SBT in the GAB. Surface temperature preferences of juveniles are combined with forecasts of sea surface temperature to predict the distribution of juveniles up to three months in advance (www.cmar.csiro.au/gab-forecasts/index.html). These forecasts aim to assist fishers in planning fishing operations in an effort to improve the economic efficiency of the fishery (Eveson et al. 2015).

The current operating models used by the CCSBT to test the management procedure for SBT are not explicitly spatial (Hillary et al. 2013); however, the CCSBT has recognized that a spatial model may be desirable to develop in the future (Anonymous 2012). Movements and distributions derived from conventional tag data may not sufficiently describe the spatial dynamics of a species at scales required for developing a fully spatial model, particularly because movement derived from these tags can be confounded with mortality, tag reporting rates, and the distribution of the fishing fleet.

The higher-resolution information on the spatial dynamics of SBT provided by electronic tags can be more informative in this regard, by reducing potential biases in estimates of natural and fishing mortality (Eveson et al. 2012).

### Knowledge Gaps on the Spatial Dynamics of Southern Bluefin Tuna

As noted earlier, relatively little is known of the spatial dynamics of age classes of SBT from direct observation, other than for those aged 2–5 years. Limited knowledge of the dispersal of larvae from the spawning grounds and of the subsequent distribution of juveniles outside of the GAB calls into question how representative GAB data is of the population as a whole (Basson et al. 2012). Further information on the importance of the GAB to the growth and survival of juvenile SBT and to their recruitment into the commercial fishery is needed, particularly for understanding the foraging behavior of individuals, key dietary requirements, and how future changes might impact these. Further understanding will also lead to important insights into how a changing environment might impact current assumptions made under frameworks for rebuilding the population and will help to ensure the ongoing sustainability of the fishery.

Knowledge of subadult movement is based on commercial catch data. Fishery-independent information is required for more details on spatial dispersal and mixing. There is also a need to better understand the proportion of young mature individuals that undertake skipped spawning. It's still not known if the proportion of skipped spawners varies between years, what the physiological determinants of skipped spawning might be, and how these might be related to the environment.

Finally, to estimate total annual fecundity as a function of length or age, overall residence time must be measured, and understanding of the spatial dynamics of adults on the spawning grounds must be improved. This is a key knowledge gap in the reproductive dynamics of SBT.

## Long-Term Recruitment-Monitoring Data for the Management-Procedure and Stock-Assessment Models

Establishing recruitment-monitoring techniques that provide reliable fishery-independent abundance indices and data on key biological processes of juveniles has been a focus in SBT research since the 1980s. These programs were motivated by concerns regarding declines in the recruitment of juveniles into the fishery and in parental biomass, contraction of juvenile surface fisheries, the lengthy time-lag on recruitment information from the stock-assessment methods used at that time, and uncertainties in key parameters of those models (Caton 1991).

Monitoring of recruitment of juveniles into the fishery is an important focus of management primarily because SBT have a slow intrinsic rate of increase (i.e., they are late in maturing), which results in a considerable lag between fisheries impacts on the juvenile year classes and their subsequent effects on the spawning component of the stock. There is, therefore, a need for early management action to mitigate the impacts of low numbers of juveniles, from poor recruitment or high juvenile fishing mortality, on spawning stock status and on stock rebuilding in the long term (Anonymous 2014b).

## Recruitment-Monitoring Research Programs

### *Conventional Tagging Programs*

Conventional tagging programs on juveniles commenced in the 1960s and were carried out, with various modifications, until the mid-2000s (Caton 1991, Anonymous 2001b, Hearn and Polacheck 2003, Davies et al. 2007). Data collected through these programs contributed to the estimates of time-varying growth, mean length at age, and fishing and natural mortality that have been integrated into SBT population dynamics models. Movement and mixing rates and age-at-length data derived from these programs have informed the design of more recent recruitment-monitoring programs.

### *National Research Institute of Far Seas Fisheries (NRIFSF) and Commonwealth Scientific and Industrial Research Organisation (CSIRO) Recruitment-Monitoring Program, 1993–2005*

In the early 1990s, Japanese and Australian scientists instigated a collaborative program aimed at developing long-term recruitment-monitoring methods, with an emphasis on fishery-independent abundance estimates of juveniles up to age 5 (Davis 2002). This collaboration was in response to high levels of fishing mortality and declining recruitment levels identified by stock assessments and tagging programs at the time (Caton 1991). Annual workshops reviewed development of experimental designs and data collected through the program. A scientific aerial survey (see the section "Recruitment Monitoring: Aerial Survey and Gene-Tagging Methods") provided an index of relative juvenile abundance that has become a key input to the CCSBT management procedure and is integrated into the SBT operating models for stock assessment. Other research included conducting acoustic surveys (Hobday et al. 2009b, Fujioka et al. 2010b), investigating residency and movement of juvenile SBT in the GAB, conducting trolling and piston line surveys for age-1 fish (Itoh and Sakai 2009, Itoh and Tsuda 2016), and collecting commercial spotting data as an additional potential indicator of juvenile abundance (Basson and Farley 2014).

*CCSBT Scientific Research Programs*

Two SRPs have been developed by the CCSBT (2001–7 and 2014–18), which have focused on improving the quality of the data available for stock assessments and on development of reliable indices to monitor future trends in abundance and fishing mortality (Anonymous 2001a, 2014b). The subprograms related to juveniles in the first SRP included a conventional tagging program (Anonymous 2001b) aimed at estimating natural and fishing mortality of juveniles aged 2–5 years (2001–7), and a scientific observer program aimed at further long-term fishery data collection and at providing estimates of tag reporting rates via comparisons between fishery logbooks and observers' data. The tagging program was ceased in 2007 owing to a lack of tag returns from some fisheries and to large uncertainties in reporting rates (Anonymous 2007, Davies et al. 2007). Gene tagging (see the section "Recruitment Monitoring: Aerial Survey and Gene-Tagging Methods") was proposed as an alternative fishery-independent method for estimating mortality or abundance (Davies at al. 2007, 2008). Advances were being made in DNA genotyping techniques and associated data analyses (Bravington et al. 2014), and the methods were becoming increasingly cost-effective and feasible for use in large-scale, genetic-based monitoring programs (Preece et al. 2015).

In the current 2014–18 SRP (Anonymous 2014b), the CCSBT has funded the following recruitment-monitoring projects (in addition to work related to adults, e.g., close-kin): the 2016 and 2017 aerial surveys (Eveson and Farley 2016), a gene-tagging design study in 2015 (Preece et al. 2015), a pilot gene-tagging program in 2016 (Preece and Bradford 2016), and subsequently, the 2017 gene-tagging project for estimating absolute abundance of age-2 fish (see the next section).

## Recruitment Monitoring: Aerial Survey and Gene-Tagging Methods

The annual relative abundance estimate of juveniles from the scientific aerial survey (Eveson and Farley 2016) is central to the current CCSBT management procedure. The aerial survey program, established in the 1990s (Cowling et al. 2003), consists of 15 north-south transect lines extending across the shelf of the GAB, which are surveyed from January to March each year. Aerial spotting of schools of juveniles (ages 2–4) is possible because juvenile SBT are resident in the GAB each summer and are visible in schools close to the ocean surface. Replicates of the survey area are flown using consistent survey methods, under specific environmental conditions, and using calibrated, expert tuna-spotters. Data are collected on the location and biomass of SBT sightings and on the environmental conditions present during the survey. Sightings and biomass data are standardized to a common set of spotters and environmental conditions to produce a time series of relative abundance of juvenile (ages 2–4) SBT that can be compared in a consistent manner across years (Eveson and Farley 2016).

The methodology of the aerial survey was reviewed after an initial period (1993–2000; Bravington 2003), and the surveys resumed in 2005. The surveys have been conducted annually since then, with the exception of 2015, when a lack of funding temporarily suspended the program. This suspension and the reliance on a diminishing number of experienced spotter pilots were catalysts for consideration within the CCSBT of cost-effective methods that could be developed to replace the aerial survey as the primary recruitment-monitoring approach (Anonymous 2014b, 2015e; Preece et al. 2015).

The CCSBT has agreed to a gene-tagging program to provide an absolute abundance estimate for use in a future management procedure. SBT gene tagging is a mark-recapture tagging method similar to conventional tagging. Fish are captured at sea (by pole and line fishing), a small tissue sample is taken (the "tag"), and the fish is quickly released back into the sea. A second set of tissue samples is collected at a later date, and the unique genotype is used to find matches in the two sets of tissue samples, analogous to recapturing a tagged fish. The number of genotype matches provides an estimate of absolute abundance of the cohort of fish that was tagged. This method has become feasible for large-scale tagging programs as costs associated with genotyping have decreased and automated methods that assist in quality control, speed, and repeatability have been developed (Bradford et al. 2015, Grewe et al. 2015, Bravington et al. 2016b). The methodology has the advantages that the tags are invisible (so a "tagged" fish cannot be identified), there is no tag shedding, the tags last the lifetime of the fish, and no reporting rates are required.

A design study for the gene-tagging program (Preece et al. 2015) calculated the costs and sample sizes required for abundance estimates (with a target coefficient of variation of 0.25), considered issues that may bias estimates, and demonstrated methods for integrating data into the SBT operating models used in the testing of management procedures. This was followed by a pilot field study, initiated in 2016, to test the feasibility of field and laboratory logistics required for large-scale implementation (Anonymous 2015d, 2016a; Preece and Bradford 2016). A long-term gene-tagging recruitment-monitoring program commenced in 2017 (Anonymous 2016b). The first estimate of abundance (of age-2 fish in 2016) from the pilot study became available in early 2018 and is being incorporated into the monitoring and management processes of CCSBT, including into a new management procedure (Preece, pers. comm.).

### Value of Recruitment-Monitoring Data in the Management Procedure

The importance of recruitment data to the scientific advice given in regard to TACs was evaluated in management strategy evaluation (MSE) simulation tests, using the management procedure in 2015 (Anonymous 2015c). Results showed that rebuilding of the SBT stock would likely be delayed and average catches reduced without a recruitment index in the management procedure (Anonymous 2015a). This is because the recruitment index provides a clear signal of future spawning stock levels, and it therefore provides time for a management procedure to anticipate these impacts and adjust recommended catches accordingly. A new management procedure scheduled for development in 2016–19 is to consider other available sources of data in combination with the gene-tagging data (Davies et al. 2016b). Once candidate management procedures are developed, they will require testing in an MSE framework to examine and refine performance. A single management procedure to set the global TAC for 2021–23 will be selected by the CCSBT, with implementation planned for 2019.

## Close-Kin Mark-Recapture

CPUE data are notoriously problematic even when appropriately standardized to obtain a relative abundance index (Maunder et al. 2006). There have long been concerns about the interpretation of CPUE data used in the SBT assessments (Davies et al. 2007) and the effects on the estimates of absolute spawning stock biomass (Polacheck and Davies 2008; Polacheck 2012a,b). The identification of more reliable, preferably fishery-independent, estimates of adult abundance has been a long-term priority (e.g., Anonymous 2001a, Davies et al. 2007). An innovative way to estimate absolute adult abundance, and other demographic parameters, known as CKMR abundance estimation, was developed based on counting genetically identified parent-offspring pairs (POPs) from tissue samples that were collected from large numbers of juveniles and mature adults (Skaug 2001, Bravington et al. 2016a). The number and pattern of pairs found are used to fit a modified mark-recapture model, based on the simple idea that each animal "marks" its two parents. One of the advantages of the method, unlike conventional mark-recapture, is that samples can come directly from dead catches. Importantly, and unlike conventional stock assessment, there is no need to use any catch-rate (or even catch) data to estimate abundance.

### Stand-Alone Abundance Estimation for Southern Bluefin Tuna (2006–2010)

To avoid biasing abundance estimates, CKMR needs to be implemented in a complex framework that includes time gaps between birth and capture, growth, fecundity, mortality rate, and selectivity. This requires a thorough knowledge of the biology of the species and access to length- and age-composition data. SBT are a good candidate for CKMR because much of the species' population biology is well documented, and monitoring systems were in place that facilitated the provision of high-quality data and tissue samples of known spawning adults and juveniles.

For SBT, 25 specifically selected microsatellite loci (Bravington et al. 2014, 2016b) were used to identify 45 POPs from approximately 14,000 samples of known spawning adults (sampled in Indonesia) and known-age juveniles (sampled in the GAB), giving 38 million usable comparisons. These were embedded in a statistical mark-recapture framework and combined with (1) length- and age-composition data from Indonesian longline catches on the spawning ground and (2) histological information on relative daily fecundity-at-size into a stand-alone stock assessment of adults ("stand alone" in that immature age classes were not modeled, and no data from those age classes were used). The model was able to estimate a time series of absolute spawning stock biomass, the effective annual fecundity-at-size, and the total mortality rate of the mature component of the population (Bravington et al. 2014, 2016b).

these strategic research programs have been central to complementary funding from individual members for domestic research and monitoring.

*Continuity of data streams.* The long-lived SBT are said to demonstrate "slow" population dynamics and thus require long-term monitoring for rigorous research and management. Research on and management of SBT have benefited from long-term data series, the collection of which has been planned with appropriate statistical design so the resulting data are informative for the intended application and can be compared over time. The collection of these data required significant resources, which were provided by member countries individually or by the CCSBT through the collaborative research programs run through the commission. A transparent review process with important guidance from the CCSBT Advisory Panel has also been essential to evaluate the data collected and to demonstrate cost-effectiveness for the CCSBT and its members. Data have been collected on and parameters have been set around length, age, maturity, recruitment indices, adult abundance indices, and estimates of fishing and natural mortality. As the dynamics of the SBT population change over time, as a result of population rebound, climate change, or changes in fishing operations, the parameter estimates and use of the data, monitoring, assessment, and management may also require updating and/or revision.

*Fishery-independent data.* Research programs on SBT have highlighted the need for fishery-independent data and a reduction of dependence on CPUE indices. Members of the CCSBT ESC have agreed that the fishery-independent methods developed since the early 1990s, including the scientific aerial survey, gene tagging, and the CKMR abundance-estimation methods described above, are essential. These are now funded predominantly through the CCSBT SRP. These methods provide the fishery-independent indices required to set management procedures and for input into the operating models used in the periodic assessments of stock status and in the evaluation and refinement of management procedures.

*Biological and spatial dynamics data.* The availability of key biological parameter estimates for SBT, such as age, growth, and maturity, has led to more accurate estimates of stock status, and information from tagging programs has provided key insights into the age-varying spatial dynamics of the population. Methods developed to estimate the annual age of SBT in the early 1990s were standardized among CCSBT member nations in the early 2000s. Other biological research has included monitoring changes in juvenile growth rates, estimating size-specific daily fecundity, a preliminary maturity schedule, and an improved understanding of spawning migrations. Information on the spatial dynamics of particular age groups has been key to the development of spatial management frameworks and has helped to improve operational efficiency of the Australian purse seine fleet.

*Continuity of expertise and collaboration.* SBT research and management has also benefited from continuity among the scientists, the CCSBT Advisory Panel, and the organizations involved in the research and monitoring programs. International scientific collaboration has led to the design and development of SBT research programs that built upon the knowledge of previous programs. The communication of information among scientists, the fishing industry, and managers has also been an essential part of the science-based management process.

The CCSBT rebuilding strategy centers on the CCSBT management procedure, which was adopted in 2011 and has been used to provide recommendations on the global TAC since 2012. The future for research and data collection to inform management of the stock looks positive, with the CCSBT providing funding to members to undertake research programs. The programs will provide fishery-independent indices of abundance for juveniles (gene tagging) and adults (close kin) that can be used in the new management procedure being developed. The next stock assessment, scheduled for 2017, will integrate new data from close-kin research and the recent positive trends (7+ years) in recruitment indices and longline CPUE, and it will provide an update on the progress made toward rebuilding the stock. As long as catches adhere to the levels recommended by the management procedure, research and monitoring are adequately funded, and members and stakeholders are well educated in the advantages of the management-procedure approach and continue to support its use, the future looks positive for rebuilding the SBT stock.

## Acknowledgments

The authors are especially grateful to the many scientists whose expertise and innovative ideas have led to the achievements and successes of the SBT research programs. They particularly thank the Recruitment Monitoring Program scientists, the CCSBT Scientific Committee members and Secretariat, and the many supporting researchers. The authors greatly appreciate the enumerators, observers, and coordinators for their assistance with the monitoring programs, and the SBT fishing industry for their continued

Cape Town. Report CCSBT-SC/0108/17, prepared for the second meeting of the Stock Assessment Group, August 19–28, Tokyo, Japan.

Farley, J.H., T.L.O. Davis, S.J. Gunn, N.P. Clear, and A.L. Preece. 2007. Demographic patterns of southern bluefin tuna, *Thunnus maccoyii*, as inferred from direct age data. *Fisheries Research* 83:151–61.

Farley, J.H., A.J. Williams, S.D. Hoyle, C.R. Davies, and S.J. Nicol. 2013. Reproductive dynamics and potential annual fecundity of South Pacific albacore tuna (*Thunnus alalunga*). *PLOS ONE* 8 (4): e60577. doi:10.1371/journal.pone.0060577.

Farley, J.H., J.P. Eveson, T.L.O. Davis, R. Andamari, C.H. Proctor, B. Nugraha, and C.R. Davies. 2014. Demographic structure, sex ratio and growth rates of southern bluefin tuna (*Thunnus maccoyii*) on the spawning ground. *PLOS ONE* 9 (5): e96392. doi:10.1371/journal.pone.0096392.

Farley, J.H., T.L.O. Davis, M.V. Bravington, R. Andamari, and C.R. Davies. 2015. Spawning dynamics and size related trends in reproductive parameters of southern bluefin tuna, *Thunnus maccoyii*. *PLOS ONE* 10 (5): e0125744. doi:10.1371/journal. pone.012 5744.

Feutry, P., O. Berry, P.M. Kyne, R.D. Pillans, R.M. Hillary, P.M. Grewe, J.R. Marthick, et al. 2017. Inferring contemporary and historical genetic connectivity from juveniles. *Molecular Ecology* 26:444–56.

Fraile, I., H. Arrizabalaga, and J.R. Rooker. 2015. Origin of Atlantic bluefin tuna (*Thunnus thynnus*) in the Bay of Biscay. *ICES Journal of Marine Science* 72 (2): 625–34.

Fujioka, K., A.J. Hobday, R. Kawabe, K. Miyashita, K. Honda, T. Itoh, and Y. Takao. 2010a. Interannual variation in summer habitat use by juvenile southern bluefin tuna (*Thunnus maccoyii*) in southern Western Australia. *Fisheries Oceanography* 19:183–95.

Fujioka, K., R. Kawabe, A.J. Hobday, Y. Takao, K. Miyashita, O. Sakai, and T. Itoh. 2010b. Spatial and temporal variation in the distribution of juvenile southern bluefin tuna *Thunnus maccoyii*: Implication for precise estimation of recruitment abundance indices. *Fisheries Science* 76:403–10.

Grewe, P.M., P. Feutry, P.L. Hill, R.M. Gunasekeral, K.M. Schaefer, D.G. Itano, D.W. Fuller, S.D. Foster, and C.R. Davies. 2015. Evidence of discrete yellowfin tuna (*Thunnus albacares*) populations demands rethink of management for this globally important resource. *Scientific Reports* 5:16916. doi:10.1038/srep16916.

Gunn, J.S., N.P. Clear, T.I. Carter, A.J. Rees, C.A. Stanley, J.H. Farley, and J.M. Kalish. 2008. Age and growth in southern bluefin tuna, *Thunnus maccoyii* (Castelnau): Direct estimation from otoliths, scales and vertebrae. *Fisheries Research* 92:207–20.

Hampton, J., J. Majkowski, and G. Murphy. 1984. The 1982 Assessment of the southern bluefin tuna (*Thunnus maccoyii*) population and the determination of catch levels which stabilize the parental biomass. CSIRO Marine Laboratories Report No. 165, Hobart, Tasmania.

Hearn, W.S., and T. Polacheck. 2003. Estimating long-term growth-rate changes of southern bluefin tuna (*Thunnus maccoyii*) from two periods of tag-return data. *Fishery Bulletin* 101:58–74.

Hillary, R., A. Preece, C. Davies, M. Bravington, P. Eveson, and M. Basson. 2012. Initial exploration of options for inclusion of the close-kin data into the SBT operating model. Report CCSBTESC/1208/21, prepared for the seventeenth meeting of the CCSBT Extended Scientific Committee, August 27–31, Tokyo, Japan.

Hillary, R., A. Preece, and C. Davies. 2013. Updates to the CCSBT operating model including new data sources, data weighting and re-sampling of the grid. Working paper CCSBT-ESC/1309/15, prepared for the eighteenth meeting of the CCSBT Extended Scientific Committee, September 2–7, 2013, Canberra, Australia.

Hillary, R., A. Preece, and C. Davies. 2014. Assessment of stock status of southern bluefin tuna in 2014 with reconditioned operating model. Report CCSBT-ESC/1409/21, prepared for the nineteenth meeting of the CCSBT Extended Scientific Committee, September 1–6, Auckland, New Zealand.

Hillary, R.M., A.L. Preece, C.R. Davies, H. Kurota, O. Sakai, T. Itoh, A.M. Parma, D.S. Butterworth, J. Ianelli, and T.A. Branch. 2016. A scientific alternative to moratoria for rebuilding depleted international tuna stocks. *Fish and Fisheries* 17 (2): 469–82.

Hobday, A.J., and K. Hartmann. 2006. Near real-time spatial management based on habitat predictions for a longline bycatch species. *Fisheries Management and Ecology* 13: 365–80.

Hobday, A.J., N. Flint, T. Stone, and J.S. Gunn. 2009a. Electronic tagging data supporting flexible spatial management in an Australian longline fishery. In *Tagging and Tracking of Marine Animals with Electronic Devices,* edited by J. Nielsen, H. Arrizabalaga, N. Fragoso, A. Hobday, M. Lutcavage, and J. Sibert, 381–404. Dordrecht, Netherlands: Springer.

Hobday, A.J., R. Kawabe, Y. Takao, K. Miyashita, and T. Itoh. 2009b. Correction factors derived from acoustic tag data for a juvenile southern bluefin tuna abundance index in southern Western Australia. In *Tagging and Tracking of Marine Animals with Electronic Devices,* edited by J. Nielsen, H. Arrizabalaga, N. Fragoso, A. Hobday, M. Lutcavage, and J. Sibert, 405–22. Dordrecht, Netherlands: Springer.

Hobday, A.J., J.R. Hartog, T. Timmis, and J. Fielding. 2010. Dynamic spatial zoning to manage southern bluefin tuna capture in a multi-species longline fishery. *Fisheries Oceanography* 19 (3): 243–53.

Hobday, A.J., K. Evans, J.P. Eveson, J.H. Farley, J.R. Hartog, M. Basson, and T.A. Patterson. 2015. Distribution and migration—southern bluefin tuna (*Thunnus maccoyii*). In *Biology and Ecology of Bluefin Tuna,* edited by T. Kitagawa and S. Kimura, 189–210. London: CRC Press.

Itoh, T., and O. Sakai. 2009. Report of the piston-line trolling survey in 2007/2008. Working paper CCSBT-ESC/0909/32, prepared for the fourteenth meeting of the CCSBT Extended Scientific Committee, September 5–11, Busan, Korea.

Itoh, T., and N. Takahashi. 2016. Update of the core vessel data and CPUE for southern bluefin tuna in 2016. Report CCSBT-ESC/1609/21, prepared for the twenty-first meeting of the CCSBT Extended Scientific Committee, September 5–10, Kaohsiung, Taiwan.

Itoh, T., and Y. Tsuda. 2016. Report of the piston-line trolling monitoring survey for the age-1 southern bluefin tuna recruitment index in 2015/2016. Report CCSBT-ESC/1609/26, prepared for the twenty-first meeting of the CCSBT Extended Scientific Committee, September 5–10, Kaohsiung, Taiwan.

Itoh, T., and S. Tsuji. 1996. Age and growth of juvenile southern bluefin tuna *Thunnus maccoyii* based on otolith microstructure. *Fisheries Science* 62:892–96.

Itoh, T., H. Kurota, N. Takahashi, and S. Tsuji. 2002. Report of 2001/2002 spawning ground surveys. Working paper CCSBT-SC/0209/20, prepared for the seventh meeting of the CCSBT Scientific Committee, September 9–11, Canberra, Australia.

Itoh, T., H. Kemps, and J. Totterdell. 2011. Diet of young southern bluefin tuna *Thunnus maccoyii* in the southwestern coastal waters of Australia in summer. *Fisheries Science* 77:337–44.

Jenkins, G.P., and T.L.O. Davis. 1990. Age, growth rate, and growth trajectory determined from otolith microstructure of southern bluefin tuna *Thunnus maccoyii* larvae. *Marine Ecology Progress Series* 63:93–104.

Josse, E., P. Bach, and L. Dagorn. 1998. Simultaneous observations of tuna movements and their prey by sonic tracking and acoustic surveys. *Hydrobiologia* 371–372:61–69.

Kalish, J.M., J.M. Johnston, J.S. Gunn, and N.P. Clear. 1996. Use of the bomb radiocarbon chronometer to determine age of southern bluefin tuna (*Thunnus maccoyii*). *Marine Ecology Progress Series* 143:1–8.

Kemps, H.A., J.A. Totterdell, T. Nishida, and H.S. Gill. 1998. Dietary comparisons of southern bluefin tuna (*Thunnus maccoyii*) of different sizes and in different locations—with suggestions and a rationale for future work. Report of the Tenth Workshop: Southern Bluefin Tuna Recruitment Monitoring and Tagging Program, September 1998. *CSIRO* pp. 14–17.

Laslett, G.M., J.P. Eveson, and T. Polacheck. 2004. Fitting growth models to length frequency data. *ICES Journal of Marine Science* 61:218–30.

Leigh, G.M., and W.S. Hearn. 2000. Changes in growth of juvenile southern bluefin tuna (*Thunnus maccoyii*): An analysis of length-frequency data from the Australian fishery. *Marine and Freshwater Research* 51:143–54.

Lin, Y.-T., and W.-N. Tzeng. 2010. Sexual dimorphism in the growth rate of southern bluefin tuna *Thunnus maccoyii* in the Indian Ocean. *Journal of the Fisheries Society of Taiwan* 37 (2): 135–51.

Luo, J., P.B. Ortner, D. Forcucci, and S.R. Cummings. 2000. Diel vertical migration of zooplankton and mesopelagic fish in the Arabian Sea. *Deep Sea Research II* 47:1451–73.

Majkowski, J., and G.J. Eckert. 1986. Exploitation of southern bluefin tuna passing through the Australian fishing grounds. Report SBFWS/86/6A, prepared for the fifth southern bluefin tuna Trilateral Scientific meeting, June 10–14, Shimizu, Japan.

Maunder, M.N., J.R. Sibert, A. Fonteneau, J. Hampton, P. Kleiber, and S.J. Harley. 2006. Interpreting catch per unit effort data to assess the status of individual stocks and communities. *ICES Journal of Marine Science* 63:1373–85.

Middleton, J.F., and J.A.T. Bye. 2007. A review of the shelf slope circulation along Australia's southern shelves: Cape Leeuwin to Portland. *Progress in Oceanography* 75:1–41.

Nielsen, A., and J.R. Sibert. 2007. State-space model for light-based tracking of marine animals. *Canadian Journal of Fisheries and Aquatic Sciences* 64:1055–68.

Nishikawa, Y., M. Honma, S. Ueyanagi, and S. Kikawa. 1985. Average distribution of larvae of oceanic species of scombrid fishes, 1956–1981. *Far Seas Fisheries Research Laboratory*, S. Series, 12:1–99.

Patterson, T.A., K. Evans, T.I. Carter, and J.S. Gunn. 2008. Movement and behaviour of large southern bluefin tuna (*Thunnus maccoyii*) in the Australian region determined using pop-up satellite archival tags. *Fisheries Oceanography* 17:352–67.

Polacheck, T. 2012a. Assessment of IUU fishing for southern bluefin tuna. *Marine Policy* 36 (5): 1150–65.

Polacheck, T. 2012b. Politics and independent scientific advice in RFMO processes: A case study of crossing boundaries. *Marine Policy* 36 (1): 132–41.

Polacheck, T., and C. Davies. 2008. Considerations of the implications of large unreported catches of southern bluefin tuna for assessments of tropical tunas, and the need for independent verification of catch and effort statistics. CSIRO Marine and Atmospheric Research Paper 023. http://pandora.nla.gov.au/tep/68165.

Polacheck, T., J.P. Eveson, and G.M. Laslett. 2004. Increase in growth rates of southern bluefin tuna (*Thunnus maccoyii*) over four decades: 1960–2000. *Canadian Journal of Fisheries and Aquatic Sciences* 61:307–22.

Preece, A., and R. Bradford. 2016. Gene-tagging 2017 work plan and research mortality allowance request. Paper CCSBT-ESC/1509/18, prepared for the twenty-first meeting of the CCSBT Extended Scientific Committee, September 5–10, Kaohsiung, Taiwan.

Preece, A., P. Eveson, C. Davies, P. Grewe, R. Hillary, and M. Bravington. 2015. Report on gene-tagging design study. Paper CCSBT-ESC/1509/18, prepared for the twentieth meeting of the CCSBT Extended Scientific Committee, September 1–5, Incheon, South Korea.

Rees, A.J., J.S. Gunn, and N.P. Clear. 1996. Age determination of juvenile southern bluefin tuna, *Thunnus maccoyii,* based on scanning electron microscopy of otolith microincrements. In *The Direct Estimation of Age in Southern Bluefin Tuna: Final Report to FRDC, Project No. 1992/42,* edited by J.S. Gunn, N.P. Clear, A.J. Rees, C. Stanley, J.H. Farley, and T. Carter, appendix. Hobart, Tasmania: CSIRO Marine Research.

Richardson, D.E., K.E. Marancik, J.R. Guyon, M.E. Lutcavage, B. Galuardi, C.H. Lam, H.J. Walsh, S. Wildes, D.A. Yates, and J.A. Hare. 2016. Discovery of a spawning ground reveals diverse migration strategies in Atlantic bluefin tuna (*Thunnus thynnus*). *Proceedings of the National Academy of Sciences* 113 (12): 3299–3304.

Ridgway, K.R., and S.A. Condie. 2004. The 5500-km-long boundary flow off western and southern Australia. *Journal of Geophysical Research* 109:C04017.

Rooker, J.R., H. Arrizabalaga, I. Fraile, D.H. Secor, D.L. Dettman, N. Abid, P. Addis, et al. 2014. Crossing the line: Migratory and homing behaviors of Atlantic bluefin tuna. *Marine Ecology Progress Series* 504: 265–76.

Schaefer, K.M. 2001. Reproductive biology of tunas. In *Tuna: Physiology, Ecology and Evolution,* edited by B.A. Block and S.E. Stevens, 225–70. San Diego, CA: Academic Press.

Schahinger, R.B. 1987. Structure of coastal upwelling vents observed off the south-east coast of South Australia during February 1983–April 1984. *Australian Journal of Marine and Freshwater Research* 38:439–59.

Shiao, J.-C., S.-K. Chang, Y.T. Lin, and W.N. Tzeng. 2008. Size and age composition of southern bluefin tuna (*Thunnus maccoyii*) in the central Indian Ocean inferred from fisheries and otolith data. *Zoology Studies* 47:158–71.

Shimose, T., and J.H. Farley. 2015. Age, growth and reproductive biology of bluefin tunas. In *Biology and Ecology of Bluefin Tuna,* edited by T. Kitagawa and S. Kimura, 47–77. London: CRC Press.

Shingu, C. 1978. Ecology and stock of southern bluefin tuna [in Japanese]. Japan Association of Fishery Resources Protection. *Fisheries Study* 31:81. English translation in CSIRO Division of Fisheries and Oceanography, *Report* 131 (1981): 79.

Sippel, T., J.P. Eveson, B. Galuardi, C. Lam, S. Hoyle, M. Maunder, P. Kleiber, et al. 2014. Using movement data from electronic tags in fisheries stock assessment: A review of models, technology and experimental design. *Fisheries Research* 163:152–60.

Skaug, H. 2001. Allele-sharing methods for estimation of population size. *Biometrics* 57: 750–56.

Thorogood, J. 1987. Age and growth rate determination of southern bluefin tuna, *Thunnus maccoyii,* using otolith banding. *Journal of Fish Biology* 30:7–14.

Ward, T.M., L.J. Mcleay, W.F. Dimmlich, P.J. Rogers, S. McClatchie, R. Matthews, J. Kampf, and P.D. Van Ruth. 2006. Pelagic ecology of a northern boundary current system: Effects of upwelling on the production and distribution of sardine (*Sardinops sagax*), anchovy (*Engraulis australis*) and southern bluefin tuna (*Thunnus maccoyii*) in the Great Australian Bight. *Fisheries Oceanography* 15:191–207.

Young, J.W., T.D. Lamb, D. Le, R.W. Bradford, and A.W. Whitelaw. 1997. Feeding ecology and interannual variations in diet of southern bluefin tuna, *Thunnus maccoyii,* in relation to coastal and oceanic waters off eastern Tasmania, Australia. *Environmental Biology of Fishes* 50:275–91.

CHAPTER 11

# Rebuilding Southern Bluefin Tuna

## Past, Present, and Future

Richard M. Hillary, Ann L. Preece, and Campbell R. Davies

## Introduction

Southern bluefin tuna (SBT) is a long-lived (to near 40 years), late-maturing (from about age 8–20), and highly migratory tuna found throughout the southern temperate oceans, except for the more easterly regions of the South Pacific (Farley et al. 2007; see also chapter 10, this volume). Commercial surface and longline fisheries for SBT began in the 1950s, with annual catches reaching a maximum of 81,750 metric tons (t) in 1961 and remaining relatively high until catch restrictions were first introduced in 1989 (see Figure 11.1). The impetus for these early management arrangements was evidence of high fishing mortality rates, estimates of declining recruitment, and the demise of the purse seine surface fishery off the southeast coast of Australia in the mid-1980s (Caton 1991).

The international SBT fishery was first managed through informal tripartite agreements among Australia, Japan, and New Zealand from the early 1980s. This became the responsibility of the Commission for the Conservation of Southern Bluefin Tuna (CCSBT) in 1993 (Anonymous 1994). The CCSBT is one of five regional fisheries management organizations (RFMOs) for tuna. These international governance bodies are constituted by nation-states involved in the fishery under the auspices of the United Nations Convention of the Law of the Sea. Typically, science is conducted by working groups acting under a scientific committee, which formulates recommendations for the CCSBT. The CCSBT is the decision-making body, with each commissioner acting on behalf of the jurisdiction they represent. CCSBT decisions are generally made by consensus and are subsequently implemented by member states through their national fisheries agency.

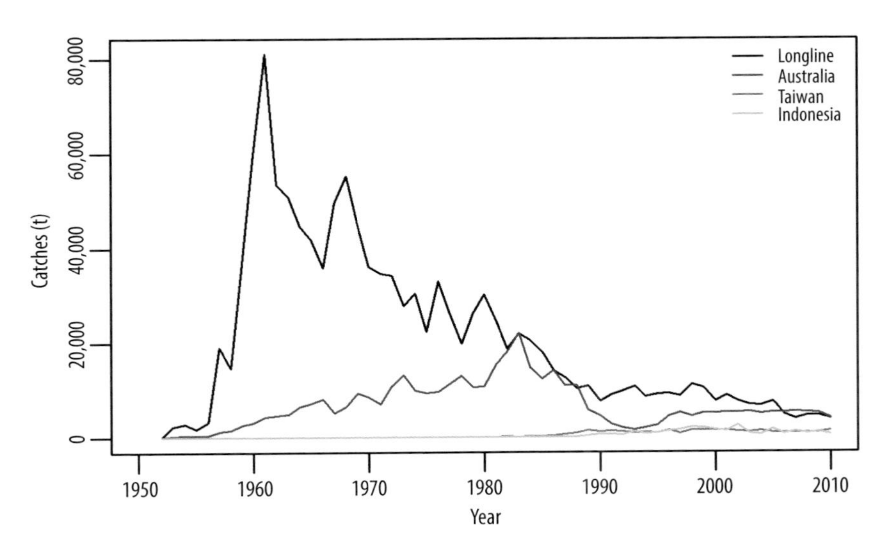


**Figure 11.1.** Southern bluefin tuna (SBT) catch history by gear and national fleet.

In the early 1980s, it became clear that historical levels of exploitation had caused the biomass of SBT to decline substantially, with increased fishing mortality rates on both mature and immature fish, which had resulted in reduced levels of recruitment. Notwithstanding this evidence, the 1990s were characterized by uncertainty within the CCSBT among members as to how *much* change had occurred. The different perceptions of stock status and productivity were driven by different stock assessment outcomes arising from the use of different models and data. Given that management via consensus is the RFMO process, this led to a lack of agreement on what—if anything—needed to be changed regarding fishing levels, and how best to deal with the considerable uncertainty in stock status and productivity. This impasse eventually led to members taking a formal dispute to the International Tribunal for the Law of the Sea (Polacheck 2002).

Outcomes from this dispute process included the appointment of independent chairs of the Stock Assessment Group and the Scientific Committee; the creation of an external advisory panel to assist in the interpretation of the science advice and, if required, to provide independent advice to the CCSBT; and an agreement on a process to develop and test formal management procedures (MPs) (see Cooke 1999, Butterworth 2007, Punt and Donovan 2007) as a basis for resolving the scientific uncertainty in stock status and productivity and to provide robust management advice. Over the following five years, this technical process led to consolidation of the available data and previous methodological approaches and underlying stock assessments into a single, unified model; and also to use of this model as an operating model (OM) to evaluate candidate MPs designed to set quotas and to rebuild the stock to an agreed level specified by the CCSBT (Kolody et al. 2008, Butterworth 2008).

Through the early 2000s, a considerable amount of OM development and testing of candidate MPs was completed (Anonymous 2005a). After several rounds of testing of a diverse range of model-based and empirical MPs, a final MP was selected and adopted by the CCSBT for implementation in 2005 (Anonymous 2005b). The MP, however, was never implemented, owing to the revelation in 2006 of substantial, long-term underreporting of longline catches (Anonymous 2006, Polacheck and Davies 2008, Polacheck 2012). Neither the OM nor the tuning of the adopted MP had allowed for underestimation in the catches of major fleets (sometimes around 100% of the reported catch), and thus the MP could not be used to provide reliable advice on catch levels that would lead to spawning stock rebuilding (Anonymous 2006). The 2006 stock assessment process also confirmed a succession of the lowest annual recruitments on record in the late 1990s and early 2000s, as well as the most pessimistic estimate of stock depletion and outcomes for a range of projected catch levels (Anonymous 2006).

It is in this context that 2006 forms such a crucial pivot point in the CCSBT science and policy history. Much good work had been completed prior to 2005 in terms of MP development, including provisional adoption of an MP by the CCSBT, but the overcatch issue derailed implementation at a time when the spawning stock was estimated to be at its lowest level historically, and with the lowest year classes on record about to make their way into the longline fisheries and eventually the spawning stock. At this crucial juncture, the CCSBT confirmed its commitment to the MP approach (Anonymous 2006) and to the rebuilding of the stock. The rest of this chapter covers the story from that point through to 2016, including the following:

- Testing, adoption and implementation, in 2011, of the CCSBT MP, which was named the Bali Procedure after the location of the meetings in which it was selected and adopted.
- Performance of the Bali Procedure during its implementation from 2011 to 2016.
- Phasing out of the Bali Procedure and the development of a new MP for the CCSBT.

## The Bali Procedure: Testing, Adoption, and Implementation

Between 2007 and 2009, the OMs used in assessing stock status and in testing and selecting MPs underwent significant development. They incorporated existing mark-recapture data with a more appropriate formulation as well as the scientific aerial survey (a relative biomass index of 2–4-year-old fish in summer in Southern Australia [see chapter 10, this volume]), and examined implications of the large-scale overcatch (i.e., impacts on estimated total catch levels and use of longline catch per unit effort [CPUE] as an index of abundance). Additionally, there were significant quota reductions in 2006 and 2009, in response to consistent evidence of four historically low cohorts between 1999 and 2004.

At the Scientific Committee meeting in 2009, member scientists were invited to initiate development and testing of candidate MPs, under the auspices of the Operating Model and Management Procedure technical group of the Scientific Committee. Candidate MPs were required to use two monitoring series as inputs: the index of recruitment from the aerial survey and the standardized CPUE for 4+-year-old fish from the Japanese longline fleet. The latter provides an index of relative abundance of the harvested age classes for the majority of catches taken by the longline fleets. The inclusion of the recruitment index reflected the ongoing concern of the Scientific Committee and of the CCSBT about the very low status of the spawning stock, the recent very weak year classes, and the substantial lag between recruitment to the fisheries (2 years old for the surface fishery and 4 years old for the longline fisheries) and recruitment to the spawning stock (minimum 8 years old, with 50% maturity at 10 to 12 years old; see chapter 10 in this volume for current estimates). The inclusion of the recruitment index was intended to provide a basis for MPs to respond to weak year classes and/or declining trends in recruitment earlier than would be the case if only longline CPUE was used.

## A Brief Tour of the CCSBT Operating Models Used to Test Candidate Management Procedures ca. 2009–2011

The SBT OMs were based on the previous stock-assessment models and were designed to incorporate the range of uncertainty in the state and dynamics of the stock and fishery. Full specification of the model can be found in Anonymous (2011b, 2014). It is a complex, integrated, statistical age-based model fitted to the following data series:

- Catch biomass for each fleet (six in total).
- Length composition of the catch of the four main longline fleets.
- Age composition of the catch of the surface-school and spawning-ground fisheries.
- Japanese longline CPUE abundance index for 4+-year-old fish.
- Aerial survey of estimates of 2–4-year-old biomass in the mid to late summer in the Great Australia Bight (GAB).
- Multicohort mark-recapture data from conventional tags from the 1990s.

An important aspect of the CCSBT approach is the use of a suite of OMs as opposed to the selection of a single base case, i.e., a single model that is judged to be the *best* relative to other plausible representations. In this case, an orthogonal cross, or grid, of influential parameters for which reliable information may be lacking (because they are poorly estimated or, in the judgment of the Scientific Committee, the available data are unable to provide reliable estimates of the parameter of interest) is used to generate a consistent suite of OMs. This approach explicitly encapsulates both the structural (i.e., model) uncertainty as well as the relevant process and observation uncertainty in the assessment of stock status, the provision of management advice, and the testing and selection of candidate MPs. Influential and uncertain parameters included in the grid are the steepness of the stock-recruit relationship ($h$), natural mortality at ages 0 and 10 ($M_0$ and $M_{10}$, respectively), alternative CPUE standardizations (w.5 and w.8, which are alternative hypotheses about effort concentration in the longline fleet and its impact on the relationship between CPUE and abundance), and the age range to which the standardized CPUE index applies (ages 4–18 or 8–12).

This array of grid permutations (240 uniquely specified models, in the case of the 2011 OMs) represents the reference set of OMs used for advice on stock status and for testing of MPs in 2010–11. From this reference set, 2,000 individual realizations are used in future projections and MP evaluations (i.e., Management Strategy Evaluation, or MSE). Table 11.1 provides the grid settings (both parameter values and resampling specifications) for the reference set of OMs against which all MPs were initially tested. Full details can be found in Anonymous (2011a).

**Table 11.1.** Settings for the southern bluefin tuna operating model grid in 2011

| | Levels | CumulN | Values | Prior | Weighting |
|---|---|---|---|---|---|
| $h$ | 5 | 5 | 0.55, 0.64, 0.73, 0.82, 0.9 | Uniform | Objective function |
| $M_0$ | 4 | 20 | 0.3, 0.35, 0.4, 0.45 | Uniform | Objective function |
| $M_{10}$ | 3 | 60 | 0.07, 0.1, 0.14 | Uniform | Objective function |
| CPUE | 2 | 120 | w.5, w.8 | Uniform | prior |
| $q$ age-range | 2 | 240 | 4–18, 8–12 | 0.67, 0.33 | prior |

*Notes*: CumulN refers to the cumulative number of levels in the grid. CPUE = catch per unit effort.

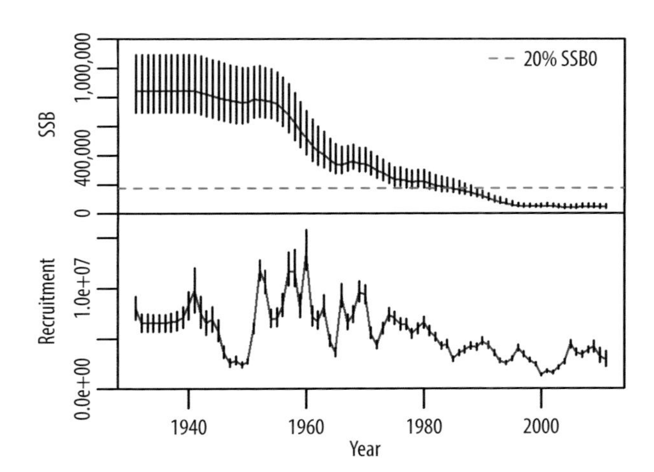


**Figure 11.2.** Median and 80% probability interval for southern bluefin tuna spawning stock biomass (SSB) (*top*) and recruitment (*bottom*) in the 2011 operating model.

To be considered as a viable MP, each candidate must have been able to meet the rebuilding targets of the CCSBT across the full suite of reference set models (i.e., from the most optimistic to the most pessimistic). Figure 11.2 shows the historical spawning stock biomass (SSB) and recruitment dynamics for the full reference set of operating models to 2011. Estimates of SSB depletion ranged from 3% to 8%, with mean recruitment decreased by ~50% relative to the unfished state and most recent $F$ levels at, or just above, maximum sustainable yield (MSY) (Anonymous 2011b).

## Final Data Sources for Input to Candidate Management Procedures

An essential step in the MSE process is agreement on what data sources may be used as inputs to candidate MPs. Initially, the data monitoring series included in candidate MPs were

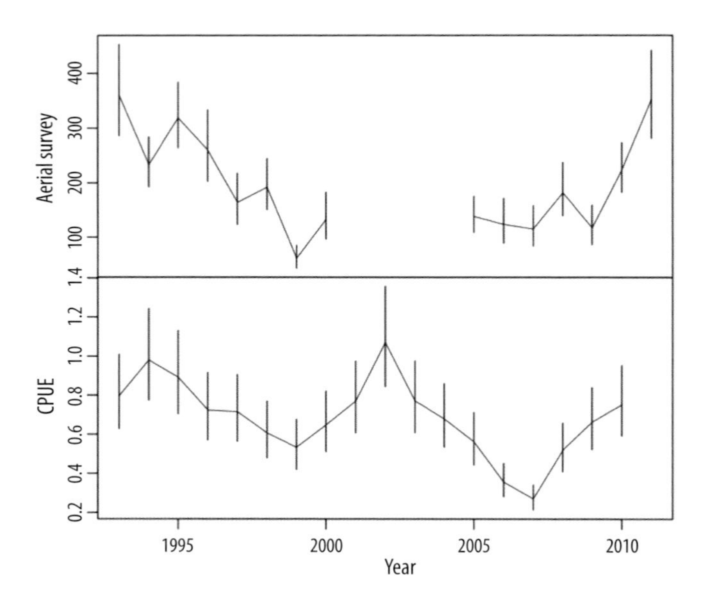


**Figure 11.3.** Median and 80% probability interval for aerial survey (*top*) and longline catch per unit effort (CPUE) (*bottom*) used in the adopted Bali Procedure in 2011.

- Japanese longline CPUE (relative abundance index of 4+-year-old fish).
- Aerial survey of juvenile SBT in the GAB in summer (relative biomass index of 2–4-year-old fish).
- Catch composition from the main longline fleet (age structure converted from length data).

In the first iteration of the evaluation process, it was agreed that any candidate MP should use *both* the CPUE and aerial survey indices (Anonymous 2010b). This eventually led to the agreement that these two indices should be the only data inputs to the MPs (Anonymous 2010a). Summaries of these two monitoring series are given in Figure 11.3.

A wide array of initial candidate MPs were reviewed and evaluated at a dedicated technical meeting in 2010 (Anonymous 2010b). There were biomass-dynamic surplus production model-based MPs (including the one originally adopted in 2005), model-based relative abundance models, more empirical-style MPs, and even one based on fuzzy logic. At that meeting, two candidate MPs were selected on the basis of initial performance and behavior for more detailed consideration at the Scientific Committee meeting later in the same year. The first, MP1, which had a model-based harvest-control rule (HCR), used the following integrated population dynamics model to smooth the aerial survey and CPUE indices:

$$\begin{aligned} B_y &= J_y + g_{y-1} B_{y-1}, \\ J_y &= \exp(\mu_J + \epsilon_y^J), \\ g_y &= \exp(\mu_g + \epsilon_y^g), \\ \epsilon_y^{J,g} &\sim N(0, \sigma_{J,g}^2). \end{aligned} \quad (1)$$

The term $B_y$ represents the adult relative biomass (including a combination of early, medium, and fully mature animals) and is assumed proportional to the longline CPUE; the term $J_y$ represents the juvenile relative biomass (covering the immature, ages 2–4), assumed proportional to the previous year's aerial survey index; the term $g_y$ represents a combination of total mortality and surplus production effects within the adult biomass range. This is a variant of a model first proposed in Trenkel (2008). The HCR for MP1 had a target-based structure for the estimated parameter, $B_y$ (which effectively specified that the longline CPUE level observed in the 1980s should be reached around the interim rebuilding target), with limit terms for the juvenile biomass, $J_y$, and adult biomass growth terms based on observed minima in the historical data.

The total allowable catch (TAC) for MP1 was defined using the TAC from the previous year and an adjusted target catch as follows:

$$TAC_{y+1}^{[MP1]} = 0.5 \times (TAC_y + C_y^{targ} \Delta_y^J) \quad (2)$$

where

$$C_y^{targ} = \begin{cases} \delta \left[ \dfrac{B_y}{B^*} \right]^{1-\varepsilon_b} & \text{for} \quad B_y \geq B^* \\ \delta \left[ \dfrac{B_y}{B^*} \right]^{1+\varepsilon_b} & \text{for} \quad B_y < B^* \end{cases} \quad (3)$$

and $\varepsilon_b \in [0,1]$ represents the degree to which the response to a biomass level above or below the target level $B^*$ is asymmetric. The target catch level $\delta$ is the tuning parameter of the HCR. The recruitment adjustment $\Delta_y^J$ is defined as follows:

$$\Delta_y^J = \begin{cases} \left[ \dfrac{\bar{J}}{\mathcal{J}} \right]^{1-\varepsilon_J} & \text{for} \quad \bar{J} \geq \mathcal{J} \\ \left[ \dfrac{\bar{J}}{\mathcal{J}} \right]^{1+\varepsilon_J} & \text{for} \quad \bar{J} < \mathcal{J} \end{cases} \quad (4)$$

and $\varepsilon$-$J$.$\in[0,1]$ is the level of asymmetry in response to the current moving (arithmetic) average recruitment levels, $J$ (now calculated using the years up to and including year $y$):

$$\bar{J}=\frac{1}{\tau_J}\sum_{i=y-\tau_J+1}^{y}J_i, \qquad (5)$$

of length $\tau_J$ relative to the average, $\mathcal{J}$, calculated over the years for which the estimates are based on observed data.

The second candidate MP (MP2) was an empirical MP with a slope-in-the-index structure for the CPUE part of the HCR:

$$TAC_{y+1}^{[MP2]}=TAC_y\times\begin{cases}1-k_1|\lambda|^{\gamma} & \text{for } \lambda<0\\ 1+k_2\lambda & \text{for } \lambda\geq 0\end{cases}, \qquad (6)$$

where $\lambda$ is the slope in the regression of $\ln B_y$ against year (from years $y-\tau_B+1$ to year $y$). MP2 also had a term that used the aerial survey in a more target-oriented format.

The 2010 Scientific Committee tested the performance of these MPs, and both, as well as their average, were recommended to the CCSBT for consideration for a range of rebuilding scenarios (Anonymous 2010a).

## Two Become One: Agreeing on a Single Management Procedure for Recommendation to the CCSBT

Although the CCSBT could not agree on an MP in 2010, it defined the following array of rebuilding objectives and constraints on scale and frequency of TAC changes:

- The interim rebuilding level of SSB is 20% of $SSB_0$, the unfished level.
- The rebuilding time frames were 2030, 2035, and 2040.
- The rebuilding probabilities were 0.6, 0.7, and 0.9.
- Minimum and maximum permitted changes in TAC were 100 t and 3,000 t or 5,000 t, respectively.
- TACs were permitted to change every three years.
- There would be no lag prior to the initial implementation of the TAC, and a one-year lag thereafter.

The 2011 Scientific Committee meeting was additionally complicated because MP2 was potentially unable to meet the tuning criteria for some of the key rebuilding objectives (specifically the set of objectives ultimately selected—to rebuild to 0.2 SSB with 0.7 probability by 2035) given a more optimistic outlook in the projected dynamics. At the meeting, member scientists agreed that there were features of both MPs that worked well on different robustness tests, so perhaps an MP that combined the two candidate MPs would perform better across more of the robustness tests. The combined MP used the model-based and target/limit-driven features of MP1 and the more trend-based features of the CPUE-driven part of the rule at the center of MP2. The TAC from the combined MP was, then, a combination of those in equations 2 and 6:

$$TAC_{y+1} = \frac{1}{2}(TAC_{y+1}^{MP1} + TAC_{y+1}^{MP2}). \qquad (7)$$

This new MP was then tuned to all the relevant rebuilding objectives and was subjected to all the robustness tests (see Hillary et al. 2016 for MP tuning specifications).

In terms of performance criteria, there was, as in almost all fisheries, a natural dichotomy between biomass rebuilding and catch-based performance measures, which often trade off with each other. In terms of rebuilding performance, the focus was firmly on the SSB. Factors such as probability of future declines, the probability that the SSB would reach the target level halfway to the interim target year, and the probability that the future SSB would drop below the minimum estimated level were also evaluated. Fishery performance was assessed through various catch-based statistics such as average interannual variation (time-averaged percentage change in catch from year to year), probability that the TAC would go down after going up over the first two and first four TAC decisions, and average catches for various periods. Figure 11.4 shows an example of the primary summary plots used to assess performance across multiple robustness trials and for a range of performance statistics. This particular plot is for the Bali Procedure for the reference set of OMs ("base") and recruitment failure (*lowR*) and catchability change (*upq2008*) robustness trials (see Table 11.2 and Anonymous 2011 for further details).

Plots such as these, while displaying a vast amount of information, are essential for the discussions (both inside and outside of the meeting process) that occur in the evaluation of an MP in an RFMO setting, where there are often large differences in the degree of understanding of the MP process within the RFMO Scientific Committee and the CCSBT. Summary

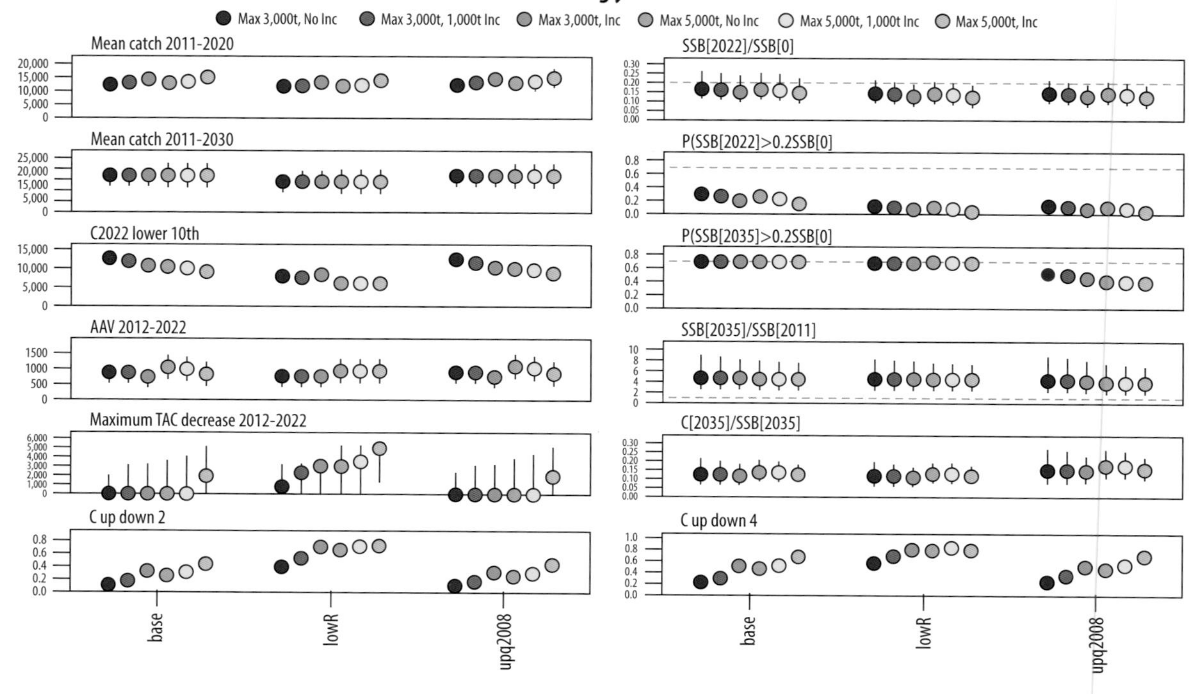


**Figure 11.4.** Performance statistics summary for the combined Bali Procedure across the reference operating model and two key robustness trials for a range of maximum total allowable catch (TAC) and initial TAC increase options. AAV = annual average variation; SSB = spawning stock biomass.

**Table 11.2.** Summary of most influential performance statistics for the Bali Procedure for the reference set (3,000 t and 5,000 t maximum total allowable catch [TAC] change) and key robustness trials, in terms of rebuilding probabilities, expected catch levels (C), and likely TAC changes

| Scenario | $p(B_{2035} > 0.2B_0)$ | $p(B_{2035} > 0.2B_0)$ | $\mathbb{E}(B_{2022}/B_{2011})$ | $\mathbb{E}(C_{2012\text{–}2022})$ | $p(C_{\uparrow\downarrow})$ |
|---|---|---|---|---|---|
| Reference set | | | | | |
| 3,000 t change | 0.19 | 0.7 | 2.76 | 15,200 | 0.49 |
| 5,000 t change | 0.14 | 0.7 | 2.65 | 15,600 | 0.71 |
| Robustness trials | | | | | |
| *lowR*, 3,000 t | 0.06 | 0.66 | 2.32 | 13,200 | 0.83 |
| *upq*, 3,000 t | 0.08 | 0.45 | 2.58 | 15,300 | 0.50 |
| *STWin*, 3,000 t | 0.01 | 0.34 | 2.39 | 12,872 | 0.81 |
| *omega75*, 3,000 t | 0.06 | 0.48 | 2.74 | 13,304 | 0.74 |

*Notes*: The $p(C_{\uparrow\downarrow})$ statistic is the probability (over all iterations) that the TAC would go down at the 2013 TAC decision (for the 2016–18 quota block) after it went up at the 2011 TAC decision (for the 2012–15 quota block). Robustness trials are a set of pessimistic scenarios designed to test how well the management procedure (MP) responded to situations not included in the reference set used to tune the MP to the rebuilding target: *lowR* is recruitment failure; *upq* is catchability change (true *q* increases); *STWin* is a more pessimistic CPUE series; and *omega75* is a sublinear catch-per-unit-effort–to-abundance relationship.

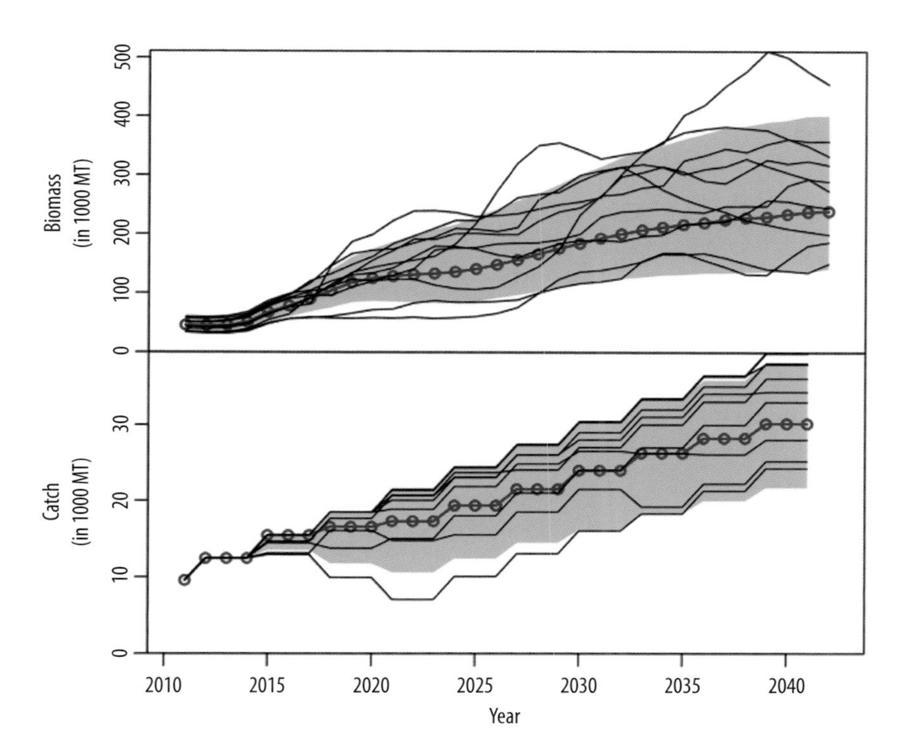


**Figure 11.5.** Projected median, 80% probability interval (*in blue*), and 10 randomly selected individual realizations (*black*) for spawning stock biomass (*top*) and total allowable catches (TACs) (*bottom*) for the selected management procedure, with a 3,000-metric-ton constraint on the maximum TAC, allowing a full increase at the first TAC decision (2012). MT = metric ton.

tables are not as effective for communicating large amounts of information on performance, but are better suited for higher-level summaries (for example, one MP and a handful of robustness trials and settings). Another very important reality to communicate to all representatives—and especially to CCSBT commissioners—is that whatever MP is chosen, the actual outcomes will *never* look like the means or medians they see in projections and statistical summaries. Figure 11.5 shows an example of a "worm" plot that is frequently used to make this point: future SSB and catch (median and 80% probability intervals) are shown alongside 10 random draws from the 2,000 projections. The more erratic black lines are representative of the expected rebuilding trajectory of the stock and the catches, not the smooth median.

Given positive signals in both the MP input indices, as well as a more optimistic reconditioned OM in 2011, relative to 2009 (Anonymous 2011b), all candidate MPs predicted an increase in the first TAC decision (2012–14 quota). A final additional constraint was added that limited the initial increase in TAC to a maximum of 1,000 t—a precautionary measure directed by the CCSBT. A very general summary of the performance of the Bali Procedure (3,000 t/5000 t maximum TAC change, 2035 target year, 1,000 t maximum initial TAC increase) is given in Table 11.2, which summarizes results for the reference set and the key robustness tests listed above.

For the recruitment failure robustness test, *lowR*, although the probability of approaching the rebuilding target by 2022 is low, the MP almost manages to reach the rebuilding target by 2035 by virtue of lower catches after the initial TAC increase when the signs of recruitment failure (artificially created in this robustness test) appear in the aerial survey and, subsequently, in CPUE indices. For the catchability change robustness test, the MP fails to reach the rebuilding target. This is driven by a slightly less optimistic starting point in 2011 and by above-average catches taken as a result of the bias in the CPUE index (which was artificially created for this robustness test). For the more pessimistic CPUE robustness test, using a spatiotemporal CPUE index, the starting point is noticeably less optimistic than for the reference set, and even with appreciably lower average catches over the projection period, the MP can attain the rebuilding target only with a probability of 0.34—less than half the target probability. For the *omega75* robustness test of a sublinear relationship between CPUE and abundance, while the starting point is below the reference set, the MP manages to attain a rebuilding probability of 0.48 with lower average catches. This performance is enhanced by the nature of the test itself: historically, the nonlinear relationship between CPUE and abundance means the decline in CPUE with respect to actual abundance is even stronger relative to the reference set; in the projections when the abundance increases, the CPUE also increases, but at a slower rate, so the MP is slower to react (i.e., increase the TAC) to the positive signals, resulting in a better level of rebuilding than if future CPUE was directly proportional to abundance. For all robustness tests, the MP managed to more than double the SSB from 2011 to 2022 (range from 2.3 to 2.8).

In 2011 the Scientific Committee of the CCSBT recommended the adoption of the MP created by the combination of MP1 and MP2. The CCSBT also had to decide on appropriate settings and TAC constraints (Anonymous 2011a). A special meeting of the CCSBT was held immediately after the Scientific Committee meeting, and a number of additional runs with different settings of the candidate MP were requested prior to the annual meeting of the CCSBT later that year. Member scientists began these additional simulations and collaborated across the various member countries to produce a final MP performance summary for the CCSBT. The CCSBT formally adopted the combined MP, known as the Bali Procedure, with the agreed target year and probability of 2035 and 0.7, a maximum change of 3,000 t (Anonymous 2011b). In terms of the implementation lag, there was zero lag for the first decision and a one-year lag thereafter, so the first three-year TAC schedule was from 2012 to 2014, and subsequently 2015–17, 2018–20, etc. More detailed information about the Bali Procedure and its evaluation and adoption can be found in Hillary et al. (2016).

## The Bali Procedure in Practice: Setting Total Allowable Catches from 2012 to 2020

Given positive trends in both of the MP input indices (see Figure 11.3), the Bali Procedure calculated an increase (less than the maximum 3,000 t permitted) in the TAC for the 2012–14 TAC block. While all indices were well within the range tested, because of the rebuilding purpose of the MP, the CCSBT decided to limit this increase to 1,000 t (Anonymous 2011a) in the first year, with a phased increase from 9,449 t between 2009 and 2011 to 12,449 t by 2014.

In 2013 the decision for the 2015–17 quota block was due. The signals in the updated input data for the MP were as follows:

- Aerial survey: two more years of data (2012 and 2013) were available. In 2012 the index was low (the second lowest in its history), but in 2013 it was well above the average. The MP acts strongly to reduce the TAC if recent (five-year moving average) aerial survey points are below the historical (1993–2011) mean level, and as of 2013, the five-year average was still above the historical mean (limit) level.
- CPUE: two more years of CPUE data (2011 and 2012) were available, both of which were above average and were above the lows of 2006 and 2007. The MP permits increases to the TAC if the seven-year log-linear trend in CPUE is positive (which it was) and if the CPUE is tracking toward the interim target level of the CPUE of the early 1980s (which it was).

Given these trends in the data, at the 2013 Scientific Committee meeting, the MP calculated an increase from 12,449 t to 14,647 t for 2015–17, the period available for increase (Anonymous 2013a), and this decision was agreed on at the subsequent CCSBT meeting (Anonymous 2013b).

In 2014 a full reconditioning of the OMs was required to assess stock status, as part of the implementation framework adopted with the MP. This included not just an update of previous models with the most recent data, but also the first formal inclusion of the close-kin parent-offspring pair (POP) mark-recapture data (Hillary et al. 2013, Anonymous 2014). The POP data provide information on absolute abundance of the spawning stock (Bravington et al. 2016a), and the independent estimate using these data had indicated a substantially larger spawning population than previously thought, based on the reference set of OMs (Bravington et al. 2016b). Work to include these data in the OM began in 2012 (Anonymous 2012, Bravington et al. 2014), and when it was completed in 2014, it was suggested that absolute spawning stock size was ~2 times larger than indicated by the reference set of OMs; depletion levels were more optimistic, at 8%–12%, than the 2011 OMs (3%–8%); and overall mortality of older reproducing fish was actually lower (Anonymous 2014).

In the conventional stock assessment context, a change of this scale would complicate how management advice might be revised, and it would potentially undermine confidence in the base-case stock assessment. However, with a fully tested MP in place as the agreed basis for management advice, it was relatively straightforward to explore the impact of including these data on the understanding of stock status and productivity (and the implications for the projected dynamics), and whether this implied a substantial departure from the conditions under which the MP was tested. These analyses indicated that for the reference set of OMs, the probability of meeting the rebuilding objective was now 0.75 instead of 0.7, and across a range of optimistic and pessimistic robustness trials, it was never less than 0.63 (Hillary et al. 2014). This represented a notable and generally positive difference from the 2011 conditions in which the MP was adopted (see Table 11.2). As a result of these analyses, the CCSBT Scientific Committee recommended, and the CCSBT agreed, that the Bali Procedure did not require retuning and thus could continue to be used to set TACs in pursuit of the CCSBT's rebuilding objective.

For the third scheduled TAC decision, in 2016, the aerial survey indices in 2014 and 2016 (the survey was not undertaken in 2015) were the second highest and the highest points in the series, respectively—and by a sizable margin for 2016. The most recent CPUE data points (2014 and 2015) were also at the highest levels since the early 1990s and were approaching the target level of the values observed in the mid-1980s. The MP calculations indicated that the full 3,000-metric-ton increase (from 14,647 t to 17,647 t) could be taken. The MP acts weakly, in terms of permitted increases in TAC, as the aerial survey index increases. So, even with such a large increase in the recent average value of the index, only a small proportion of the recommended increase in TAC was attributable to the aerial survey. The large majority of the contribution to the TAC increase was due to the sustained positive trend in the CPUE index over recent years *and* the fact that the actual CPUE levels were very close to the target level specified in the HCR of the MP. For more detail on how all these factors were reviewed and considered by the Scientific Committee, see Anonymous (2016). The Scientific Committee recommended, and the CCSBT agreed to, the full TAC increase, less an amount for estimated catches by nonmembers (Anonymous 2016).

## Phasing Out the Bali Procedure and Designing the Next-Generation Management Procedure

For a number of reasons, including cost of operation and distribution of this cost among members, the aerial survey was not undertaken in 2015, and the general consensus was that it was not likely to continue after 2017. The 2015 Scientific Committee scientists completed an in-depth analysis of the impact of a recruitment-monitoring index on the likely performance of an MP, versus an MP based solely on the longline CPUE index. The committee made an unequivocal recommendation that a recruitment-monitoring index provided a clear benefit to an MP that uses the longline CPUE. Since the aerial survey index was very likely to be discontinued, the current Bali Procedure could not be used beyond the current TAC block (i.e., the 2016 decision above); the committee recommended that any future MPs consider some form of recruitment-monitoring index (Anonymous 2015a).

For a number of years, researchers investigated the concept of using genetic tagging (GT) as an alternative recruitment-monitoring index focused on 2–3-year-old fish (Davies et al. 2008; Preece et al. 2013; Bradford et al. 2016; chapter 10, this volume). Previous tagging programs had been both successful and informative (in the sense of improving the OMs), but the most recent had been undermined by high uncertainty in reporting rates and unusual mixing characteristics for 2-year-old releases (Anonymous 2007). It was determined that moving to a GT approach and a sensible release-and-recapture strategy would have a high probability of ameliorating both these issues (Preece et al. 2013, Preece et al. 2015). This was recommended to the CCSBT by the Scientific Committee as the most promising recruitment-monitoring index (Anonymous 2015a). The CCSBT requested an initial pilot study of the GT approach, in addition to the ongoing tissue collection and genotyping for close-kin mark-recapture studies (Anonymous 2015b).

By this time, the close-kin mark-recapture approach had developed to the point where not just POPs, but also half-sibling pairs (HSPs), among juvenile comparisons, were possible (Bravington et al. 2016b; Davies et al. 2016). The HSP data are informative for both adult abundance *and* adult mortality, relative to the POPs, which primarily inform adult abundance and reproductive potential-at-age (Bravington et al. 2016b). Member scientists reviewed how these new data sources, both GT and close-kin (POPs and HSPs), might be used in a more empirical MP framework (Davies et al. 2016, Hillary et al. 2016, Anonymous 2016). For GT data, and for the sample sizes agreed on by the CCSBT, a relative recruitment index of age-2 fish derived from the GT data was found to correlate very strongly with the true underlying recruitment trends over time. Empirical indices based on the POP and HSP data—again for agreed sample sizes—correlated linearly with the true spawning population at a level comparable to an unbiased survey with a CV of around 25% (Hillary et al. 2016). While a survey of this nature is not logistically possible for a highly migratory species such as SBT, it served to underscore the potential value of the close-kin data as an input monitoring series for future MPs, in addition to their value to OMs for stock assessment. As a result, MP development over the next two years will consider, for adoption in 2019, MPs that use a combination of GT, CPUE, and close-kin data. The selected MPs would then be used to set the TAC for 2021–23 (Anonymous 2016).

## Discussion

It is no exaggeration to say that the decision by the CCSBT to commit to the simulation-tested MP approach as a basis for management advice (as opposed to the best assessment/negotiated TAC process common to most tuna RFMOs) has transformed how this international fishery is managed and how science is done. Given the relatively slow dynamics of the stock, it is too early to say whether the benefits, in terms of rebuilding of the spawning stock, have been realized, but the early signs are promising. Clear, how-

Punt, A.E., and G.P. Donovan. 2007. Developing management procedures that are robust to uncertainty: Lessons from the international whaling commission. *ICES Journal of Marine Science* 64:603–12.
Trenkel, V. 2008. A two-stage biomass random effects model for stock assessment without catches: What can be estimated using only biomass survey indices? *Canadian Journal of Fisheries and Aquatic Sciences* 65:1024–35.

CHAPTER 12

# Bluefin Tunas in a Changing Ocean

Alistair J. Hobday, Barbara A. Muhling, Elliott L. Hazen,
Haritz Arrizabalaga, J. Paige Eveson, Mitchell A. Roffer,
and Jason R. Hartog

## Introduction

All three species of bluefin tuna, Pacific (*Thunnus orientalis,* PBT), Atlantic (*T. thynnus,* ABT), and southern (*T. maccoyii,* SBT), have been depleted to low levels because of overexploitation (SBT: CCSBT 2015a, PBT: PBTWG 2015, ABT: SCRS 2015). For SBT, spawning biomass over the past decade has been estimated to be below recognized reference limits, leading to the development of a scientifically tested management procedure implemented by the Commission for the Conservation of Southern Bluefin Tuna (CCSBT) as part of its plan to rebuild the stock (Hillary et al. 2016). In the case of PBT, high market value has kept fishing pressure high and resulted in continued declines in the stock, making new management approaches particularly important (Pons et al. 2017). Thus, while some populations are recovering as a result of improved data quality and stock assessments, quota setting, and compliance, there remains a strong need to carefully manage populations to enable recovery while they are still being fished sustainably. In addition to conventional quota-based management approaches, which specify how many fish can be captured by targeted fisheries, spatial management tools have a role to play in aiding recovery and in minimizing interaction with other fisheries. Although commercial fishing is still permitted on all three bluefin species, there is a recognized need to avoid unwanted interactions in some regions, for some age classes, and at particular times of year. These interactions can occur even when targeted fisheries seek other species, such as yellowfin tuna in eastern Australia (Hartog et al. 2011) and the Gulf of Mexico (GOM) (e.g., Armsworth et al. 2010) and swordfish in the Mediterranean (e.g., Garibaldi 2015).

Fisheries managers can use spatial tools to separate fishing activities from bluefin in areas of concern and have implemented static time-area closures in some regions (e.g., GOM: NOAA NMFS 2014) based on historical fisheries data. In addition to fisheries management, fishers can also benefit from information on spatiotemporal distributions to voluntarily avoid one species (e.g., Howell et al. 2008) or to more efficiently catch their quota (Dunn et al. 2011, Eveson et al. 2015). Given the high reliance on one market (Japan) for bluefin species and a growing supply of cultured bluefin (Sawada et al. 2005, Benetti et al. 2016), in cases where fishing is sustainable, spatial approaches that lead to reduced bluefin fishing costs and thus sustain fishery profits are also desired by fishers and managers (Eveson et al. 2015).

The bluefin species are wide ranging, with individual movements often spanning the range of the species (e.g., SBT: Hobday et al. 2015b, ABT: Block et al. 2001, PBT: Itoh et al. 2003). The nature of these movements differs over the lifetime of the species and includes ontogenetic migrations and seasonal spawning and feeding migrations. Changes in movement patterns and species distribution have occurred in response to both short-term (Briscoe et al. 2017) and long-term changes in the environment (e.g., Fromentin et al. 2014, Mackenzie et al. 2014). These complex dynamics result in considerable spatiotemporal variation in the distribution of individuals over annual to decadal timescales; however, many patterns are still somewhat predictable based on the time of year, preferred environmental conditions, and life stage.

These semipredictable distribution patterns, common to many wide-ranging species, have been important in the relatively new field of dynamic ocean management (Hobday et al. 2014, Lewison et al. 2015, Maxwell et al. 2015). Dynamic ocean management is defined as management that rapidly changes in space and time in response to changes in the ocean and its users through the integration of near-real-time biological, oceanographic, social, and/or economic data (Maxwell et al. 2015). Dynamic management can refine the temporal and spatial scale of managed areas and achieve a balance between ecological (conservation) and economic (fishery) objectives. For species with regular patterns or long-term trends in spatial distribution, such as bluefin, static spatial management may not offer the best solution, as fishing activities may be restricted when bluefin are not present or the size of the managed area may be too small when bluefin are present

(Maxwell et al. 2015, Dunn et al. 2016). Dynamic management, based on habitat models conditioned with current or forecasted environmental conditions, can also aid managers and fishers in anomalous seasons when tuna distribution may change substantially (e.g., Mackenzie et al. 2014). Given the high value of bluefin fisheries, early warning of changed distribution can assist planning of adaptation strategies by managers, fishers, fleets, and countries, such that overexploitation does not occur in the new regions and fishers can adapt and operate more efficiently. For bluefin species that are managed under sustainable quotas, knowing their distribution in advance should not automatically increase exploitation levels, but rather provide for even more flexible and effective management and fishing strategies.

Understanding the spatiotemporal patterns in bluefin distribution has been aided in recent decades by the widespread use of electronic tags (see reviews by Gunn and Block 2001, Block et al. 2003, Nielsen et al. 2009, and references therein). The tags gather information on location, depth, and in some cases environmental conditions such as temperature. Subsequent matching of fish location to additional environmental conditions has further expanded the available data and underpinned the development of habitat models (Teo and Block 2010, Hazen et al. 2013, Hobday et al. 2014). Habitat models have also been developed from fishery-dependent catch data (Teo and Block 2010; Druon et al. 2011, 2016), and fishery-independent survey data (Muhling et al. 2010, Druon et al. 2016). These models have been used to understand environmentally driven variation in fish distribution, and they represent an important step in predicting fish distribution on a range of timescales (Hobday et al. 2014).

Habitat predictions, based on habitat models, require information about future environmental conditions, which can come from a range of ocean models. Depending on available environmental data, forecasts can be developed for a range of timescales, including the present (referred to as nowcasts), short term (days to weeks), seasonal (weeks to months), multiyear (years to decade), and climate scale (many decades). While much attention has been paid to projecting bluefin distributions at centennial timescales in response to climate change (e.g., Hobday 2010, Hartog et al. 2011, Hazen et al. 2013, Muhling et al. 2015), and these studies have been important in informing responses to climate policy (Bell et al. 2013), habitat models can also inform management on a range of shorter timescales (Hobday et al. 2016). In fact, most tactical decisions for both managers and fishers are made at subannual timescales (Hobday et al. 2016). The typical decisions made by fishers over short timescales are about where to go fishing, whereas decisions to invest in expensive equipment to adapt to new situations require information over longer timescales. Managers need to decide where to allow fishing at short timescales, or how to allocate quotas between different fishing sectors over longer periods.

Habitat predictions can also be used to guide noncommercial activities, such as ecological research. For the most part, fishery-independent surveys are designed on the basis of fixed grids (Royer et al. 2004, Kitagawa et al. 2010, Lyczkowski-Shultz et al. 2013) and are not adjusted in response to local environmental conditions that might influence the distribution of the surveyed organisms. There is potential for the use of nowcasts and short-term forecasts to significantly increase sampling efficiency in some instances, particularly in cases where bluefin tuna are rare or sparsely distributed. For example, sampling for larval ABT has been aided by environmental data in the Mediterranean Sea (Alemany et al. 2010) and the Gulf of Mexico (see Case Study 3, below). Habitat models have also been proposed as tools to increase the efficiency of shipboard or aerial surveys in other regions (Brill and Lutcavage 2001, Royer et al. 2004); however, there are few operational examples of this approach.

Here, we focus on the development and application of bluefin habitat nowcasts and seasonal forecasts, describing several case studies in which these approaches have been implemented. In each case, we note the management or scientific issue that has been addressed, the forecasting approach, and if applicable, the delivery method and the uptake and benefit of the approach. We then consider forecasting prospects for bluefin species in other regions or situations and discuss the specific technical, implementation, and regulatory challenges that may limit these applications. We conclude with a synthesis of the existing and prospective approaches, comment on forecasting limitations, and offer some prospects for bluefin forecasts on longer (multiyear and climate) timescales.

## The Case Studies and Development of New Applications

### Case Study 1: Forecasting Juvenile Southern Bluefin Tuna Habitat in the Great Australian Bight

Large numbers of juvenile SBT are found in the Great Australian Bight (GAB) during the austral summer (December–April) and are caught in a purse seine fishery worth ~\$60 million AUD annually. In recent fishing seasons, changes in the distribution of SBT affected the timing and location of fishing activity and contributed to economic pressure. In particular, in the 2011–12 season, SBT moved through the GAB quickly and were distributed farther east than in the previous two decades. This resulted in less than 15% of purse seine catches being taken from traditional fishing grounds; a similar unusual distribution occurred in the following season (2012–13) that again impacted the fishery. The bluefin industry association then sought scientific support to improve the operational planning in this fishery. As many decisions central to planning fishing operations need to be made weeks to months in advance, a seasonal forecasting system was seen as a useful approach, and an industry-researcher partnership formed.

First, a habitat model for juvenile SBT in the GAB was developed on the basis of fish location data collected over many years from electronic tags. The ocean conditions where fish were found were compared with the conditions available to them throughout the GAB in summer to see which conditions they tended to "prefer" (Eveson et al. 2014, 2015). For example, if 10% of fish locations occur in 19°C–20°C water, but only 5% of the GAB contains temperatures in that range, then that temperature bin is assigned a preference value of 10/5 = 2; thus, values greater than 1 indicate preferred conditions. Sea surface temperature (SST) was found to have the greatest influence on fish distribution, with temperatures in the range of 19°C–22°C preferred (Figure 12.1a); however, chlorophyll *a* was also influential, with the preferred range depending on the value of SST, but generally from 0.1 to 0.3 mg/m$^3$ (Figure 12.1b). The validity of the habitat model was tested using independent sightings data from an annual aerial survey of SBT in the GAB (Farley et al. 2016; chapter 10, this volume). The habitat model with SST alone was found to be informative in predicting sighting locations; however, the addition of chlorophyll improved its predictive performance. More specifically, when chlorophyll was added to the model, the area deemed to be preferred habitat became smaller while still containing the majority of sightings (Figure 12.1c,d; see Eveson et al. 2014 for details).

To predict locations of preferred SBT habitat in the future, habitat preferences were coupled with an environmental forecasting model developed at the Bureau of Meteorology, the Predictive Ocean Atmosphere Model for Australia (POAMA) (Spillman et al. 2012). Because chlorophyll is not a forecasted variable, only the habitat model with SST could be used in forecasting. POAMA's ability to predict SST in the GAB during the summer months was evaluated using historical data and was found to have sufficient

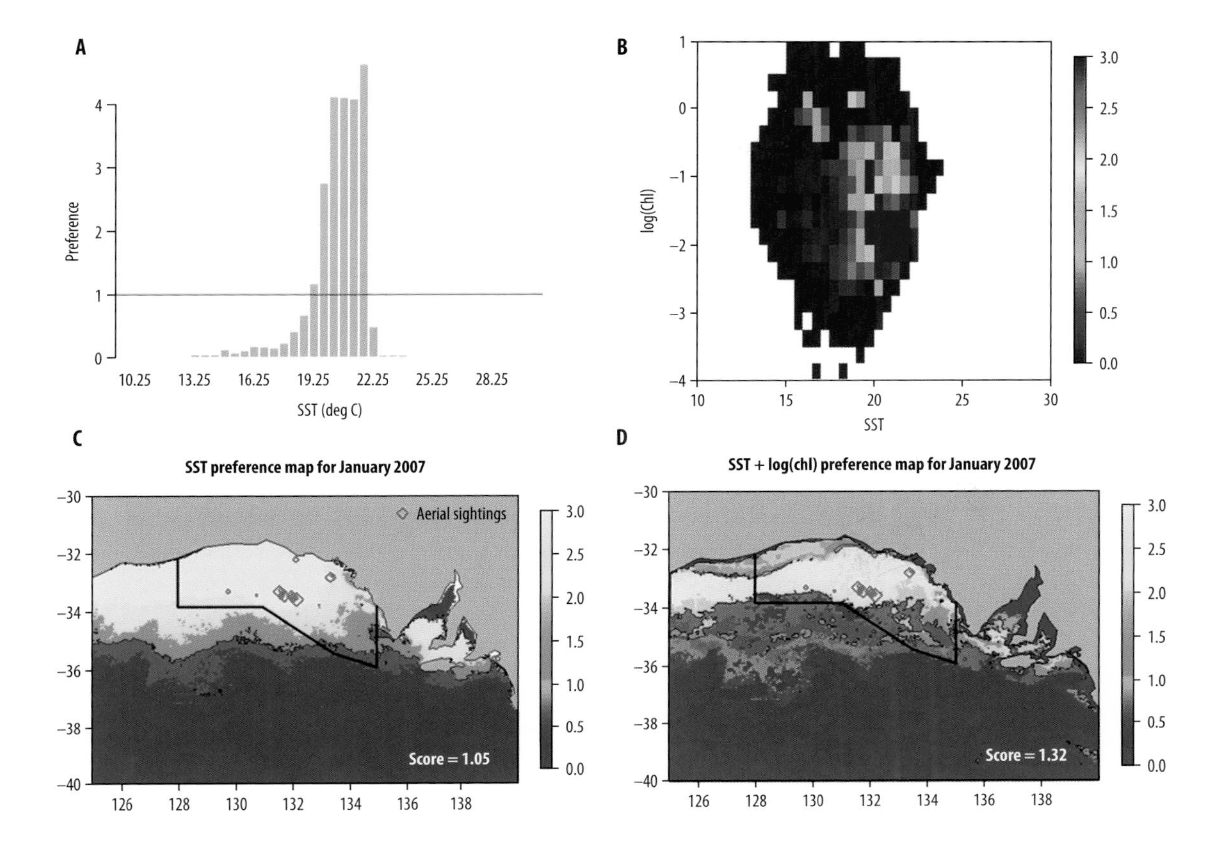


**Figure 12.1.** (*a*) Habitat preference curve for southern bluefin tuna (SBT) in the Great Australian Bight (GAB) based on sea surface temperature (SST) data only. Preference values were derived by dividing the distribution of SST values at all locations where fish were found in the GAB during January–March of 1998–2009 (based on archival tag data) by the distribution of all SST values present in the GAB during that time period. Values > 1 indicate preferred temperatures (i.e., temperatures found in greater proportion in the fish distribution than in the ocean distribution). (*b*) Same as (*a*), but based on SST and chlorophyll combined, resulting in a two-dimensional habitat preference surface. (*c*) Map showing areas of preferred habitat for January 2010 based on SST only, and (*d*) based on SST and chlorophyll combined. Pink diamonds represent aerial sightings of SBT made in January 2010 as part of the scientific aerial survey (symbol size proportional to biomass of sighting); the solid black line delineates the survey area. The score gives the proportion of sightings within preferred habitat divided by the proportion of the survey area deemed preferred habitat; higher scores indicate better model predictive performance.

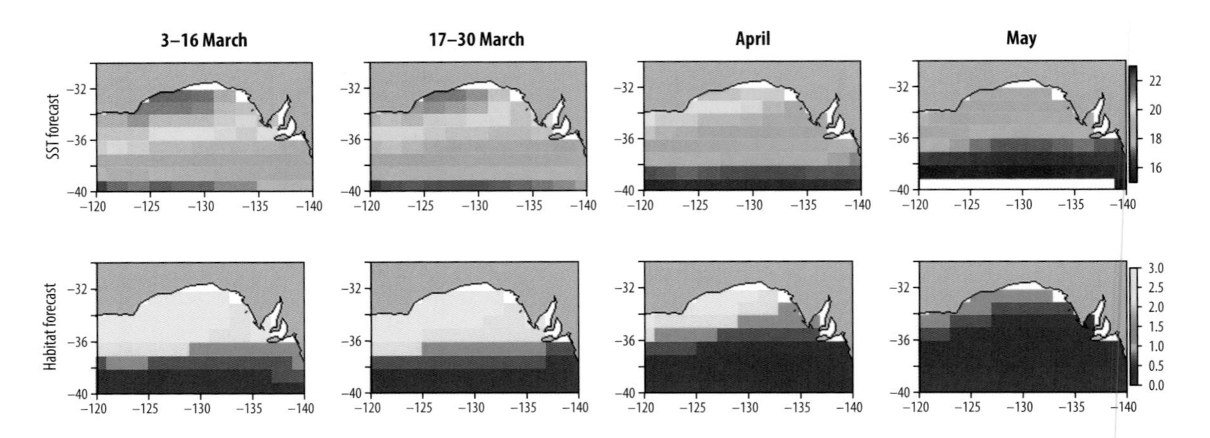


**Figure 12.2.** Great Australian Bight environment and southern bluefin tuna (SBT) habitat forecasts. Forecasts of sea surface temperature (SST) (°C) (*top row*) and preferred SBT habitat (*bottom row*) issued on March 3, 2016, for the next two fortnights and the next two calendar months. Preference values greater than 1 indicate preferred habitat.

skill for lead times up to two months (Eveson et al. 2015). Forecasts of SST in the GAB and corresponding areas of preferred SBT habitat for the next two fortnights and the next two calendar months (Figure 12.2) were delivered to fishers via a daily-updating website (www.cmar.csiro.au/gab-forecasts). Industry feedback showed the forecasts have been a valuable tool for fishers making decisions such as when and where to position vessels and to conduct fishing operations (Eveson et al. 2015). Because the SBT fishery is quota-managed, the forecasting approach will not lead to increased catches (and thus impact sustainability) but should enable fishers to catch their quota more efficiently, thereby increasing profitability.

### Case Study 2: Forecasting Southern Bluefin Tuna Distribution in Eastern Australia

Larger SBT (subadult and adult) occur seasonally on the east coast of Australia (Patterson et al. 2008, Hobday et al. 2015b), where they are captured in the Australian eastern tuna and billfish fishery. This fishery targets a range of pelagic species, including yellowfin tuna, and the catch of SBT is restricted by quota. Capture of SBT was initially limited by a static zoning arrangement, but in 2003 a dynamic spatial management approach was initiated to limit nonquota capture of SBT (Hobday and Hartmann 2006). Initially, a habitat nowcast was utilized along with other data inputs (such as recent fishing catch rates) by fishery managers during the fishing season to restrict access to regions predicted to have a high density of SBT and to allocate observer coverage to vessels fishing the restricted region to estimate bluefin capture. The nowcast was based on a habitat model conditioned with temperature-at-depth data obtained from pop-up satellite archival tags deployed on SBT, coupled with an ocean model (Hobday and Hartmann 2006, Hobday et al. 2010). The habitat preferences of SBT in this region are determined by calculating, within each defined depth strata from 0 to 200 m, the percentage of time fish are found in temperatures colder than the observed ocean value in each spatial grid cell, then taking a weighted average of these percentages across depth strata using the time spent in each depth strata as the weight (refer to Hobday et al. 2010 for details). The dynamic nowcast approach reduced the need for large area closures while still meeting the management goal of reducing the likelihood of vessels without quota capturing SBT. It also required fishing operators to develop more flexible fishing strategies, including planning vessel movements, home-port selection, and quota purchase (Hobday et al. 2010).

In 2011 the habitat model was used to develop a seasonal forecast of SBT habitat distribution (Hobday et al. 2011). This required replacing the ocean model component with ocean temperature information derived from POAMA (see Case Study 1). The nowcast and forecast habitat maps were emailed to managers every two weeks and showed probabilistic zones of tuna distribution coded as OK (unlikely to encounter SBT), Buffer (likely to encounter SBT), and Core (very likely to encounter SBT) (Figure 12.3). The forecast has useful skill out to several months (Hobday et al. 2011) and has informed and encouraged managers and fishers to think about decisions on longer timescales, although no formal management actions were implemented in response to forecasts at the time; managers preferred to continue to use the nowcasts to set the zones (Hobday et al. 2016). In 2015 the fishery management agency initiated video monitoring of every fishing set in the fleet, which reduced the need for spatial management approaches, as each captured SBT could now be verified against quota (which could be purchased retrospectively if needed to cover unanticipated capture of SBT). Although the dynamic spatial management approach could still be used to minimize any bycatch and prevent early closure of the overlapping yellowfin fishing region when all the quota is used, this has not occurred at the time of writing.

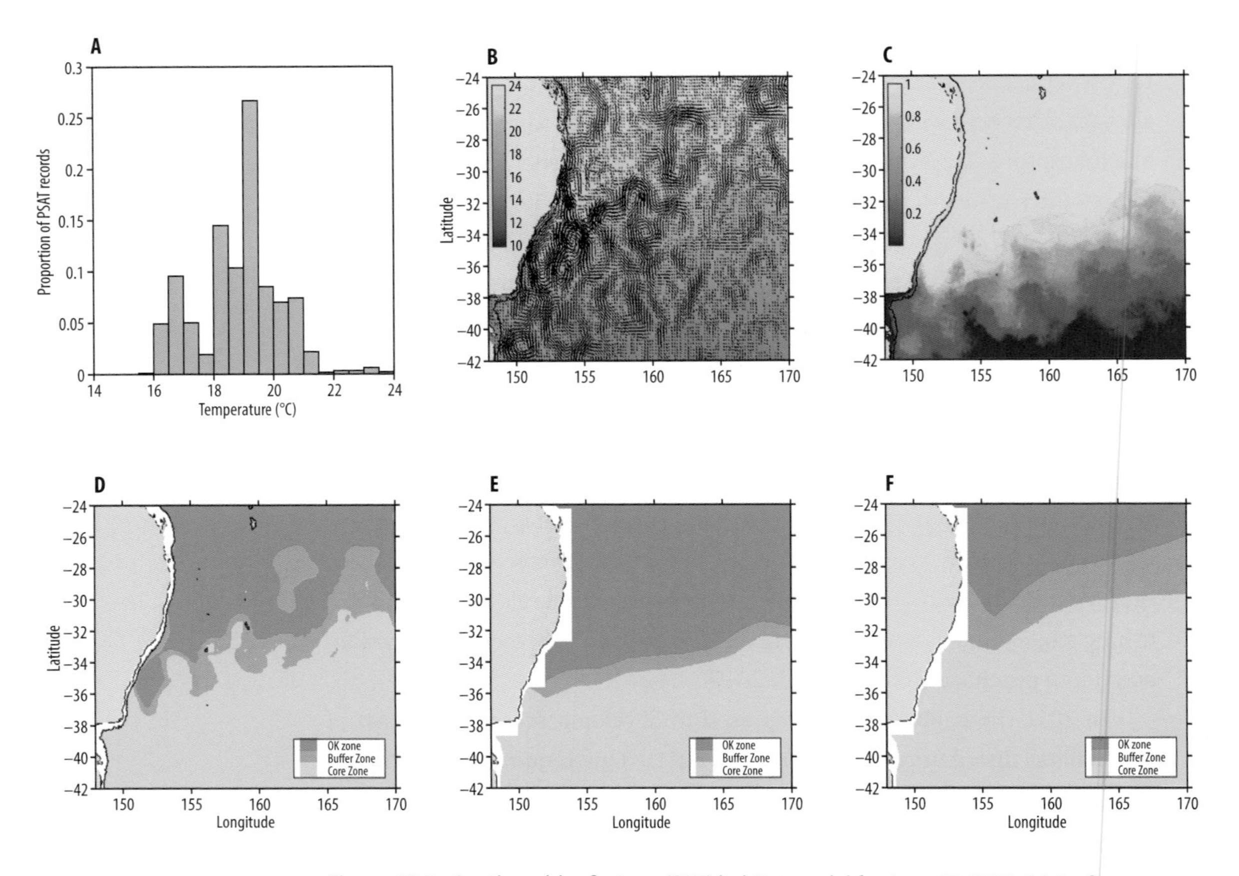


**Figure 12.3.** Southern bluefin tuna (SBT) habitat model for June 12, 2016. (*a*) Surface temperature preference of tagged SBT within 70 days of the analysis day of the year (thus tagged in any year). (*b*) Sea surface temperature for June 12, 2016, along with the surface currents. (*c*) Probability of SBT preferring temperatures that are colder than the value in each pixel on the analysis date. This process is repeated for each of the subsurface layers available in the model to a depth of 200 m. (*d*) Probability maps for each depth layer are combined and categorized to form the habitat nowcast. (*e*) Habitat forecast for the next two weeks using POAMA seasonal forecasts of ocean temperature. (*f*) Habitat forecast for August 1, 2016, a lead time of two months.

## Case Study 3: Using Real-Time Satellite Data and Ocean Model Outputs in Exploratory Larval Atlantic Bluefin Tuna Cruises

The current stock assessment for western ABT includes only one fishery-independent index of spawning stock biomass: the GOM larval index (Scott et al. 1993, Ingram et al. 2010). The index is derived using larval abundances from more than 30 years of spring surveys in the northern GOM. A length-age curve is used to model annual abundance of larvae with a first daily increment formed per 1,000 $m^2$ of seawater sampled from bongo net gear. Bongo nets are 60 cm in diameter with 333 μm mesh and are towed obliquely to 200 m depth (or within 10 m of the bottom in shallower water) (Habtes et al. 2014). Time of day, date, area of the GOM, and year are also included as covariates (Ingram et al. 2010). However, the usefulness of the index as a proxy for spawning stock biomass could be compromised if significant spawning events are occurring outside the northern GOM. To investigate this possibility, exploratory cruises have been completed in adjacent areas of the western central Atlantic since 2009.

Larval ABT are associated with particular environmental conditions in the GOM, including a relatively narrow SST range (~24°C–28°C), low surface chlorophyll (< 0.2 mg/$m^3$), and low-moderate surface currents (0.1–0.5 m/s). These relationships can be summarized using multivariate predictive habitat models (Muhling et al. 2010, 2012; Domingues et al. 2016). A habitat model incorporating satellite data and ocean model outputs that gave good results in the northern GOM was adapted for the southern GOM, Caribbean Sea, and Bahamas regions, and candidate areas for further investigation were identified as those with theoretically favorable larval habitat during spring. Environmental variables included were SST, temperature at 200 m depth, surface chlorophyll, sea surface height, and surface current magnitude. The predictive habitat model was a multilayer perceptron artificial neural network, and it was trained on larval catch data from annual spring surveys in the northern GOM (Muhling et al. 2012, Lyczkowski-Shultz et al. 2013, Habtes et al. 2014).

Once a broader region of interest was selected for each cruise, real-time infrared and ocean-color satellite imagery and nowcasts from ocean models were used daily to guide sampling while at sea. Areas with high probabilities of occurrence from the habitat model were identified and targeted. Analysis of sequential daily satellite images and daily ocean model outputs showed the development and dynamics of favorable spawning areas, which could then be followed and sampled for several days (Figure 12.4). Satellite imagery was also used to highlight frontal features and eddies, which can be associated with higher larval abundances (Lindo-Atichati et al. 2012) and therefore facilitated sampling in a wider range of water masses. In addition, sampling across finer-scale oceanographic features and boundaries led to the development of larval ecology studies, with comparisons of larval growth and feeding success across different planktonic foraging environments (Llopiz et al. 2015, Malca 2014).

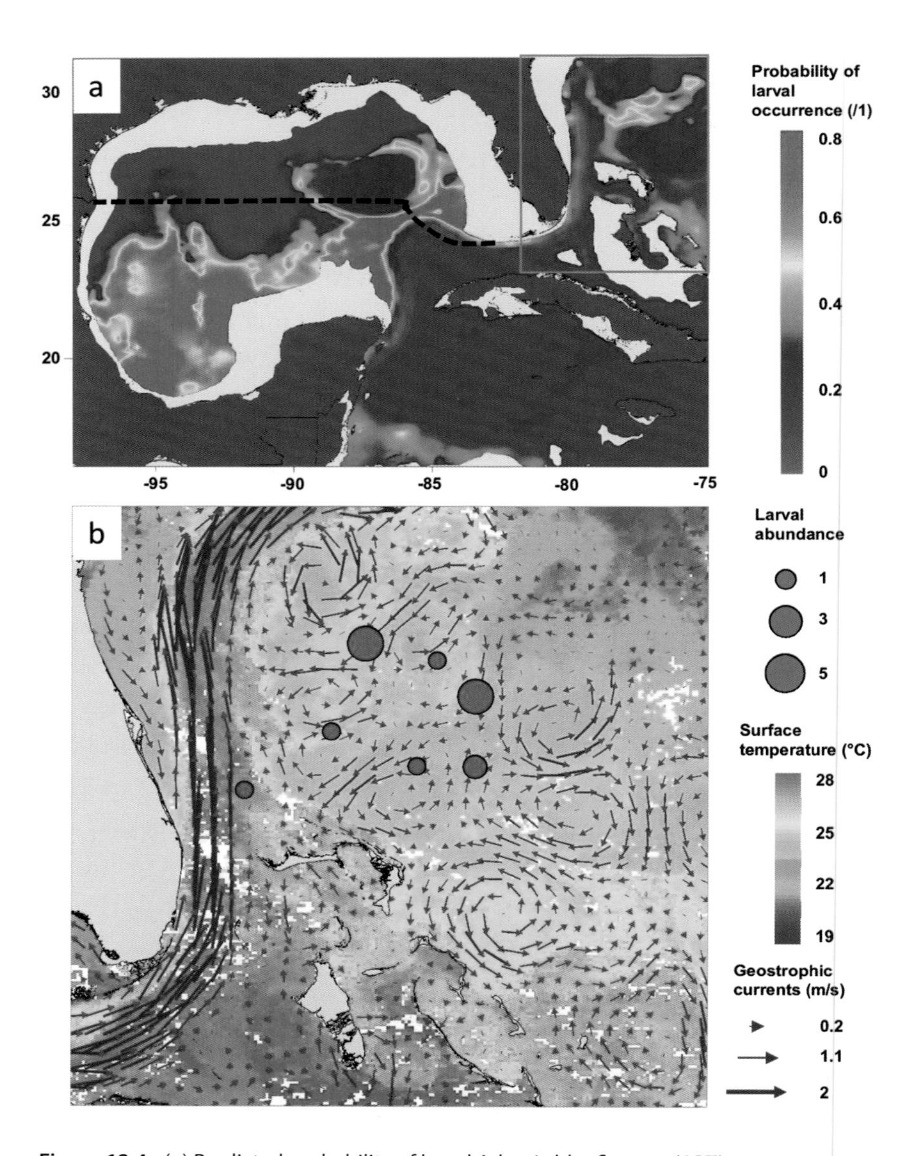


**Figure 12.4.** (*a*) Predicted probability of larval Atlantic bluefin tuna (ABT) occurrence from an artificial neural network habitat model, April 14–20, 2013. The area occupied by the annual spring survey (mid-April to the end of May) is north of the black dashed line. Note that habitat suitability is low in the northern Gulf of Mexico because the dates shown are early in the spawning season in a cooler-than-usual year. However, the analysis shows potentially suitable habitat developing north of the Bahamas, which was then targeted on a sampling cruise. (*b*) Bahamas region (pink box in panel *a*) showing mean sea surface temperature from MODIS Aqua, geostrophic currents from absolute dynamic topography (satellite altimetry), and catches of larval ABT from all sampling gears, May 15–20, 2013.

Cruise results showed ABT spawning activity in several regions outside the northern GOM, including the southwest GOM, the western Caribbean, and an area north of the Bahamas. However, the relative importance of these areas to recruitment remains unresolved. Other recent sampling from the Mid-Atlantic Bight by other researchers has also collected larval ABT, within similar temperature ranges to those in the GOM (Richardson et al. 2016). Overall, these results have cast uncertainty on assumptions of stock productivity and spawning behaviors, and they have implications for the validity of life history parameters used in the current stock-assessment model (Richardson et al. 2016). Future studies will continue to use real-time oceanographic data to sample known and potential ABT spawning grounds. Results will be used to address key knowledge gaps that contribute uncertainty to stock-assessment models, and to inform subsequent management decisions.

### Development of New Bluefin Forecast Applications

*Southern Bluefin Tuna on Their Spawning Ground*

Habitat nowcasts and forecasts for SBT have been implemented in the two major Australian fishing regions (Case Studies 1 and 2); however, other important areas have not been considered. For example, as for ABT (Case Study 3), models of larval distribution could be developed and used to guide surveys in the large putative spawning ground south of Indonesia and to reduce bycatch of spawning SBT in the tropical tuna fishery in the region. Information from tagged animals that are on the spawning ground could also be used to develop adult habitat models. In contrast to the coastal Australian regions where the Australian fisheries operate, access by researchers to the offshore spawning area between Australia and Indonesia is difficult. Ship-based larval surveys have been rare (Nishikawa et al. 1985, Davis et al. 1991), and so the ability to generate a habitat model based on occurrence of larvae is constrained. Likewise, large fish tagged with pop-up satellite tags in southern and eastern Australia have rarely traveled to the spawning grounds during their relatively short deployment times (Patterson et al. 2008), limiting the data available for constructing a habitat model. A dedicated tagging program directed at fish approaching the spawning ground may be needed to fill the data gap. Alternatively, information from commercial longline fisheries could be used (Davis and Farley 2001); however, there may be biases in the locations where fishers are active, and the quality of location information is often low, limiting the matching of synoptic environmental variables to putative fish capture locations.

Whereas SST is the most important variable influencing SBT distribution elsewhere, on the spawning ground, where surface temperatures are high, mixed layer depth and temperatures at the base of the thermocline may be more important. These variables are not well represented in existing forecast models used for SBT, such as POAMA. Thus, even if habitat models could be developed, the important variables may not be forecasted with sufficient spatial resolution or accuracy to be useful, in which case applications would be restricted to nowcasts. Finally, even with development of a habitat forecast model in this region, the regulatory capacity to make use of forecasts to regulate bycatch does not currently exist, such that enforcement on the SBT spawning ground of Indonesian longline fleet activity is probably unrealistic at present, and the vessels operating in this area may not be able to reliably access information to manage interactions. Thus, effort should be spent understanding the decision context and the needs of fishers and managers in this region before a forecasting system is developed.

*Pacific Bluefin Tuna in the Eastern Pacific*

There is a small fishery for PBT in the United States, limited to 425 metric tons in 2015 and 2016, although more are caught in international waters (PBTWG 2015), and some of the juvenile stock are bycatch in the California drift gillnet fishery (DGN) (Hanan et al. 1993). Current management approaches for the DGN fleet have focused on reducing sea turtle and marine mammal bycatch, with less attention on PBT, although holistic dynamic management approaches that consider multiple species are being developed (Lewison et al. 2015, Maxwell et al. 2015). Habitat models have been developed for PBT habitat and foraging success (Whitlock et al. 2015) and to examine the response of PBT (and a suite of other top predators) to climate change (Hazen et al. 2013), but neither approach has been run in forecast mode. Habitat forecasts could be used to analyze the risk-return trade-off when considering the likelihood of target species catch, bycatch of PBT, and bycatch of protected species. Two delivery methods are also under development: a mobile application-based tool that also allows input of catch data and the bycatch species of concern while at sea, and a web-based tool for managers and fishers to enter data back on land. These delivery approaches would allow fishers to optimize maps as they pertain to their fishing goals while at sea and would allow managers to create maps that accurately reflect the bycatch limits put in place by fishery management council regulations. These approaches would support both nowcasts and, when coupled with a seasonal forecasting model, forecasts to reduce PBT bycatch in this region.

*Atlantic Bluefin Tuna in the East and West Atlantic Ocean*

There are several habitat models for western ABT that could be developed as part of forecasting applications. Teo et al. (2007), Teo and Block (2010), and Hazen et al. (2016) used tagging data to relate ABT distributions on the GOM spawning ground to oceanographic variables. This work highlighted environmentally driven spatial habitat use and raised the possibility of future dynamic closures in the GOM and elsewhere, on the basis of mesoscale oceanographic features. However, avoidance of ABT in the GOM by yellowfin tuna and swordfish fishing fleets using habitat models may be difficult to implement. Yellowfin tuna and ABT habitats in the GOM appear to be best separated in space by their use of different mesoscale oceanographic features (Teo and Block 2010). These are more ephemeral and spatially complex than the strong north-south temperature gradient used in dynamic management to reduce bycatch of SBT off eastern Australia (Case Study 2) (Hobday and Hartmann 2006), although eddies may also be identified and tracked (Hobday and Hartog 2014). The simplest variable available for separating ABT from other species in the GOM is time of year, as ABT are most abundant from March to May and are less common from June to February. In addition, ABT are usually restricted to offshore waters, whereas yellowfin tuna and swordfish are found on the continental shelf. As a result, the recently implemented management areas in the GOM are closed to fishing from April to May, and the closures encompass offshore regions of high historical interactions with ABT (NOAA NMFS 2014). These closed areas are fixed in space and do not take environmental variability into account.

Another management issue with western ABT that could be addressed using habitat models is the current disparity between several catch-per-unit-effort (CPUE) indices from the northwest Atlantic. Fishery-dependent abundance indices from the Gulf of Saint Lawrence have increased strongly in recent years, while those from southwest Nova Scotia and the Gulf of Maine have not (Vanderlaan et al. 2014, Hanke et al. 2015). The primary mechanisms for a northward shift in ABT distribution remain unresolved, but they are likely to involve both prey dynamics (Golet et al. 2013, Hanke et al. 2015) and water temperature (Vanderlaan et al. 2014). Besides resolving the conflicting CPUE patterns that might cause confusion in stock assessment, an ability to forecast distributions of ABT in the northwest Atlantic several months ahead could allow advance negotiations for quota shares between the two countries primarily affected—Canada and the United States—and pave the way for future climate change adaptation. However, models with good predictive skill for subsurface temperature and biological variables would be required. Forecasts of these parameters are generally more complex and difficult than forecasting SST and are also limited by a lack of historical observations with which to validate forecast skill (Stock et al. 2015).

Several habitat models have been developed for the eastern stock of ABT, which spawns in the Mediterranean and mixes throughout the whole Atlantic with the western stock (Fromentin and Powers 2005, Rooker et al. 2014), but to our knowledge, none have been used for nowcast or forecast applications. Fromentin et al. (2014) and Arrizabalaga et al. (2015) used coarse-resolution fishery data to investigate habitat preferences and to predict spatial distribution of adult ABT, whereas Druon et al. (2011, 2016) used a wider range of finer-resolution data to predict the feeding and spawning habitats of juveniles and adults across the North Atlantic. Biological occurrence data were obtained from a variety of sources such as tagging, surveys, and fishing records, and environmental variables were sourced from satellites and ocean models. These authors all noted the potential for the planning of marine protected areas and development of dynamic spatial management approaches using their habitat models. In addition, the spawning habitat around the Balearic Islands has also been studied using larval survey data (e.g., Alemany et al. 2010), and Alvarez-Berástegui et al. (2016) recently used operational oceanography to produce near-real-time habitat maps for ABT spawning habitat, which could also be used for management applications such as the design of dynamic marine protected areas or to guide scientific surveys (as in Case Study 3).

As for the eastern ABT stock, the development of habitat nowcast and forecast capabilities could have additional management and fishery applications. The International Commission for the Conservation of Atlantic Tunas (ICCAT) has implemented a strict multi-annual stock recovery plan. The implemented measures include, among others, a reduced total allowable catch (TAC), fleet-specific fishing seasons, a minimum size, and bycatch limits for fleets not targeting ABT. ICCAT also requested that the Scientific Committee identify spawning grounds as precisely as possible and provide advice on the creation of sanctuaries (ICCAT 2014). During low TAC periods, it is important to avoid, to the extent possible, bycatch by fleets that do not have a quota share. Forecasts of multispecies habitat models would allow regulation of fishing areas to avoid ABT bycatch while catching target species (e.g., swordfish). Depending on the habitat overlap between ABT (the bycatch species in this case) and the target species, there might be more or less opportunity to minimize ABT bycatch using dynamic management tools. This information could also be useful for managers in allocating ABT quota between fleets targeting other species. For example, managers could allocate less ABT quota to fleets that can more easily avoid ABT bycatch based on nowcasts and forecasts. As the stock rebuilds, different needs might arise, because TACs are expected to increase, and catching the allocated quotas might require substantially more effort. In this situation, the fleets might benefit from using nowcast and forecast information to decide where to fish or, in the case of tuna farming, where to place tuna cages to make operations more profitable.

## Discussion

The case studies described here have shown how habitat predictions have supported decision-making by bluefin tuna fishers (Case Study 1), managers (Case Study 2), and scientists (Case Study 3). The underlying habitat models have all been based on many years of data, collected through electronic tagging studies and scientific cruises. In developing forecasting capability, nowcast development preceded the delivery of forecasts in eastern Australia and in the emerging applications for other species, while in the GAB (Case Study 1), need for a forecast was the initial driver. There is clearly interest and need for forecasts at a range of timescales, depending on the application. For fishers and scientists planning upcoming fishing or survey activities, forecasts of fish distribution over the upcoming weeks to months are most relevant, whereas for fishing and survey activities currently under way, nowcasts or very-short-term forecasts are of greater relevance (Hobday et al. 2016). In the Australian SBT case studies, codevelopment of the forecasting applications with industry proved particularly valuable and led to rapid uptake of information (Eveson et al. 2015). In fact, industry and management engagement is critical in helping researchers to understand which decisions are made at lead times relevant to seasonal forecasts and to design effective forecast delivery systems (Hobday et al. 2016).

Notably, forecasting of tuna distribution, and other species in general, is limited by the output variables from the physical forecast model. Although temperature, the most common forecasted variable, can be an important predictor of tuna distribution, it is not the sole important variable. For example, chlorophyll was found to be important in the habitat model for SBT in the GAB (Case Study 1, Figure 12.1), but it cannot be included in the habitat forecasts because it is not an output of the environmental forecasting model. Nowcasts can be based on a more complete set of explanatory variables than forecasts, and as a result they may be more reliable. Forecast models also have variable skill in different ocean regions and for different environmental variables in the same region. Thus, assessment of model skill is an important step in creating both nowcasts and forecasts (Hobday et al. 2016). The spatial resolution of environmental data also tends to decline from nowcasts, which can use satellite observations, to forecasts, which use medium-resolution ocean models. This can limit the skill of the habitat model for some applications (Scales et al. 2016). Some improvement in spatial resolution can be delivered via downscaling (Vanhatalo et al. 2016), by coupling to a fine-resolution model (Kaplan et al. 2016, Tommasi et al. 2017), and by increased primary model resolution.

While SST and chlorophyll are often used as proxies when describing tuna habitat, aggregating features and prey density are likely more proximate drivers of distribution. For example, much debate remains over the drivers of PBT migration in the California Current: Do tuna seek to optimize physiological scope (environmental driver) or to maximize foraging potential (prey driver)? Upwelling-driven productivity in the northern California Current occurs in parallel with warming waters in the southern California Current concurrent with northward migration (Boustany et al. 2010, Block et al. 2011), confounding examination of these competing hypotheses. Existing habitat models for PBT currently lack explicit information on productive frontal dynamics (Scales et al. 2014) and on bluefin prey distribution, which could increase predictive accuracy. In fact, the development and integration of predictive prey models could improve understanding of migration and distribution of bluefin in all locations. In the California Current, for example, habitat models for one tuna prey (sardines) exist for the southern California Bight and for the coast of Washington but currently operate in isolation. Multispecies prey models are likely to be needed for tuna, which utilize a range of prey, depending on local abundance (e.g., Itoh et al. 2011). Mechanistic models can also elucidate drivers of prey abundance, which ultimately could inform dynamics of tuna migrations and be incorporated in forecast models. End-to-end models for tropical tuna that include prey dynamics have already been developed and used to make climate projections (Lehodey et al. 2010, 2013, 2015), and with refinement and extension to bluefin, these models could also provide seasonal forecasts.

Use of forecasts is increasingly important in a changing ocean, where considerable interannual variation exists, or when there is long-term change in the environment (Hobday et al. 2016, Tommasi et al. 2017). Here, we have described how nowcasts and short-term habitat forecasts can be used in decision-making applications by scientists, fishers, and managers, and we give examples of such applications for two bluefin species. A range of additional opportunities exist for different species and life history stages, and they will be addressed over the coming years, particularly for ABT and PBT where clear management needs have been identified. The use of habitat forecasts can be particularly relevant in bluefin tuna fisheries where strict quotas and regulations apply. For example, for stocks managed under a strict quota, catching more fish is not an option; instead, fish must be caught more efficiently for the fishery to remain economically viable. Habitat forecasts can increase catching efficiency by providing information about expected fish distribution (as demonstrated in Case Study 1). Additionally, when quotas are strict, finding methods to reduce bycatch of the quota-limited species in other fisheries becomes an important management issue; habitat forecasts for both the target and bycatch species can be very useful in this regard (see Case Study 2).

Habitat forecasts can also inform conservation planning and fisheries management at decadal and longer timescales. Climate-scale forecasts can provide policymakers and countries with information to aid access negotiations and to ensure access to fish for food (Bell et al. 2013). For example, as biomass increases, ABT might colonize new feeding areas to reduce density-dependent competition. In fact, MacKenzie et al. (2014) report the novel occurrence of ABT off Greenland, which is likely due to a combination of warm temperatures that are physiologically more tolerable and immigration of an important prey species to the region. ABT has also been observed recently in other northern countries such as Norway (SCRS 2015). These movements may require new negotiations between catching nations, as has been noted for other species (Miller 2007, Miller et al. 2013). Managers might need to negotiate new quota allocations between ICCAT contracting parties, and short-, medium-, and even longer-term forecasts might inform new bilateral fishing agreements between countries with historical fishing rights (and knowledge) and the new countries. The additional management complexity due to distributional changes of species in response to climate change can be aided by greater use of early warning systems (Hobday et al. 2015a). Overall, we anticipate that the use of habitat forecasts will continue to play a role in helping all three bluefin species recover from overexploitation and in supporting sustainable harvest into the future.

## Acknowledgments

Alistair J. Hobday, J. Paige Eveson, and Jason R. Hartog acknowledge funding support from the Australian Southern Bluefin Tuna Industry Association, the Australian Fisheries Management Authority, and the Fisheries Research and Development Corporation. Elliott L. Hazen acknowledges funding support from National Oceanic and Atmospheric Administration's Integrated Ecosystem Assessment program. Barbara A. Muhling and Mitchell A. Roffer would like to thank W. Turner and M. Estes (NASA) for project management and support, and J. Roberts (Duke University) for the MGET Toolbox. G. Gawlikowski and M. Upton (ROFFS™), J. Olascoaga (University of Miami), and L. Fiorentino (NOAA NDBC) provided oceanographic analyses and discussions for cruises. The authors also acknowledge J. Lamkin, T. Gerard, E. Malca, S. Privoznik, A. Shiroza, and the rest of the ELH/FORCES group, as well as J. Lyczkowski-Shultz, A. Millet, G. Zapfe, R. Nero, and J. Walter (NOAA SEFSC). The SEAMAP program and captains /crew of NOAA vessels are thanked for provision of historical survey data. Muhling and Roffer were partially supported by NASA grants NNX11 AP76G S07 and NNX08AL06G. Altimetry products were produced by Ssalto/Duacs and distributed by Aviso, with support from CNES. Funding for the development of HYCOM was provided by the National Ocean Partnership Program and the Office of Naval Research. Data assimilative products using HYCOM were funded by the US Navy. Review by Craig Proctor and Jessica Farley is appreciated. This is a contribution from the CLIOTOP task team on seasonal forecasting.

## REFERENCES


Alemany, F., L. Quintanilla, P. Velez-Belchi, A. Garcia, D. Cortes, J.M. Rodriguez, M.L. Fernandez de Puelles, C. González-Pola, and J.L. Lopez-Jurado. 2010. Characterization of the spawning habitat of Atlantic bluefin tuna and related species in the Balearic Sea (western Mediterranean). *Progress in Oceanography* 86:21–38.

Alvarez-Berástegui, D., M. Hidalgo, M.P. Tugores, P. Reglero, A. Aparicio-González, L. Ciannelli, M. Juza, et al. 2016. Pelagic seascape ecology for operational fisheries oceanography: Modelling and predicting spawning distribution of Atlantic bluefin tuna in western Mediterranean. *ICES Journal of Marine Science* 73 (7): 1851–62. doi:10.1093/icesjms/fsw041.

Armsworth, P.R., B.A. Block, J. Eagle, and J.E. Roughgarden. 2010. The economic efficiency of a time–area closure to protect spawning bluefin tuna. *Journal of Applied Ecology* 47:36–46.

Arrizabalaga, H., F. Dufour, L.T. Kell, G. Merino, L. Ibaibarriaga, G. Chust, X. Irigoien, et al. 2015. Global habitat preferences of commercially valuable tuna. *Deep Sea Research II* 113:102–12.

Bell, J.D., A. Ganachaud, P.C. Gehrke, S.P. Griffiths, A.J. Hobday, O. Hoegh-Guldberg, J.E. Johnson, et al. 2013. Mixed responses of tropical Pacific fisheries and aquaculture to climate change. *Nature Climate Change* 3 (6): 591–99. doi:10.1038/NCLIMATE1838.

Benetti, D.D., G.J. Partridge, and A. Buentello. 2016. *Advances in Tuna Aquaculture*. Cambridge, MA: Academic Press.

Block, B.A., H. Dewar, S.B. Blackwell, T.D. Williams, E.D. Prince, C.J. Farwell, A. Boustany, et al.. 2001. Migratory movements, depth preferences, and thermal biology of Atlantic bluefin tuna. *Science* 293:1310–14.

Block, B.A., D.P. Costa, G.W. Boehlert, and R.E. Kochevar. 2003. Revealing pelagic habitat use: The tagging of Pacific pelagics program. *Oceanologica Acta* 5 (5): 255–66.

Block, B.A., I.D. Jonsen, S.J. Jorgensen, A.J. Winship, S.A. Shaffer, S.J. Bograd, E.L. Hazen, et al. 2011. Tracking apex marine predator movements in a dynamic ocean. *Nature* 475:86–90. doi:10.1038/nature10082.

Boustany, A., R. Matteson, M.R. Castleton, C.J. Farwell, and B.A. Block. 2010. Movements of pacific bluefin tuna (*Thunnus orientalis*) in the Eastern North Pacific revealed with archival tags. *Progress in Oceanography* 86:94–104.

Brill, R.W., and M.E. Lutcavage. 2001. Understanding environmental influences on movements and depth distributions of tunas and billfishes can significantly improve population assessments. *American Fisheries Society Symposium* 25:179–98.

Briscoe, D.K., A.J. Hobday, A. Carlisle, K. Scales, J.P. Eveson, H. Arrizabalaga, J.N. Druon, and J.-M. Fromentin. 2017. Ecological bridges and barriers in pelagic ecosystems. *Deep Sea Research II*, http://doi.org/10.1016/j.dsr2.2016.11.004.

CCSBT (Commission for the Conservation of Southern Bluefin Tuna). 2015a. Report of the Twentieth Meeting of the Scientific Committee. CCSBT-20th Meeting of the Scientific Committee Incorporating the Extended Scientific Committee, September 5, Incheon, South Korea. https://www.ccsbt.org/sites/ccsbt.org/files/userfiles/file/docs_english/meetings/meeting_reports/ccsbt_22/report_of_SC20.pdf.

Davis, T.L.O., and J.H. Farley. 2001. Size distribution of southern bluefin tuna (*Thunnus maccoyii*) by depth on their spawning ground. *Fishery Bulletin* 99:381–86.

Davis, T.L.O., V. Lyne, and G.P. Jenkins. 1991. Advection, dispersion and mortality of a patch of southern bluefin tuna larvae *Thunnus maccoyii* in the East Indian Ocean. *Marine Ecology Progress Series* 73 (1): 33–45.

Domingues, R., G. Goni, F. Bringas, B. Muhling, D. Lindo-Atichati, and J.F. Walter. 2016. Variability of preferred environmental conditions for Atlantic bluefin tuna (*Thunnus thynnus*) larvae in the Gulf of Mexico during 1993–2011. *Fisheries Oceanography* 25:320–36.

Druon, J.N., J.M. Fromentin, F. Aulanier, and J. Keikkonen. 2011. Potential feeding and spawning habitats of Atlantic bluefin tuna in the Mediterranean Sea. *Marine Ecology Progress Series* 439:223–40.

Druon, J.-N., J.-M. Fromentin, A.R. Hanke, H. Arrizabalaga, D. Damalas, V. Tièina, G. Quílez-Badia, et al. 2016. Habitat suitability of the Atlantic bluefin tuna by size class: An ecological niche approach. *Progress in Oceanography* 142:30–46. http://dx.doi.org/10.1016/j.pocean.2016.01.002.

Dunn, D.C., A.M. Boustany, and P.N. Halpin. 2011. Spatio-temporal management of fisheries to reduce by-catch and increase fishing selectivity. *Fish and Fisheries* 12:110–19.

Dunn, D.C., S.M. Maxwell, A.M. Boustany, and P.N. Halpin. 2016. Dynamic ocean management increases the efficiency and efficacy of fisheries management. *Proceedings of the National Academy of Sciences USA* 113 (3): 668–73. www.pnas.org/cgi/doi/10.1073/pnas.1513626113.

Eveson, J.P., A.J. Hobday, J.R. Hartog, C.M. Spillman, and K.R. Rough. 2014. Forecasting spatial distribution of southern bluefin tuna habitat in the Great Australian Bight. Fisheries Research and Development Corporation, FRDC Project No. 2012/239, Australia. http://frdc.com.au/research/Final_reports/2012-239-DLD.pdf.

Eveson, J.P., A.J. Hobday, J.R. Hartog, C.M. Spillman, and K.M. Rough. 2015. Seasonal forecasting of tuna habitat in the Great Australian Bight. *Fisheries Research* 170:39–49.

Farley, J.H., A.L. Preece, M.V. Bravington, J.P. Eveson, C.R. Davies, K. Evans, T.A. Patterson, et al. 2016. Strategic research and long-term monitoring: keys to advancing the management of southern bluefin tuna. Bluefin Futures Symposium, January 18–20. Monterey, CA.

Fiechter, J., K.A. Rose, E.N. Curchitser, and K.S. Hedstrom. 2015. The role of environmental controls in determining sardine and anchovy population cycles in the California Current: Analysis of an end-to-end model. *Progress in Oceanography* 138:381–98.

Fromentin, J.-M., and J.E. Powers. 2005. Atlantic bluefin tuna: Population dynamics, ecology, fisheries and management. *Fish and Fisheries* 6:281–306.

Fromentin, J.-M., G. Reygondeau, S. Bonhomeau, and G. Beaugrand. 2014. Oceanographic changes and exploitation drive the spatiotemporal dynamics of Atlantic bluefin tuna (*Thunnus thynnus*). *Fisheries Oceanography* 23 (2): 147–56.

Garibaldi, F. 2015. By-catch in the mesopelagic swordfish longline fishery in the Ligurian Sea (Western Mediterranean). *ICCAT Collective Volume of Scientific Papers* 71 (3): 1495–98.

Golet, W.J., B. Galuardi, A.B. Cooper, and M.E. Lutcavage. 2013. Changes in the distribution of Atlantic bluefin tuna (*Thunnus thynnus*) in the Gulf of Maine 1979–2005. *PLOS ONE* 8:e75480. doi:10.1371/journal.pone.0075480.

Gunn, J., and B. Block. 2001. Advances in acoustic, archival, and satellite tagging of tunas. In *Tuna: Physiology, Ecology and Evolution,* edited by B.A. Block and E.D. Stevens, 167–224. San Diego, CA: Academic Press.

Habtes, S., F.E. Muller-Karger, M.A. Roffer, J.T. Lamkin, and B.A. Muhling. 2014. A comparison of sampling methods for larvae of medium and large epipelagic fish species during spring SEAMAP ichthyoplankton surveys in the Gulf of Mexico. *Limnology and Oceanography: Methods* 12:86–101.

Hanan, D.A., D.B. Holts, and A.L. Coan. 1993. The California drift gill net fishery for sharks and swordfish, 1981–82 through 1990–91. Fish Bulletin 175, State of California, Resources Agency, Department of Fish and Game.

Hanke, A.R., I. Andrushchenko, and C. Whelan. 2015. Indices of stock status from the Canadian bluefin tuna fishery: 1981 to 2013. *ICCAT Collective Volume of Scientific Papers* 71:983–1017.

Hartog, J., A.J. Hobday, R. Matear, and M. Feng. 2011. Habitat overlap of southern bluefin tuna and yellowfin tuna in the east coast longline fishery—implications for present and future spatial management. *Deep Sea Research II* 58:746–52.

Hazen, E.L., S.J. Jorgensen, R.R. Rykaczewski, S.J. Bograd, D.G. Foley, I.D. Jonsen, S.A. Shaffer, et al. 2013. Predicted habitat shifts of Pacific top predators in a changing climate. *Nature Climate Change* 3: 234–38. doi:10.1038/NCLIMATE1686.

Hazen, E.L., A. Carlisle, S.G. Wilson, J.E. Ganong, M.R. Castleton, S.J. Bograd, and B.A. Block. 2016. Impacts of the Deepwater Horizon oil spill on bluefin tuna spawning habitat in the Gulf of Mexico. *Scientific Reports* 6:33824. doi:10.1038/srep33824.

Hillary, R.M., A.L. Preece, C.R. Davies, H. Kurota, O. Sakai, T. Itoh, A.M. Parma, et al. 2016. A scientific alternative to moratoria for rebuilding depleted international tuna stocks. *Aquatic Sciences and Fisheries* 17 (2): 469–82.

Hobday, A.J. 2010. Ensemble analysis of the future distribution of large pelagic fishes in Australia. *Progress in Oceanography* 86 (1–2): 291–301. doi:10.1016/j.pocean.2010.04.023.

Hobday, A.J., and K. Hartmann. 2006. Near real-time spatial management based on habitat predictions for a longline bycatch species. *Fisheries Management and Ecology* 13 (6): 365–80.

Hobday, A.J. and J.R. Hartog. 2014. Dynamic ocean features for use in ocean management. *Oceanography* 27 (4): 134–45.

Hobday, A.J., J.R. Hartog, T. Timmis, and J. Fielding. 2010. Dynamic spatial zoning to manage southern bluefin tuna capture in a multi-species longline fishery. *Fisheries Oceanography* 19 (3): 243–53.

Hobday, A.J., J. Hartog, C. Spillman, and O. Alves. 2011. Seasonal forecasting of tuna habitat for dynamic spatial management. *Canadian Journal of Fisheries and Aquatic Sciences* 68:898–911.

Hobday, A.J., S.M. Maxwell, J. Forgie, J. McDonald, M. Darby, K. Seto, H. Bailey, et al. 2014. Dynamic ocean management: Integrating scientific and technological capacity with law, policy and management. *Stanford Environmental Law Journal* 33 (2): 125–65.

Hobday, A.J., J.D. Bell, T.R. Cook, M.A. Gasalla, and K.C. Weng. 2015a. Reconciling conflicts in pelagic fisheries under climate change. *Deep Sea Research II* 113:291–300.

Hobday, A.J., K. Evans, J.P. Eveson, J.H. Farley, J.R. Hartog, M. Basson, and T.A. Patterson. 2015b. Distribution and migration—southern bluefin tuna (*Thunnus maccoyii*).

In *Biology and Ecology of Bluefin Tuna*, edited by T. Kitagawa and S. Kimura, 189–210. London: CRC Press.

Hobday, A.J., C.M. Spillman, J.P. Eveson, and J.R. Hartog. 2016. Seasonal forecasting for decision support in marine fisheries and aquaculture. *Fisheries Oceanography* 25 (S1): 45–56.

Howell, E.A., D.R. Kobayashi, D.M. Parker, G.H. Balazs, and J.J. Polovina. 2008. TurtleWatch: A tool to aid in the bycatch reduction of loggerhead turtles Caretta caretta in the Hawaii-based pelagic longline fishery. *Endangered Species Research* 5:267–78. doi:10.3354/esr00096.

ICCAT (International Commission for the Conservation of Atlantic Tunas). 2014. Recommendation 14-04 by ICCAT amending the Recommendation 13-07 by ICCAT to establish a multi-annual recovery plan for Bluefin tuna in the eastern Atlantic and Mediterranean. http://iccat.int/Documents/Recs/compendiopdf-e/2014-04-e.pdf.

Ingram, G.W., Jr., W.J. Richards, J.T. Lamkin, and B. Muhling. 2010. Annual indices of Atlantic bluefin tuna (*Thunnus thynnus*) larvae in the Gulf of Mexico developed using delta-lognormal and multivariate models. *Aquatic Living Resources* 23:35–47.

Itoh, T., S. Tsuji, and A. Nitta. 2003. Migration patterns of young Pacific bluefin tuna (*Thunnus orientalis*) determined with archival tags. *Fishery Bulletin* 101 (3): 514–34.

Itoh, T., H. Kemps, and J. Totterdell. 2011. Diet of young southern bluefin tuna *Thunnus maccoyii* in the southwestern coastal waters of Australia in summer. *Fisheries Science* 77:337–44.

Kaplan, I.C., G.D. Williams, N.A. Bond, A.J. Hermann, and S. Siedlecki. 2016. Cloudy with a chance of sardines: Forecasting sardine distributions using regional climate models. *Fisheries Oceanography* 25 (1): 15–27.

Kitagawa, T., Y. Kato, M.J. Miller, Y. Sasai, H. Sasaki, and S. Kimura. 2010. The restricted spawning area and season of Pacific bluefin tuna facilitate use of nursery areas: A modeling approach to larval and juvenile dispersal processes. *Journal of Experimental Marine Biology and Ecology* 393:23–31.

Lehodey, P., I. Senina, J. Sibert, L. Bopp, B. Calmettes, J. Hampton, and R. Murtugudde. 2010. Preliminary forecasts of Pacific bigeye tuna population trends under the A2 IPCC scenario. *Progress in Oceanography* 86:302–15.

Lehodey, P., I. Senina, B. Calmettes, J. Hampton, and S.J. Nicol. 2013. Modelling the impact of climate change on Pacific skipjack tuna population and fisheries. *Climatic Change* 119:95–109. doi:10.1007/s10584-012-0595-1.

Lehodey, P., I. Senina, S. Nicol, and J. Hampton. 2015. Modelling the impact of climate change on South Pacific albacore tuna. *Deep Sea Research II* 113:246–59.

Lewison, R.L., A.J. Hobday, S.M. Maxwell, E.L. Hazen, J.R. Hartog, D.C. Dunn, D.K. Briscoe, et al. 2015. Dynamic ocean management: Identifying the critical ingredients of dynamic approaches to ocean resource management. *Bioscience* 65:486–98.

Lindo-Atichati, D., F. Bringas, G. Goni, B. Muhling, F.E. Muller-Karger, and S. Habtes. 2012. Varying mesoscale structures influence larval fish distribution in the northern Gulf of Mexico. *Marine Ecology Progress Series* 463:245–57.

Llopiz, J.K., B.A. Muhling, and J.T. Lamkin. 2015. Feeding dynamics of Atlantic bluefin tuna (*Thunnus thynnus*) larvae in the Gulf of Mexico. *ICCAT Collective Volume of Scientific Papers* 71:1710–15.

# AQUACULTURE

weigh more than 0.1 g and is able to eat several dozens, even hundreds, of ABT larvae every day (Figure 13.5).

## Weaning

In general, the larvae are taken from the larval rearing tanks at 24–26 dph, when average weight is ~0.1–0.2 g. After counting and removing the small and large juveniles, the remaining fish are split into different tanks (20–55 $m^3$) at a density of 100–200 fingerlings/$m^3$. The natural temperature of seawater at this time ranges between 24 and 28°C. Although the fish are exposed to the natural photoperiod, a low-intensity light is placed above the tank during the night.

The success of weaning the BFT juveniles onto an inert diet or baitfish has been mixed; it resulted in only 15% survival in 2011. Consequently, bait fish (*Ammodytes* sp.) has been used since. Nevertheless, dry diet performance is improving. Figure 13.6 shows little difference in survival with age when weaning with baitfish and two dry diets—a commercial diet (Magokoro, by Marubeni Nisshin Feed Co. Ltd.) and an experimental diet made by Skretting.

In 2015 the dry diet Magokoro was used with promising results: the survival rate was close to 40% during the weaning process in large tanks (Ortega, unpublished data).

During the first days of weaning, YSL are also supplied, but from 30 dph onward, when tunas weigh between 0.5 and 1 g, only inert diet is provided.

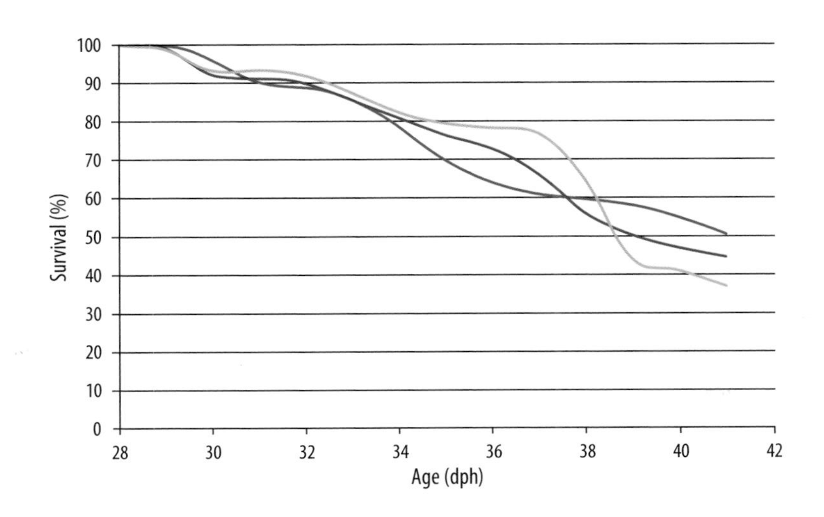


**Figure 13.6.** Survival rate (%) in the weaning process obtained in an experiment carried out in 2014. Blue line shows survival with bait fish, and red and green lines show survival with a couple of dry diets. dph = days post hatching.

Tuna juveniles grow very fast, and by 35–38 dph, they must be moved to the cages. During the time leading up to transfer, the problems that became more prevalent were diseases and mortality by wall collision (Ortega et al. 2014).

## Ongrowing

Tuna juveniles are transported to cages when their weight is ~3 g (35 to 38 dph). First trials showed that tunas grown beyond this to more than 5 g in the tanks exhibited increasing mortality from wall collision, and the fish suffered considerably more from handling during transport. Moreover, when the sea conditions are not optimal during the first days in the cages, mortality can continue to occur (de la Gándara et al. 2012b, Ortega 2015).

The transfer to sea cages takes a number of hours. To diminish handling, small tanks (1,000 l) are usually moved by crane directly from truck to boat, and then from the boat to the cages. Mortality has been reduced during the past few years (Figure 13.7) and stood in 2016 at approximately 15%–25%, including the three days following the cage stocking.

Tuna mortality was high during the first stages in the net cage (Figure 13.8), particularly during the first month but also as long as tunas were lighter than a half kilogram. During this period, total mortality ranged from 60% to 90%. From the fifth month, the mortality rate decreased (Figure 13.9) to less than 2% monthly (Ortega et al. 2014).

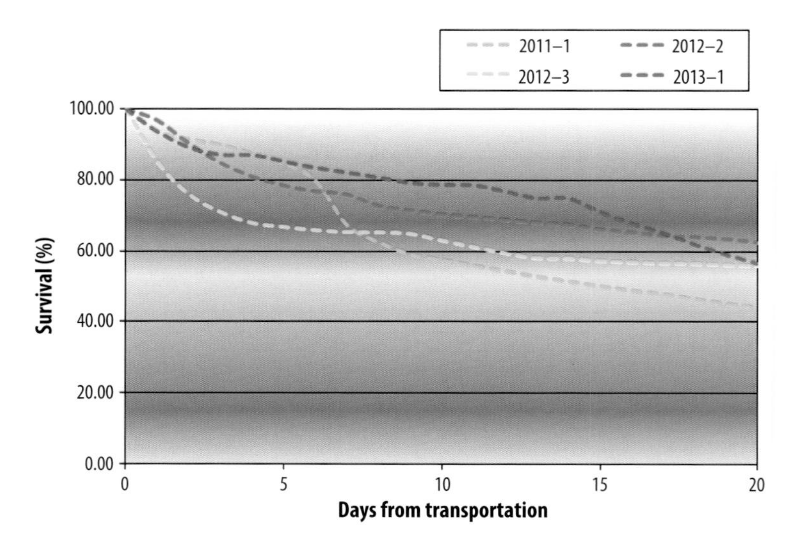


**Figure 13.7.** Survival rate (%) during transport from land-based facilities to the cages, 2011–13.

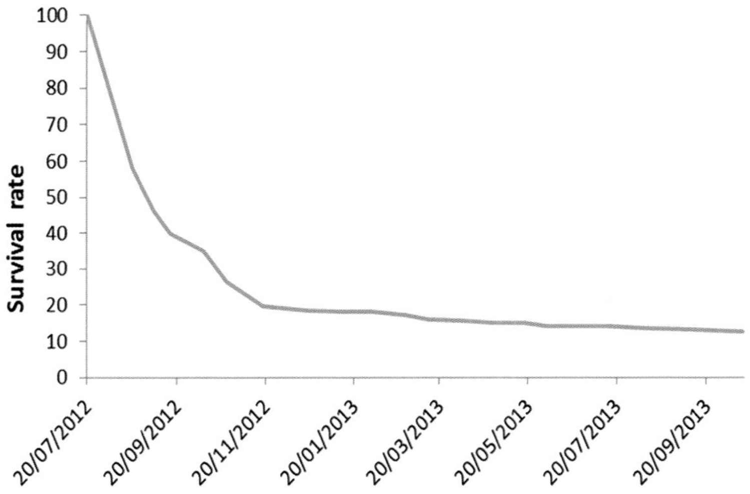

Figure 13.8. Survival rate (%) of bluefin tuna in the cages after transport.

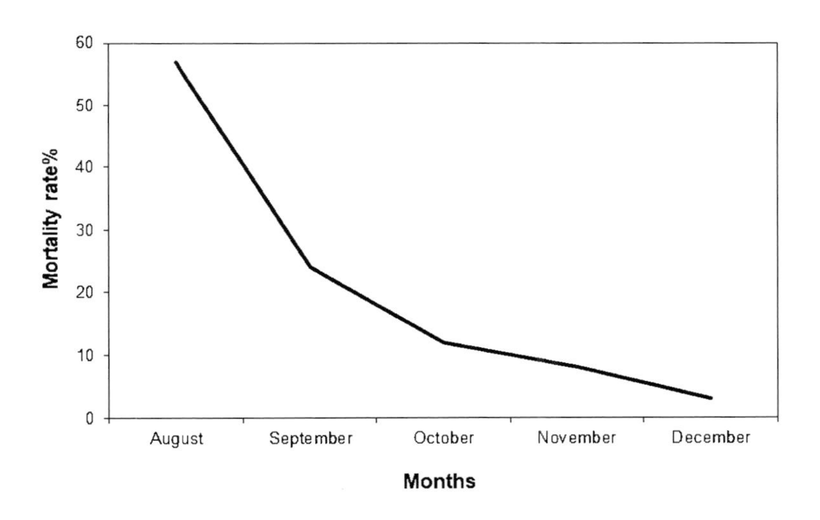


Figure 13.9. Monthly mortality rate (%) of the Atlantic bluefin tuna in the cages.

In the cages, tunas were fed to satiation several times per day (Ortega et al. 2014). Food consisted of defrosted bait. At the beginning, the first feeding was sand eel (*Ammodytes* sp.), which was replaced later, when ABT fingerlings were > 100 grams, by another species such as European pilchard (*Sardina pilchardus*), round sardinella (*Sardinella aurita*), mackerel (*Scomber scombrus*), and/or herring (*Clupea harengus*).

ABT fingerlings showed rapid growth (Figure 13.10) during the first months in the cages (de la Gándara et al. 2016). After two months, they reached 0.5 kg. They gained an additional 1 kg in the third month, and they reached ~2 kg after five months (Figure 13.11). During this time the specific growth rate remained above 4, with growth slowing down during the

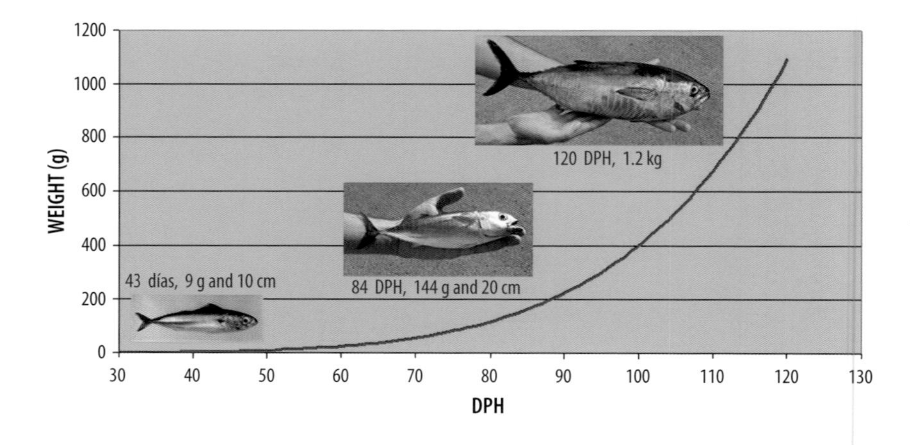


**Figure 13.10.** Growth of juvenile Atlantic bluefin tuna in the cages. DPH = days post hatching.

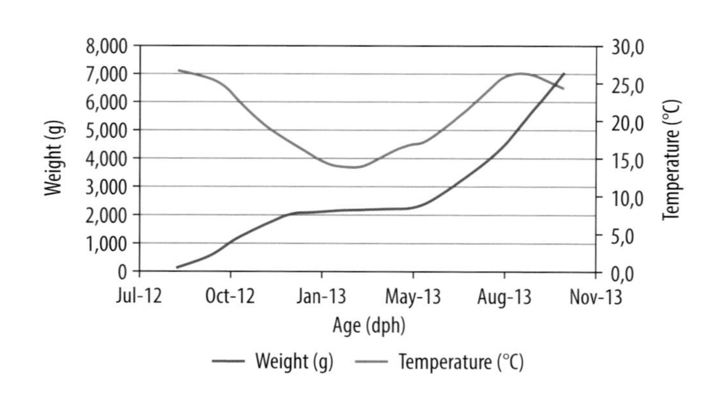


**Figure 13.11.** Growth of juvenile Atlantic bluefin tuna in oceanic cages and temperature profiles in southeastern Spain.

winter when temperatures were below 18°C, and stopping altogether when temperatures were below 16°C (Ortega et al. 2014).

Improving the success of grow-out, understanding the nutritional requirements at different stages of growth, and developing a suitable dry diet are essential for industrial ABT aquaculture to advance and to be able to close the life cycle in captivity.

## Land-Based Facility to Control the Reproduction of Atlantic Bluefin Tuna

A new land-based facility to control ABT reproduction in tanks, Infrastructure for Controlling the Reproduction of Atlantic bluefin tuna (ICRA), was built by IEO in 2015 (Ortega 2015), and the broodstock was introduced during 2016.

The ICRA (Figure 13.12) was 80% funded by the European Regional Development Fund (FEDER), and the IEO contributed the remaining 20%.

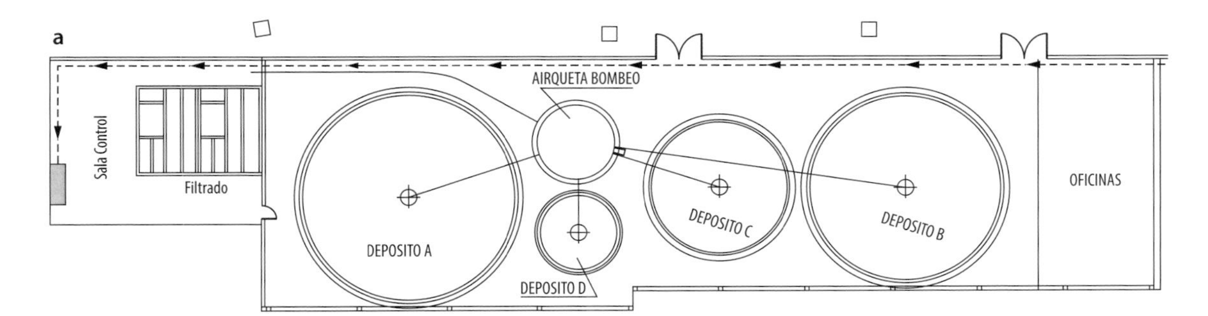

**Figure 13.12.** Land-based Atlantic bluefin tuna aquaculture facility in Murcia, in the southeast of Spain. (*a*, *b*) General design and (*c*) interior of one of the big tanks.

This was enabled by an agreement between the IEO, the Spanish Science Ministry, and the Regional Government of Murcia. The facility consists of two large tanks for broodstock (A and B tanks have volumes of 3,500 and 2,600 $m^3$, respectively) and two smaller tanks for ongrowing juvenile tunas (C and D tanks have volumes of 900 and 150 $m^3$, respectively). To maintain constant seawater parameters and good water quality, a recirculation system (mechanical and biological filters, a heating and chilling system, skimmers, and ozone, ultraviolet, and oxygen injection) was implemented and is completely automated.

In June 2016 we achieved, for the first time, the closing of the life cycle of ABT in captivity. A batch of 30 tunas from the 2011–12 season, weighing an average of 35 kg each, together with another batch of ~70 tunas born in 2013 and 2014 and weighing ~12 kg each, were moved in late May to 28 m diameter cages and then transported from El Gorguel, Cartagena, to another facility in San Pedro del Pinatar (both in southeastern Spain). Tunas were fed with minced fish (herring and mackerel), and ten days after the tuna arrived in San Pedro, a tarpaulin egg collector system was placed around the cage perimeter. A batch of 50,000 fertilized eggs was collected on June 18. The eggs were incubated and hatched ~30 hours later. Then they were moved to a tank to carry out the larval rearing. Twenty-five days later, ~300 larvae were moved to a larger tank to start the weaning process. This broodstock spawned only a few times more, and only low numbers of eggs were collected.

The reason for the low number of eggs obtained could be the youth of the broodstock. This was their first spawning event, and possibly only one or two females contributed spawning eggs. This fact should have been confirmed in the next spawning season, but unfortunately a great storm broke the cages where the broodstock was located, and all escaped.

Fingerlings produced from these eggs as well as those produced the following years will set up broodstock lots for ICRA that could contribute to production and offer ABT eggs for research projects and/or for commercial production purposes in the years to come.

To even consider the fattening of captured ABT as a highly profitable commercial activity is a new perspective gained from the recent progress in tuna aquaculture; it was brought about by fully closing the cycle of ABT, producing seeds and fish up to marketable sizes, similar to what is occurring with other cultured marine fish.

## REFERENCES


Abate, T.G., R. Nielsen, M. Nielsen, G. Drillet, P.M. Jepsen, and B.W. Hansen. 2015. Economic feasibility of copepod production for commercial use: Result from a prototype production facility. *Aquaculture* 436:72–79.

de la Gándara, F., and A. Ortega. 2008. Eight years of research on bluefin tuna (*Thunnus thynnus*) culture at the Spanish Institute of Oceanography (IEO). *Proceedings of the EAS 2008, Krakow, Poland,* pp. 185–86. http://hdl.handle.net/10508/5766.

de la Gándara, F., C.C. Mylonas, D. Coves, and C.R. Bridges. 2010. SELFDOTT report 2009. http://hdl.handle.net/10508/356.

de la Gándara, F., C.C. Mylonas, D. Coves, and C.R. Bridges. 2012a. SELFDOTT final report (abstract). http://hdl.handle.net/10508/1119.

de la Gándara, F., C.C. Mylonas, D. Coves, and C.R. Bridges. 2012b. SELFDOTT report 2010–2011. http://hdl.handle.net/10508/1118.

de la Gándara, F., A. Ortega, and A. Buentello. 2016. Tuna aquaculture in Europe. In *Advances in Tuna Aquaculture: From Hatchery to Market,* edited by D.D. Benetti, G.J. Partridge, and A. Buentello, 115–57. New York: Elsevier Academic Press.

Fromentin, J.M. 2009. The key importance of the underlying stock-recruitment assumption when evaluating the potential of management regulations of Atlantic bluefin tuna. *ICCAT Collective Volume of Scientific Papers* 64 (2): 558–67.

García-Gómez, A. 2007. REPRO-DOTT final report. Reproduction of the Bluefin Tuna in captivity—feasibility study for the domestication of *Thunnus thynnus.* Spanish Institute of Oceanography (IEO), Government of Spain. Contract number: Q5RS-2002-0153. http://hdl.handle.net/10508/1010.

Honryo, T., T. Tanaka, A. Guillen, J.B. Wexler, A. Cano, D. Margulies, V.P. Scholey, M.S. Stein, and Y. Sawada. 2016. Effect of water surface condition on survival, growth and swim bladder inflation of yellowfin tuna, *Thunnus albacares* (Temminck and Schlegel), larvae. *Aquaculture Research* 47 (6): 1–9. doi:10.1111/are.12641.

Kurata, M., M. Seoka, Y. Ishibashi, T. Honryo, S. Katayama, K. Takii, H. Kumai, S. Miyashita, and Y. Sawad. 2015. Timing to promote initial swimbladder inflation by surface film removal in Pacific bluefin tuna, *Thunnus orientalis* (Temminck and Schlegel), larvae. *Aquaculture Research* 46 (5): 1–11. doi:10.1111/are.12277.

Ortega, A. 2015. Cultivo Integral de dos especies de escómbridos: Atún rojo del Atlántico (*Thunnus thynnus,* L. 1758) y Bonito Atlántico (Sarda sarda, Bloch 1793). PhD diss., Universidad de Murcia, Cartagena, Spain. http://hdl.handle.net/10508/10066.

Ortega, A., J. Viguri, J.R. Prieto, A. Belmonte, D. Martinez, M. Velazquez, F. de la Gándara, and M. Seoka. 2014. First results on ongrowing of hatchery reared Atlantic bluefin tuna, *Thunnus thynnus,* kept in sea cages. *Proceedings of the EAS Aquaculture Europe 2014, San Sebastián, Spain, October 14–17,* pp. 931–32. http://hdl.handle.net/10508/2757.

Ortega, A., P. Reglero, E. Blanco, and F. de la Gándara. 2015. Efecto de la dieta basada en nauplios del copépodo *Acartia tonsa,* del rotifero *Brachionus plicatilis* enriquecido en acidos grasos, y de una mezcla de ambos, sobre el crecimiento y supervivencia de larvas de atun rojo (*Thunnus thynnus*). *Actas del XV Congreso Nacional de Acuicultura, Huelva, Spain, October 13–16,* pp. 404–5. http://hdl.handle.net/10508/9872.

Ottolenghi, F. 2008. Capture-based aquaculture of bluefin tuna. *FAO Fisheries Technical Paper* 508:169–82.

CHAPTER 14

# The Resource and Environmental Intensity of Bluefin Tuna Aquaculture

Dane H. Klinger and Nicolas Mendoza

## Introduction

Demand for seafood is increasing because of a growing global population and an expanding middle class (World Bank 2013). Production from capture fisheries has stagnated or declined, and there is limited capacity for rebuilt fisheries to satisfy growing demand (Sumaila et al. 2012, Costello et al. 2016). Aquaculture is one of the fastest growing food-production sectors and has great potential to increase seafood production to meet rising demand (FAO 2016b). However, there is substantial diversity within the aquaculture sector, and different production species and systems often result in a range of environmental impacts, resource intensities, and growth potentials (Klinger and Naylor 2012).

The ability of the aquaculture sector to increase seafood production, wealth generation, and employment, without harming the environment or wasting natural resources, depends largely on the types of aquaculture that are utilized (Le François et al. 2010, Tidwell 2012). As the aquaculture industry continues to grow and overtakes capture fisheries in production tonnage, a key question is whether the sector can learn from mistakes made by the agriculture industry as it grew and supplanted hunting (Tilman 1999, Mazoyer and Roudart 2006). To evaluate the relative potential benefits and costs to society of a specific aquaculture species, it is helpful to compare the species' current role in aquaculture (including the tonnage and value of production), typical production systems, resource use efficiency, demonstrated environmental impacts, and market attributes. In this chapter, we consider these factors in regard to bluefin tuna in the context of other common aquaculture species.

## Aquaculture Production—Tonnage and Value

The aquaculture industry is a diversified and dynamic sector. Whereas terrestrial animal agriculture focuses primarily on only a few species from different genera, aquaculture includes numerous species from different classes and phyla (Duarte et al. 2007, Tidwell 2012). Nearly 600 aquatic species have been farmed, including 362 species of finfish (FAO 2016b). From 2010 to 2014, the global aquaculture industry produced an average of 90 million metric tons (mt) of seafood annually, of which 47% was freshwater and diadromous fish, 26% was aquatic plants, 17% was mollusks, 8% was crustaceans, and 2% was marine fish. The majority of production by tonnage was concentrated in Asia, with 60% in China, 12% in Indonesia, 5% in India, 3% in Vietnam, and 20% in all other countries combined (FAO 2016a).

By reported first sale value, the aquaculture industry as a whole generated on average $147 billion USD annually from 2010 to 2014, of which 54% was from freshwater and diadromous fish, 22% was from crustaceans, 11% was from mollusks, 9% was from marine fish, and 4% was from aquatic plants (FAO 2016a). The reported first sale price (value/weight) for all aquaculture products combined was $1.64 per kg, and the average price for finfish was $2.03 per kg (FAO 2016a).

In contrast, between 2010 and 2014, the average annual global production of farmed bluefin tuna was 22,385 mt and represented an average annual value of $445 million dollars. The average annual reported first sale price was $19.28 per kg. Bluefin are farmed primarily in Japan (43% of production by tonnage), Australia (26%), Mexico (16%), and several countries around the Mediterranean, including Malta, Croatia, Spain, Turkey, Tunisia, Greece, Morocco, and Italy (15% combined), calculated from average annual production between 2010 and 2014 (FAO 2016a).

Bluefin aquaculture is a minor player in global aquaculture by tonnage (0.06% of global aquaculture finfish production) and value (0.50% of global aquaculture finfish value), but its price point is 1,181% higher than aquaculture in general and 957% higher than the average price for farmed finfish. Additionally, the majority of all aquaculture is concentrated in Asia, whereas bluefin production is distributed in four primary locations around the world (Figure 14.1).

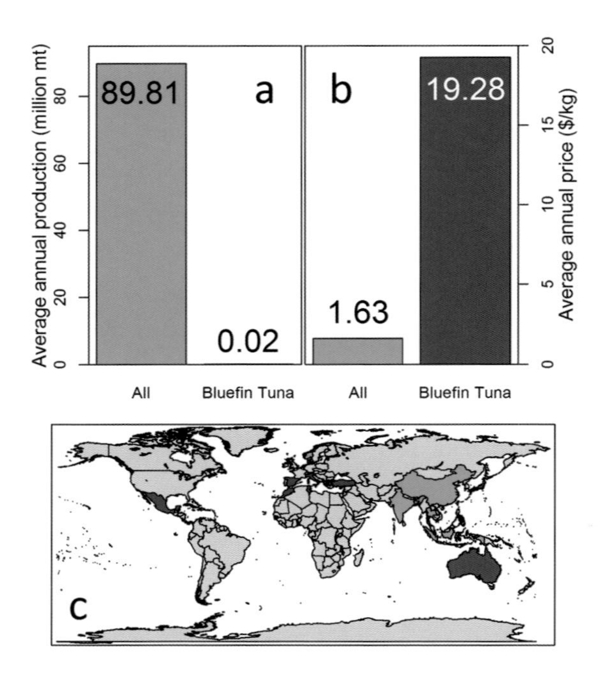


**Figure 14.1.** Production tonnages, values, and geographies of global aquaculture (including aquatic plants) and bluefin tuna. (*a*) Average annual aquaculture production tonnage of all species (*green*) and of bluefin tuna (*blue*), 2010–14; (*b*) average annual price of all aquaculture species (*green*) and of bluefin tuna (*blue*), 2010–14; and (*c*) major production countries for aquaculture as a whole (top 88% of production; *green*) and for all bluefin tuna farming countries (*blue*) for the years 2010–14. Data from FAO 2016.

## Production Systems

Globally, seafood is farmed in a broad variety of production systems, ranging from extensive to intensive (Stickney 1994). The majority of production takes place in extensive systems (e.g., unfed ponds) in which farmers introduce broodstock or juveniles and then rely largely on natural ecological productivity to feed a low-density stock, and semi-intensive systems (e.g., fed ponds or lagoons), in which farmers use fertilization, supplemental feeding, and/or other inputs to enhance productivity, allowing higher stocking densities (Bostock et al. 2010). Species commonly farmed in extensive and semi-intensive systems include carp, tilapia, shrimp, oysters, and seaweed. Intensive systems, in which farmers supply additional inputs (e.g., compound feeds and water-quality management), allowing higher stocking densities (e.g., 12–18 kg/m$^3$, or up to 50 kg/m$^3$ with oxygen injection), are becoming increasingly common (Tidwell 2012). Examples of intensive aquaculture systems include land-based recirculating aquaculture systems and large-scale marine net pens.

The goal of most aquaculture is domestication, in which the life cycle of a species is closed in captivity such that full control of reproduction exists across successive generations, independent of eggs, larvae, juveniles, and broodstock sourced from wild populations (Teletchea and Fontaine 2014). Domestication of an aquaculture species requires significant time and research investment, as it necessitates an understanding of, and an ability to meet, the biological and environmental needs of a species throughout its entire life cycle (Gjedrem 2005). When working with new species, aquaculture operations often rely heavily on wild-sourced broodstock, larvae, and/or fingerlings while the species is being domesticated (Klinger et al. 2013).

Bluefin tuna are farmed in highly intensive systems relative to other types of aquaculture. However, the bluefin life cycle has yet to be closed for most of bluefin production, contributing to the sector's production inefficiencies.

Bluefin tuna aquaculture began as a capture-based endeavor, in which wild-caught tuna are maintained in intensive, nearshore net pens and are raised to market size on a diet of forage fish (Ottolenghi 2008). There are several economic incentives to capture-based farming, which is sometimes called ranching or fattening. These include the higher market price for fish with a higher fat content and the ability of producers to time harvests with the market to maximize the value of each fish sold (Shamshak 2011). Through the 1990s, consistently high demand for bluefin tuna meat (primarily in Japan) and the global decline of wild bluefin populations due to overexploitation created the economic incentive for industrial-scale capture-based operations in Japan, Mexico, Australia, and across the Mediterranean (Metian et al. 2014). Despite significant research investment since the 1970s, efforts to close the life cycle of bluefin tuna have had limited success until very recently, and today most farm production comes from capture-based operations, with a few notable exceptions (Benetti et al. 2015).

The unique physiology and life history of bluefin tuna present unique challenges for domestication. Bluefin tuna reproduction is temperature dependent, meaning broodstock maintained in the ocean will spawn only seasonally in most temperate regions. Maintaining tuna broodstock in land-based facilities, where water temperature can be controlled to continually meet the reproductive needs of farmed fish, is complicated by the fact that female bluefin tuna are large and long lived; they reach sexual maturity between 5 and 10 years of age, when they weigh 100 kg or more (Corriero et al. 2005, Bubner et al. 2012, Gardner et al. 2012). In addition, there are obstacles to the larval rearing and nursery phases of hatchery production. Like other carnivorous marine fish, tuna larvae are dependent on a supply of live prey (such as rotifers, artemia, and the yolk-sac larvae of other fish) that can be difficult, time-consuming, and expensive for a hatchery operation to culture. Other persistent issues include early "floating death" and "sinking mortality" (Nakagawa et al. 2011, Kurata et al. 2014), cannibalism and collision deaths at the post-flexion stage (Sawada et al. 2005, Masuma et al. 2011), and mass mortality when transferring fish from land-based hatcheries to sea cages (Ishibashi et al. 2009, Higuchi et al. 2014). Collectively, these problems have resulted in average larval survival rates of 0.01% to 4.5% (Masuma et al. 2008).

Despite the challenges of domesticating bluefin tuna, Sawada et al. (2005) reported the first closing of the Pacific bluefin tuna life cycle in 2002, after 6- and 7-year-old laboratory-produced fish successfully spawned at Kinki University (now Kindai University) in Japan. R&D consortiums in the Mediterranean and groups in Australia also made important advancements with Atlantic and southern bluefin tuna. These advances include the observation of spontaneous spawning events in captive bluefin, improved broodstock management techniques, protocols for hormone-induced spawning, the development of improved larval rearing technologies, and progress in developing formulated diets (Chen et al. 2015, de la Gándara et al. 2015). The high price of bluefin and the limited supply of wild-caught fish create substantial economic incentives for continued research into bluefin tuna farming. In the future, refinement of closed-life-cycle bluefin farming has the potential to produce continuous performance gains through the use of selective breeding programs, which average 10%–20% increases in growth per generation for other fish species (Gjedrem and Baranski 2009) and can select for behavioral attributes that are more amenable to conditions in aquaculture operations (e.g., reduced stress through decreased fear of divers and diminished antipredator responses) (Teletchea and Fontaine 2014).

## Species Selection and Physiology

The species under cultivation has a large influence on the performance of an aquaculture operation, including the resource efficiency, environmental intensity, and profitability (Naylor et al. 2009, Le François et al. 2010). Historically, aquaculturists have chosen which species to culture on the basis of numerous economic, geographic, technological, regulatory, cultural, biological, and environmental factors (Nash 2010, Stickney and Treece 2012). The desired production system also plays a key role in determining which species are most appropriate (Tidwell 2012). Aquaculturists must balance the trade-offs and synergies between factors and choose the species that best achieves the enterprise's goals, be they subsistence, profit, or something else.

The majority of aquacultured finfish, by tonnage, have physiological adaptations that are beneficial for growth and survival under culture conditions. For example, carp, tilapia, and many other fish use their buccal and opercular pumps to move oxygen-rich water across their gills for respiration, a relatively energy-efficient means of respiration (Brainerd and Ferry-Graham 2005). Additionally, several commonly cultured species, including *Pangasius*, are able to breathe air when the dissolved oxygen concentration in water becomes critically low (Lefevre et al. 2014), which can occur during mechanical failures of aeration equipment or when stock densities are too high. Most aquaculture finfish species are also ectotherms, meaning they rely on ambient water temperatures for body heat and do not expend large amounts of energy on heat generation (Angilletta 2009).

There are several aspects of bluefin tuna physiology that are markedly different from most aquaculture species (Block and Stevens 2001). Bluefin are obligate ram ventilators, which means they must perpetually swim to move oxygen-rich water across their gills for respiration (Magnuson 1978, Brill 1996). This required movement increases the routine metabolic rate of bluefin tuna, meaning there is less dietary energy available for growth relative to nonram ventilators. Additionally, bluefin tuna are regionally endothermic, meaning they maintain parts of their body at temperatures that are warmer than the surrounding water temperature (Stevens et al. 2000). Regional endothermy allows bluefin to actively hunt in cool waters that reduce the physiological performance of ectothermic prey, but it is energetically expensive to maintain, and it increases routine metabolic rates. These physiological adaptations are beneficial to bluefin in the wild, but they necessitate a large amount of feed and result in greater environmental impacts under farmed conditions.

## Resource Intensity Metrics

The resource intensity of bluefin tuna aquaculture, including feed use and growth rates, also provides valuable points of comparison with other types of aquaculture.

### Feed

Approximately 50% of global aquaculture output by tonnage comes from nonfed species, including seaweeds, filter-feeding carp, and filter-feeding mollusks. The other 50% of aquaculture production requires feed inputs, ranging from diets based on plant ingredients to diets with high concentrations of marine ingredients, such as fishmeal and fish oil (FAO 2016b). Diets rich in fishmeal and fish oil are expensive and often resource inefficient. The prices of fishmeal and fish oil are both substantially higher than the prices of terrestrially derived proteins and lipids (Asche et al. 2013), and the limited capacity for increased yields in the reduction fisheries that provide these proteins and lipids means that prices will likely remain high unless effective alternatives are developed (Tacon et al. 2011). From a resource perspective, there is a large opportunity cost to global human health when highly nutritious fish (e.g., anchoveta) are reduced to fishmeal and fish oil and then fed to carnivorous fish, resulting in a net loss of marine protein and lipid (Naylor et al. 2000). These concerns are compounded by the fact that fed aquaculture sectors, especially aquaculture of carnivorous species, are increasing at faster rates than other types of aquaculture (FAO 2016b).

Two common metrics used to describe the feed efficiency (including feeding practices and nutrient uptake) of a species or operation are feed conversion ratio (FCR) and fish-in to fish-out ratio (FI:FO). FCR measures the conversion efficiency of a species or system and describes the amount of feed required to produce a unit of farmed product. Large FCRs indicate an inefficient use of feed. At a global scale, species tend to have lower FCRs when there is widespread use of standardized, optimized production practices and formulated feeds as opposed to farm-made feeds or whole fish diets (e.g., catfish [1.4 FCR] and salmon and trout [1.3 FCR] for export markets). Species tend to have higher FCRs when they are farmed in majority small-scale, domestic-market-oriented operations (e.g., fed carp in China [1.7 FCR]) or are relatively new to commercial culture (e.g., most marine fish [1.7 FCR]), as feeds and production practices have not been optimized for the biological needs of those species (Tacon and Metian 2015). The FCR for bluefin tuna operations are substantially higher than for most common aquaculture operations and range from 10 to 15 for smaller fish and 15 to 25 for larger fish (Zertuche-González et al. 2008, Mourente and Tocher 2009, Estess et al. 2014).

Comparison of FCRs across species can be misleading because different species are often fed different feeds, and the nutritional composition of the feed is not reflected in the metric. Feeds differ in the ratio of macronutrients (carbohydrates, lipids, and proteins) and other ingredients, each of which has uses and value outside of aquaculture (Halver and Hardy 2002). FCR does not reveal the relative efficiencies at an ingredient level. When individual ingredients or groups of ingredients are of concern, individualized metrics are used. For example, FI:FO ratios describe the amount of whole fish required to produce a unit of farmed product. If FI:FO is greater than 1, there is a net loss of fish to society.

Inclusion rates of fishmeal and fish oil in aquafeeds have been declining for several decades in most species. For example, FI:FO ratios decreased between 1995 and 2006 for salmon (7.5 to 4.5), marine fish in general (3.0 to 2.2), and tilapia (0.9 to 0.4) (Tacon and Metian 2008). During the same period, the FI:FO for all fed fish species, averaged by production volume, decreased from 1.0 to 0.6 (Naylor et al. 2009). Bluefin operations typically use whole fish as feed, meaning FCRs are equal to FI:FO ratios and range from 10:1 to 25:1 (Buentello et al. 2015, de la Gándara et al. 2015). As such, bluefin tuna aquaculture requires substantially more whole fish per unit of production than other types of aquaculture.

Feed efficiency can also be compared at the macronutrient level (Glencross et al. 2007). Carbohydrates are generally an inexpensive source of energy in animal diets, whereas protein and lipid sources are more expensive (Halver and Hardy 2002). In general, freshwater fish in warm regions are able to utilize higher levels of carbohydrates than freshwater fish from cold regions and marine fish (Wilson 1994). Bluefin tuna, similar to other diadromous and marine fish, require higher absolute levels of lipid and protein than cultured freshwater fish in warm environments. As such, bluefin diets require higher levels of more expensive ingredients.

To improve the resource efficiency of bluefin tuna farming and reduce the cost of feeds, bluefin tuna operations can work with scientists and feed manufacturers to develop and deploy compound feeds and improved feeding practices. Additional research to enhance the understanding of tuna dietary requirements for proteins and specific amino acids, lipids and specific fatty

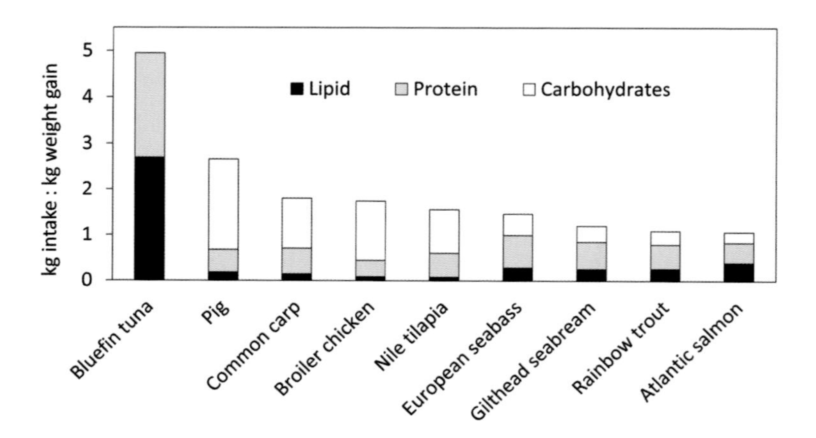


**Figure 14.2.** Macronutrient intake of bluefin tuna, relative to other terrestrial and aquatic species (modified and expanded from Naylor et al. 2009). The macronutrient intake for bluefin tuna is calculated on the basis of a whole sardine diet and a feed conversion ratio of 15:1.

acids, carbohydrates, minerals, vitamins, and other ingredients can also aid the development of commercially viable compound feeds for tuna (Buentello et al. 2015). Advances in feeds and feeding practices for other aquatic and terrestrial species have greatly reduced the amount of lipid and protein required for production of those species (Figure 14.2).

## Growth Rate

The ideal species for aquaculture attains the desired market size in the shortest amount of time. Longer growout requires greater investments in feed and other inputs and exposes the operation to risks, such as disease outbreaks and equipment failure. The growout period for most aquaculture species is less than a year, including catfish (6 months), tilapia (3 months), and shrimp (3 months) (FAO 2017). Generally, growth rates for these species are high and the market size is small (<2 kg). Although bluefin have high growth rates relative to other marine fish (Figure 14.3), the desired market size for bluefin is also high (>25 kg), meaning the growout period for closed-life-cycle bluefin can be greater than two years. If bluefin tuna operations are able to sell smaller bluefin in the future, the growout time, and subsequent risk exposure, could decrease to levels similar to that of other types of aquaculture. Selective breeding programs on closed-life-cycle bluefin could also increase growth rates and decrease time to market size (Sawada and Agawa 2015).

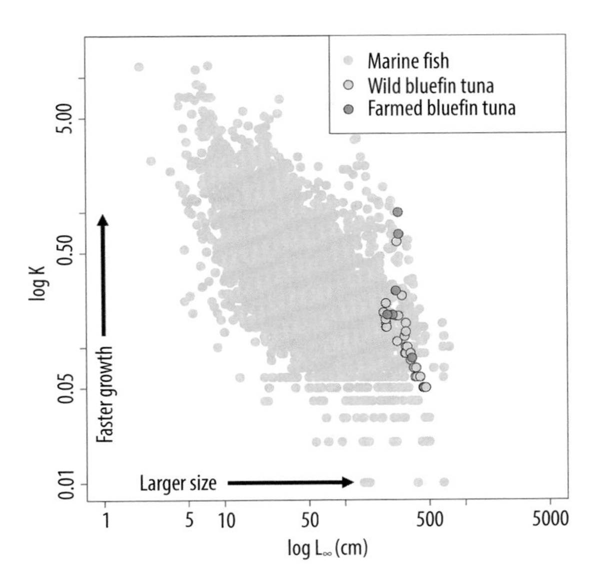


**Figure 14.3.** Von Bertalanffy growth parameters of bluefin tuna (wild, *light gray dots with black outlines;* farmed, *dark gray dots with black outlines*) relative to those of other marine fish species (*gray dots with no outlines*). The parameter $L_\infty$ describes the mean length of an old fish of a given species, and the parameter K describes how quickly the species reaches $L_\infty$. Data include more than 7,000 growth curves for marine fish from Fishbase.com.

## Environmental Intensity Metrics

The environmental impacts associated with marine aquaculture have been reviewed extensively and include conflicts with other coastal users, release of effluent and chemicals into the environment, escaped fish that can interbreed with wild conspecifics or compete with wild species, disease and parasite transmission to wild populations, farm-predator interactions, and the destruction of coastal habitat to build aquaculture infrastructure (Primavera 2006, Bert 2007, Diana 2009, Sarà et al. 2011, Klinger and Naylor 2012). Subsequently, the expansion of marine aquaculture has been met with criticism by environmental groups, fishing communities, and other stakeholders concerned about the negative impacts that farms can have on coastal and pelagic environments (Knapp and Rubino 2016, Froehlich et al. 2017). Bluefin tuna farming presents many of the same environmental concerns that have been cited for other carnivorous marine species such as salmonids, amberjacks, and seabass (see Holmer 2010). However, certain attributes of bluefin tuna physiology, culture, and economics present unique challenges, as well as opportunities, for the production of these species with reduced harm to the environment. Here, we review several negative environmental impacts associated with marine aquaculture and how bluefin tuna farming is similar or different.

### User Conflicts

Owing to the high concentration of human populations and activities along coastlines, marine aquaculture has long competed with other stakeholders for use of the coastal environment. Among the user conflicts that arise are those involving recreational and commercial fishers, environmentalists, and groups who object to aquaculture on the basis of aesthetics, particularly in developed nations (Stickney and McVey 2002). Bluefin tuna farming operations create similar conflicts. In the Aegean Sea, clustering of bluefin operations has occasionally led to conflicts with the tourism sector (Karakulak et al. 2015), and in Croatia there have been reports of waste from feedings and harvests washing up on beaches and causing foul odors, eliciting negative reactions in major tourist destinations (Katavic et al. 2003).

### Effluent and Chemical Release

Effluents from aquaculture can promote eutrophication, limit the environmental carrying capacity, and reduce biodiversity and ecosystem functioning (Naylor et al. 2000, Primavera 2006). Chemicals and therapeutants are commonly used in marine aquaculture, but the use of such compounds in tuna aquaculture appears to be very limited, perhaps because of the short growing cycle of fattening operations (~1–5 months) and an associated low incidence of disease on farms (Mylonas et al. 2010). However, recent studies on the ecological impacts of bluefin farming operations have shown that effluent from tuna ranching can negatively affect marine ecosystems, though results are mixed as to the extent and permanence of measurable impacts (Vita et al. 2004, Zertuche-González et al. 2008, Forrestal et al. 2012, Kružić et al. 2014, de la Gándara et al. 2015, Mangion et al. 2017, Mangion et al. 2018). As the industry moves from wild-capture fattening of large fish to wild-capture fattening of small fish and closed-life-cycle farming, growout duration will increase, increasing the potential for negative impacts on local ecosystems.

## Escapes and Invasiveness

Invasive species are a global threat to biodiversity, ecosystem functioning, animal health, and human economies (Bax et al. 2003, Molnar et al. 2008). Historically, the aquaculture sector has been a major source of introduced and potentially invasive species, including species that are nonnative to a region, captively bred organisms that are genetically and behaviorally different from existing wild populations, and nonnative pathogens (Naylor et al. 2001, Arthur et al. 2010). However, bluefin tuna farming has some stark differences from other types of marine aquaculture. Most bluefin aquaculture is capture based, meaning bluefin tuna are farmed in areas where they can be readily caught and transported to fattening pens (Benetti et al. 2015). Additionally, the dominant use of wild fish in operations means that no genetic differences exist between farmed fish and the wild population (Sawada and Agawa 2015). Even Pacific bluefin tuna produced through the novel Kindai University breeding program likely maintain the same genetic diversity of natural populations, as the program is in the early stages (Benetti et al. 2015). Nonetheless, bluefin have been known to escape from farms via both spawning and incidental release of individuals, and the bluefin tuna's status as a highly migratory pelagic species (perhaps the only one subject to intensive farming) leaves questions as to the effects of the farm environment on bluefin behavior (Arechavala-Lopez et al. 2015, Džoić et al. 2017). For example, escaped fish sometimes demonstrate an affinity to farms, remaining near the operation where there is constant feed availability (Arechavala-Lopez et al. 2015). As such, fattening operations could affect the migratory and schooling behavior of escaped or released bluefin, potentially altering the tuna's homing accuracy and resulting in new coastal migratory routes, resident populations, or the inability of fish to locate spawning and feeding grounds (Sarà et al. 2011). Additional research is necessary to understand the influence of offshore fish farms on the biology and ecology of bluefin, particularly as the life cycle is closed and selective breeding programs genetically differentiate farmed and wild populations (Buentello et al. 2015).

## Wildlife and Predator Interactions

Coastal and open-ocean fish farms act as fish aggregation devices, providing wild fish with structure and nutrients in the form of uneaten feed and fish waste (Dempster et al. 2004). Although there are few studies on the

**Figure 14.4.** Observation of an interaction between a farmed Pacific bluefin tuna and a California sea lion off the Pacific coast of Mexico. *Photo by Robert Schallert/TRCC.*

positive and negative impacts that bluefin farms have on wild fish assemblages, one recent study makes the case for semi-offshore tuna farms as functional marine protected areas, given the designation of farms as no-fishing zones and the subsequent enhanced fitness of aggregated wild fish (Stagličić et al. 2017). Predators such as seabirds, pelagic fish, sharks, and marine mammals congregate around operations, feeding on both captive fish and the extensive marine community attracted to the area (Figure 14.4; Holmer 2010). This raises concerns about the potential negative impacts of predator-farm interactions (e.g., Sepúlveda and Oliva 2005, Sanchez-Jerez et al. 2008). Bluefin tuna farms are no different from other marine aquaculture operations in this regard. In the Mediterranean, Australia, and Mexico, protected species, such as sea lions (*Zalophus californianus*) and white sharks (*Carcharodon carcharias*), have been killed after becoming entangled or trapped in bluefin tuna pens (Ellis 2008, Arechavala-Lopez et al. 2015). Although some farms in Mexico deploy an outer 150–200 mm mesh around the inner net to act as a barrier against predators (Zertuche-González et al. 2008), very little data are available on technologies and methods used across the industry. Ultimately, the impact of bluefin tuna farms on marine wildlife and predators remains largely unmeasured, necessitating continued study of the ecological and economic impacts of these interactions, as well as the development of technologies with the potential to minimize any risks.

## Biosecurity and Disease

Aquaculture's history is one of continual struggle with pathogens and disease—including outbreaks, solutions and victories, and subsequent new challenges (Lafferty et al. 2015). Disease outbreaks on marine fish farms are caused by parasites, bacteria, viruses, and fungi and can result in heavy mortality, decreased production, and harm to wild communities (Hayward et al. 2007). Disease losses cost the aquaculture sector many billions of dollars annually (Lafferty et al. 2015). Additionally, marine aquaculture operations can be net exporters of infectious agents, amplifying pathogens obtained from wild populations and retransmitting them (Lafferty et al. 2015).

Although a number of parasites of bluefin tuna have been identified, particularly in southern bluefin (as reviewed in de la Gándara et al. 2015, Ellis and Kiessling 2015), as of yet, global bluefin tuna aquaculture appears to be more resilient to disease than most marine aquaculture (Karakulak et al. 2015), perhaps owing to bluefin tuna's unique culture methods and physiology. Unlike other marine finfish reared in hatcheries, bluefin tuna juveniles and adults are primarily stocked from wild populations, where natural selection has favored the survival of the hardiest individuals (Hayward et al. 2007). These juvenile and adult fish also have well-developed immune systems in comparison to the typical hatchery fry stocked in marine net pens (Mylonas et al. 2010). Also, stocking densities prevalent in tuna aquaculture are substantially lower than with other marine species (reported at 2–4 kg m–3 in Australia, Mexico, and the Mediterranean [Hayward et al. 2007, Mylonas et al. 2010, Karakulak et al. 2015], as opposed to typical densities of 10–20 kg m–3 for other Mediterranean finfish species and 10–34 kg m–3 for Atlantic salmon worldwide [Turnbull et al. 2005]). It has been postulated that regional endothermy in bluefin contributes to immunocompetency throughout the year (Mladineo et al. 2008), unlike poikilothermic species that experience inhibition of immune responses during the colder winter farming period (Watts et al. 2002). It should be noted, however, that most aquaculture production systems undergo intensification as technology improves and the industry matures, increasing the long-term risk of infectious disease outbreaks in bluefin tuna farming (de la Gándara et al. 2015). Additionally, the continued use of fresh and frozen baitfish as feed presents a constant biological risk for the propagation of new pathogens on farms (Mladineo et al. 2008). Research into formulated feeds for bluefin tuna must be a focus for reasons of biosecurity and sustainability.

Additional research is also required to ensure disease detection, treatment, and prevention protocols are available and effective for both existing and emergent pathogens.

## Reducing Negative Environmental Impacts

Several changes can help reduce the negative environmental impacts of bluefin tuna farming while increasing or maintaining current production levels, including appropriate siting, locating farms farther offshore, and feed improvement. Despite a lack of consensus regarding the extent of localized negative environmental impacts from tuna aquaculture effluent, the industry would benefit from appropriate farm siting guidelines (Katavic et al. 2003). Additionally, moving operations farther offshore, away from nearshore users and fragile coastal ecosystems, can help reduce conflict with other stakeholders and minimize environmental impacts. Offshore siting also offers the dual benefits of increased growth rate and enhanced fish health (Kirchhoff et al. 2011).

Challenges to offshore farm development include the risk and uncertainty of operating in the high-energy, offshore environment (Karakulak et al. 2015). Pair this with the high startup cost, and it becomes difficult to attract investors (Atkins 2009). The development of cost-effective formulated feeds for widespread use across the bluefin tuna aquaculture sector would also help reduce the resource intensity and environmental burden of tuna aquaculture. The difficulty of working with bluefin tuna increases the cost of most types of research on tuna digestion and feeds. However, in recent years there have been substantial gains in the scientific understanding of bluefin tuna nutritional requirements (for comprehensive reviews, see Mourente and Tocher 2009, Buentello et al. 2015). The high cost of feed and large feed requirements of bluefin tuna will likely incentivize further research and the eventual development of effective formulated feeds that help increase feed efficiency and reduce farm effluent and harm to the local environment (de la Gándara et al. 2015).

## Bluefin as a Luxury Good

Bluefin tuna meat has increased in value from near worthlessness in the 1950s to being one of the most expensive animal-derived foods today (Ellis 2008). Most bluefin reaches end consumers as sushi or sashimi, and it is largely seen as a luxury good (Longo and Clausen 2011, Guillotreau et al. 2016), meaning the demand for bluefin increases as one's income rises. Bluefin is likely a specific type of luxury good, called a Veblen good, where consumption of bluefin is used primarily as a tool to display wealth or status to one's peers (Eswaran and Oxoby 2008). While status as a Veblen good has incentivized overfishing and poor management of aquaculture in the past (e.g., Safina and Klinger 2008), it also presents a unique opportunity for the bluefin industry.

Brands in other industries have used their Veblen-like status to combine luxury products with innovations to inspire change in a whole sector (Kapferer 2010). For example, Tiffany & Company, an American jewelry company, uses its position as a luxury brand to improve supply chain traceability for precious metals and gemstones (Doval et al. 2013). Valentino, an Italian fashion brand, has moved to eliminate the use of hazardous chemicals in its manufacturing, which is a chronic problem in the textile industry (King and Hower 2013). Tesla Motors, an American automobile company, is an example of a luxury brand that has developed a more sustainable product without compromising the quality associated with luxury brands (Kapferer 2010). These brands are helping to make both large and incremental improvements in their individual sectors through technology development, shifts in corporate social responsibility, and changes in consumer expectations.

Given the position of bluefin tuna as a Veblen good, there are substantial opportunities for bluefin tuna farming companies to similarly leverage their position to help improve the larger seafood industry. Innovations in production could easily inspire or be transferred to other aquaculture and seafood enterprises to make similar improvements, including development and adoption of novel technologies for traceability, operational best-management practices, and resource-efficient feeds for marine carnivores. Further, these types of collaborations and technology transfers within and between individual seafood sectors are becoming increasingly common, reducing historical barriers to precompetitive collaboration (Österblom et al. 2017).

## Conclusion

Bluefin tuna are truly distinct among aquaculture species for their unique physiology and status as a high-priced global luxury commodity. Many species in aquaculture are chosen for traits that make them resource-efficient to produce or otherwise amenable to captive conditions. Bluefin tuna farming, on the other hand, emerged as highly environmentally and resource-intensive capture-based operations, sustained by strong demand and high profit margins. As with other aquaculture species, decades of scientific research and industry investment into bluefin tuna farming technology are reaping benefits. Significant advances have been made in the areas of larval culture, nutrition, and techniques for closed-life-cycle production, incrementally reducing the footprint of the industry. But the industry continues to face a variety of environmental sustainability and resource-efficiency challenges that far exceed those of other types of aquaculture. The future of bluefin tuna aquaculture will be defined by the species' unique physiology and its status as a luxury good, which serves as an incentive for continued innovation and a guarantee that bluefin will continue to play a unique role in the global seafood industry.

## Acknowledgments

This work is a deliverable of the project Green Growth Based on Marine Resources: Ecological and Socio-Economic Constraints (GreenMAR), which is funded by Nordforsk.

### REFERENCES

Angilletta, M.J. 2009. *Thermal Adaptation: A Theoretical and Empirical Synthesis*. Oxford, UK: Oxford University Press.

Arechavala-Lopez, P., J.A. Borg, T. Šegvić-Bubić, P. Tomassetti, A. Özgül, and P. Sanchez-Jerez. 2015. Aggregations of wild Atlantic Bluefin Tuna (*Thunnus thynnus* L.) at Mediterranean offshore fish farm sites: Environmental and management considerations. *Fisheries Research* 164:178–84.

Arthur, R.I., K. Lorenzen, P. Homekingkeo, K. Sidavong, B. Sengvilaikham, and C.J. Garaway. 2010. Assessing impacts of introduced aquaculture species on native fish communities: Nile tilapia and major carps in SE Asian freshwaters. *Aquaculture* 299:81–88.

Asche, F., A. Oglend, and S. Tveteras. 2013. Regime shifts in the fish meal/soybean meal price ratio. *Journal of Agricultural Economics* 64:97–111.

Atkins, C.A. 2009. A management tool and strategy for aquaculture development in offshore United States coastal waters. USDA SBIR Phase 1 Project Final Technical Report: 1–35.

Bax, N., A. Williamson, M. Aguero, E. Gonzalez, and W. Geeves. 2003. Marine invasive alien species: A threat to global biodiversity. *Marine Policy* 27:313–23.

Benetti, D.D., G.J. Partridge, and J. Stieglitz. 2015. Overview on status and technological advances in tuna aquaculture around the world. In *Advances in Tuna Aquaculture: From Hatchery to Market,* edited by D. Benetti, G. Partridge, and A. Buentello, 1–19. Amsterdam, Netherlands: Elsevier.

Bert, T.M. 2007. Environmentally responsible aquaculture—a work in progress. In *Ecological and Genetic Implications of Aquaculture Activities,* edited by T. M. Bert, 1–31. Dordrecht, Netherlands: Springer.

Block, B.A., and E.D. Stevens, eds. 2001. *Tuna: Physiology, Ecology, and Evolution.* San Diego, CA: Academic Press.

Bostock, J., B. McAndrew, R. Richards, K. Jauncey, T. Telfer, K. Lorenzen, D. Little, et al. 2010. Aquaculture: Global status and trends. *Philosophical Transactions of the Royal Society B: Biological Sciences* 365:2897–2912.

Brainerd, E.L., and L.A. Ferry-Graham. 2005. Mechanics of respiratory pumps. In *Fish Biomechanics,* edited by R. Shadwick and G. V. Lauder, 1–28. Cambridge, MA: Academic Press.

Brill, R.W. 1996. Selective advantages conferred by the high performance physiology of tunas, billfishes, and dolphin fish. *Comparative Biochemistry and Physiology Part A: Physiology* 113:3–15.

Bubner, E., J. Farley, P. Thomas, T. Bolton, and A. Elizur. 2012. Assessment of reproductive maturation of southern bluefin tuna (*Thunnus maccoyii*) in captivity. *Aquaculture* 364–365:82–95.

Buentello, A., M. Seoka, K. Kato, and G.J. Partridge. 2015. Tuna farming in Japan and Mexico. In *Advances in Tuna Aquaculture: From Hatchery to Market,* edited by D. Benetti, G. Partridge, and A. Buentello, 189–215. Amsterdam, Netherlands: Elsevier.

Chen, B.N., W. Hutchinson, and C. Foster. 2015. Southern bluefin tuna captive breeding in Australia. In *Advances in Tuna Aquaculture: From Hatchery to Market,* edited by D. Benetti, G. Partridge, and A. Buentello, 233–52. Amsterdam, Netherlands: Elsevier.

Corriero, A., S. Karakulak, N. Santamaria, M. Deflorio, D. Spedicato, P. Addis, S. Desantis, et al. 2005. Size and age at sexual maturity of female bluefin tuna (*Thunnus thynnus* L. 1758) from the Mediterranean Sea. *Journal of Applied Ichthyology* 21: 483–86.

Costello, C., D. Ovando, T. Clavelle, C.K. Strauss, R. Hilborn, M.C. Melnychuk, T.A. Branch, et al. 2016. Global fishery prospects under contrasting management regimes. *Proceedings of the National Academy of Sciences* 113:5125–29.

de la Gándara, F., A. Ortega, and A. Buentello. 2015. Tuna aquaculture in Europe. In *Advances in Tuna Aquaculture: From Hatchery to Market,* edited by D. Benetti, G. Partridge, and A. Buentello, 115–57. Amsterdam, Netherlands: Elsevier.

Dempster, T., P. Sanchez-Jerez, J. Bayle-Sempere, and M. Kingsford. 2004. Extensive aggregations of wild fish at coastal sea-cage fish farms. *Hydrobiologia* 525:245–48.

Diana, J.S. 2009. Aquaculture production and biodiversity conservation. *Bioscience* 59:27–38.

Doval, J., E.P. Singh, and G.S. Batra. 2013. Green buzz in luxury brands. *Review of Management* 3:5–14.

Duarte, C.M., N. Marbá, and M. Holmer. 2007. Rapid domestication of marine species. *Science* 316:382.

Džoić, T., G. Beg Paklar, B. Grbec, S. Ivatek-Šahdan, B. Zorica, T. Šegvić-Bubić, V. Čikeš Keč, et al. 2017. Spillover of the Atlantic bluefin tuna offspring from cages in the Adriatic Sea: A multidisciplinary approach and assessment. *PLOS ONE* 12:e0188956.

Ellis, D., and I. Kiessling. 2015. Ranching of southern bluefin tuna in Australia. In *Advances in Tuna Aquaculture: From Hatchery to Market,* edited by D. Benetti, G. Partridge, and A. Buentello, 217–32. Amsterdam, Netherlands: Elsevier.

Ellis, R. 2008. *Tuna: A Love Story.* New York: Vintage.

Estess, E.E., D.M. Coffey, T. Shimose, A.C. Seitz, L. Rodriguez, A. Norton, B. Block, and C. Farwell. 2014. Bioenergetics of captive Pacific bluefin tuna (*Thunnus orientalis*). *Aquaculture* 434:137–44.

Eswaran, M., and R. Oxoby. 2008. Intertemporal discounting with Veblen preferences: Theory and evidence. Working Paper, University of British Columbia and University of Calgary.

FAO (Food and Agriculture Organization of the United Nations). 2016a. Fishery and Aquaculture Statistics. *FishStatJ.* Updated July 21, 2016. www.fao.org/fishery/statistics/software/fishstatj/en.

FAO (Food and Agriculture Organization of the United Nations). 2016b. The state of world fisheries and aquaculture (SOFIA) 2016. www.fao.org/3/a-i5555e.pdf.

FAO (Food and Agriculture Organization of the United Nations). 2017. Aquaculture fact sheets. www.fao.org/fishery/culturedspecies/search/en.

Forrestal, F., M. Coll, D.J. Die, and V. Christensen. 2012. Ecosystem effects of bluefin tuna *Thunnus thynnus thynnus* aquaculture in the NW Mediterranean Sea. *Marine Ecology Progress Series* 456:215–31.

Froehlich, H.E., R.R. Gentry, M.B. Rust, D. Grimm, and B.S. Halpern. 2017. Public perceptions of aquaculture: Evaluating spatiotemporal patterns of sentiment around the world. *PLOS ONE* 12:e0169281.

Gardner, L.D., N. Jayasundara, P.C. Castilho, and B. Block. 2012. Microarray gene expression profiles from mature gonad tissues of Atlantic bluefin tuna, *Thunnus thynnus* in the Gulf of Mexico. BMC *Genomics* 13:530.

Gjedrem, T., ed. 2005. *Selection and Breeding Programs in Aquaculture.* New York: Springer.

Gjedrem, T., and M. Baranski. 2009. Initiating breeding programs. In *Selective Breeding in Aquaculture: An Introduction,* edited by T. Gjedrem, 63–85. New York: Springer.

Glencross, B.D., M. Booth, and G.L. Allan. 2007. A feed is only as good as its ingredients—a review of ingredient evaluation strategies for aquaculture feeds. *Aquaculture Nutrition* 13:17–34.

Guillotreau, P., D. Squires, J. Sun, and G.A. Compeán. 2016. Local, regional and global markets: What drives the tuna fisheries? *Reviews in Fish Biology and Fisheries* 27:909–29.

Halver, J.E., and R.W. Hardy, eds. 2002. *Fish Nutrition.* Cambridge, MA: Academic Press.

Hayward, C.J., H.M. Aiken, and B.F. Nowak. 2007. Metazoan parasites on gills of southern bluefin tuna (*Thunnus maccoyii*) do not rapidly proliferate after transfer to sea cages. *Aquaculture* 262:10–16.

Higuchi, K., Y. Tanaka, T. Eba, A. Nishi, K. Kumon, H. Nikaido, and S. Shiozawa. 2014. Causes of heavy mortality of hatchery-reared Pacific bluefin tuna Thunnus orientalis juveniles in sea cages. *Aquaculture* 424–25:140–45.

Holmer, M. 2010. Environmental issues of fish farming in offshore waters: Perspectives, concerns and research needs. *Aquaculture Environment Interactions* 1:57–70.

Ishibashi, Y., T. Honryo, K. Saida, A. Hagiwara, S. Miyashita, Y. Sawada, T. Okada, and M. Kurata. 2009. Artificial lighting prevents high night-time mortality of juvenile Pacific bluefin tuna, *Thunnus orientalis,* caused by poor scotopic vision. *Aquaculture* 293:157–63.

Kapferer, J.-N. 2010. All that glitters is not green: The challenge of sustainable luxury. *European Business Review,* November–December: 40–45.

Karakulak, F.S., B. Ozturk, and T. Yıldız. 2015. From ocean to farm: Capture-based aquaculture of bluefin tuna in the eastern Mediterranean Sea. In *Advances in Tuna Aquaculture: From Hatchery to Market,* edited by D. Benetti, G. Partridge, and A. Buentello, 59–76. Amsterdam, Netherlands: Elsevier.

Katavic, I., V. Franicevic, and V. Ticina. 2003. Bluefin tuna (*Thunnus thynnus* L.) farming on the Croatian coast of the Adriatic Sea: Present stage and future plans. *Cahiers Options Méditerranéennes (CIHEAM)* 101–6.

King, B., and M. Hower. 2013. Valentino trumps Louis Vuitton in Greenpeace fashion rankings. Available at Sustainable Brands, www.sustainablebrands.com/news_and_views/articles/valentino-trumps-louis-vuitton-greenpeace-fashion-rankings.

Kirchhoff, N.T., K.M. Rough, and B.F. Nowak. 2011. Moving cages further offshore: Effects on southern bluefin tuna, T. maccoyii, parasites, health and performance. *PLOS ONE* 6:e23705.

Klinger, D.H., and R. Naylor. 2012. Searching for solutions in aquaculture: Charting a sustainable course. *Annual Review of Environment and Resources* 37:247–76.

Klinger, D.H., M. Turnipseed, J.L. Anderson, F. Asche, L.B. Crowder, A.G. Guttormsen, B.S. Halpern, et al. 2013. Moving beyond the fished or farmed dichotomy. *Marine Policy* 38:369–74.

Knapp, G., and M.C. Rubino. 2016. The political economics of marine aquaculture in the United States. *Reviews in Fisheries Science and Aquaculture* 24:213–29.

Kružić, P., V. Vojvodić, and E. Bura-Nakić. 2014. Inshore capture-based tuna aquaculture impact on Posidonia oceanica meadows in the eastern part of the Adriatic Sea. *Marine Pollution Bulletin* 86:174–85.

Kurata, M., Y. Ishibashi, K. Takii, H. Kumai, S. Miyashita, and Y. Sawada. 2014. Influence of initial swimbladder inflation failure on survival of Pacific bluefin tuna, *Thunnus orientalis* (Temminck and Schlegel), larvae. *Aquaculture Research* 45:882–92.

Lafferty, K.D., C.D. Harvell, J.M. Conrad, C.S. Friedman, M.L. Kent, A.M. Kuris, E.N. Powell, D. Rondeau, and S.M. Saksida. 2015. Infectious diseases affect marine fisheries and aquaculture economics. *Annual Review of Marine Science* 7:471–96.

Le François, N.R., M. Jobling, and C. Carter, eds. 2010. *Finfish Aquaculture Diversification.* Oxfordshire, UK: Centre for Agriculture and Biosciences International (CABI).

Lefevre, S., T. Wang, A. Jensen, N.V. Cong, D.T.T. Huong, N.T. Phuong, and M. Bayley. 2014. Air-breathing fishes in aquaculture: What can we learn from physiology? *Journal of Fish Biology* 84:705–31.

Longo, S.B., and R. Clausen. 2011. The tragedy of the commodity. *Organization and Environment* 24:312–28.

Magnuson, J.J. 1978. Locomotion by scombrid fishes: Hydromechanics, morphology, and behavior. In *Fish Physiology*, edited by W. S. Hoar and D. J. Randall, 239–313. Cambridge, MA: Academic Press.

Mangion, M., J.A. Borg, P.J. Schembri, and P. Sanchez-Jerez. 2017. Assessment of benthic biological indicators for evaluating the environmental impact of tuna farming. *Aquaculture Research* 48:5797–5811.

Mangion, M., J.A. Borg, and P. Sanchez-Jerez. 2018. Differences in magnitude and spatial extent of impact of tuna farming on benthic macroinvertebrate assemblages. *Regional Studies in Marine Science* 18:197–207.

Masuma, S., S. Miyashita, H. Yamamoto, and H. Kumai. 2008. Status of bluefin tuna farming, broodstock management, breeding and fingerling production in Japan. *Reviews in Fisheries Science* 16:385–90.

Masuma, S., T. Takebe, and Y. Sakakura. 2011. A review of the broodstock management and larviculture of the Pacific northern bluefin tuna in Japan. *Aquaculture* 315:2–8.

Mazoyer, M., and L. Roudart. 2006. *A History of World Agriculture: From the Neolithic Age to the Current Crisis.* New York: Monthly Review Press.

Metian, M., S. Pouil, A. Boustany, and M. Troell. 2014. Farming of bluefin tuna—reconsidering global estimates and sustainability concerns. *Reviews in Fisheries Science and Aquaculture* 22:184–92.

Mladineo, I., J. Žilić, and M. ČAnković. 2008. Health survey of Atlantic bluefin tuna, *Thunnus thynnus* (Linnaeus, 1758), reared in Adriatic cages from 2003 to 2006. *Journal of the World Aquaculture Society* 39:281–89.

Molnar, J.L., R.L. Gamboa, C. Revenga, and M.D. Spalding. 2008. Assessing the global threat of invasive species to marine biodiversity. *Frontiers in Ecology and the Environment* 6:485–92.

Mourente, G., and D.R. Tocher. 2009. Tuna nutrition and feeds: Current status and future perspectives. *Reviews in Fisheries Science* 17:373–90.

Mylonas, C.C., F. de La Gándara, A. Corriero, and A.B. Ríos. 2010. Atlantic bluefin tuna (*Thunnus thynnus*) farming and fattening in the Mediterranean Sea. *Reviews in Fisheries Science* 18:266–80.

Nakagawa, Y., M. Kurata, Y. Sawada, W. Sakamoto, and S. Miyashita. 2011. Enhancement of survival rate of Pacific bluefin tuna (*Thunnus orientalis*) larvae by aeration control in rearing tank. *Aquatic Living Resources* 24:403–10.

Nash, C. 2010. *The History of Aquaculture.* Hoboken, NJ: John Wiley.

Naylor, R.L., R.J. Goldburg, J.H. Primavera, N. Kautsky, M.C. Beveridge, J. Clay, C. Folke, J. Lubchenco, H. Mooney, and M. Troell. 2000. Effect of aquaculture on world fish supplies. *Nature* 405:1017–24.

Naylor, R.L., S.L. Williams, and D.R. Strong. 2001. Aquaculture—a gateway for exotic species. *Science* 294:1655–56.

Naylor, R.L., R.W. Hardy, D.P. Bureau, A. Chiu, M. Elliott, A.P. Farrell, I. Forster, et al. 2009. Feeding aquaculture in an era of finite resources. *Proceedings of the National Academy of Sciences* 106:15103–10.

Österblom, H., J.-B. Jouffray, C. Folke, and J. Rockström. 2017. Emergence of a global science-business initiative for ocean stewardship. *Proceedings of the National Academy of Sciences* 114:9038.

Ottolenghi, F. 2008. Capture-based aquaculture of bluefin tuna. *FAO Fisheries Technical Paper* 508:169–82.

Primavera, J. 2006. Overcoming the impacts of aquaculture on the coastal zone. *Ocean and Coastal Management* 49:531–45.

Safina, C., and D.H. Klinger. 2008. Collapse of bluefin tuna in the western Atlantic. *Conservation Biology* 22 (2): 243–46.

Sanchez-Jerez, P., D. Fernandez-Jover, J. Bayle-Sempere, C. Valle, T. Dempster, F. Tuya, and F. Juanes. 2008. Interactions between bluefish *Pomatomus saltatrix* (L.) and coastal sea-cage farms in the Mediterranean Sea. *Aquaculture* 282:61–67.

Sarà, G., M.L. Martire, M. Sanfilippo, G. Pulicanò, G. Cortese, A. Mazzola, A. Manganaro, and A. Pusceddu. 2011. Impacts of marine aquaculture at large spatial scales: Evidences from N and P catchment loading and phytoplankton biomass. *Marine Environmental Research* 71:317–24.

Sawada, Y., and Y. Agawa. 2015. Genetics in tuna aquaculture. In *Advances in Tuna Aquaculture: From Hatchery to Market*, edited by D. Benetti, G. Partridge, and A. Buentello, 323–32. Amsterdam, Netherlands: Elsevier.

Sawada, Y., T. Okada, S. Miyashita, O. Murata, and H. Kumai. 2005. Completion of the Pacific bluefin tuna *Thunnus orientalis* (Temminck et Schlegel) life cycle. *Aquaculture Research* 36:413–21.

Sepúlveda, M., and D. Oliva. 2005. Interactions between South American sea lions Otaria flavescens (Shaw) and salmon farms in southern Chile. *Aquaculture Research* 36: 1062–68.

Shamshak, G.L. 2011. Economic evaluation of capture-based bluefin tuna aquaculture on the US East Coast. *Marine Resource Economics* 26:309–28.

Stagličić, N., T. Šegvić-Bubić, P. Ugarković, I. Talijančić, I. Žužul, V. Tičina, and L. Grubišić. 2017. Ecological role of bluefin tuna (*Thunnus thynnus*) fish farms for associated wild fish assemblages in the Mediterranean Sea. *Marine Environmental Research* 132:79–93.

Stevens, E.D., J.W. Kanwisher, and F.G. Carey. 2000. Muscle temperature in free-swimming giant Atlantic bluefin tuna (*Thunnus thynnus* L.). *Journal of Thermal Biology* 25:419–23.

Stickney, R.R. 1994. *Principles of Aquaculture*. New York: John Wiley.

Stickney, R.R., and J.P. McVey. 2002. Responsible Marine Aquaculture. Oxon, UK: Centre for Agriculture and Biosciences International (CABI).

Stickney, R.R., and G.D. Treece. 2012. History of aquaculture. In *Aquaculture Production Systems*, edited by J. H. Tidwell, 15–50. Hoboken, NJ: Wiley-Blackwell.

Sumaila, U.R., W. Cheung, A. Dyck, K. Gueye, L. Huang, V. Lam, D. Pauly, et al. 2012. Benefits of rebuilding global marine fisheries outweigh costs. *PLOS ONE* 7: e40542.

Tacon, A.G.J., and M. Metian. 2008. Global overview on the use of fish meal and fish oil in industrially compounded aquafeeds: Trends and future prospects. *Aquaculture* 285:146–58.

Tacon, A.G., M.R. Hasan, and M. Metian. 2011. Demand and supply of feed ingredients for farmed fish and crustaceans: Trends and prospects. *FAO Fisheries and Aquaculture Technical Paper* 546:1–87.

Tacon, A.G.J., and M. Metian. 2015. Feed matters: Satisfying the feed demand of aquaculture. *Reviews in Fisheries Science and Aquaculture* 23:1–10.

Teletchea, F., and P. Fontaine. 2014. Levels of domestication in fish: Implications for the sustainable future of aquaculture. *Fish and Fisheries* 15:181–95.

Tidwell, J.H., ed. 2012. *Aquaculture Production Systems.* Hoboken, NJ: Wiley-Blackwell.

Tilman, D. 1999. Global environmental impacts of agricultural expansion: The need for sustainable and efficient practices. *Proceedings of the National Academy of Sciences* 96: 5995–6000.

Turnbull, J., A. Bell, C. Adams, J. Bron, and F. Huntingford. 2005. Stocking density and welfare of cage farmed Atlantic salmon: Application of a multivariate analysis. *Aquaculture* 243:121–32.

Vita, R., A. Marin, B. Jiménez-Brinquis, A. Cesar, L. Marín-Guirao, and M. Borredat. 2004. Aquaculture of bluefin tuna in the Mediterranean: Evaluation of organic particulate wastes. *Aquaculture Research* 35:1384–87.

Watts, M., B. Munday, and C. Burke. 2002. Investigation of humoral immune factors from selected groups of southern bluefin tuna, *Thunnus maccoyii* (Castelnau): Implications for aquaculture. *Journal of Fish Diseases* 25:191–200.

Wilson, R.P. 1994. Utilization of dietary carbohydrate by fish. *Aquaculture* 124:67–80.

World Bank. 2013. Fish to 2030: Prospect for fisheries and aquaculture. Working Paper 83177, World Bank. http://documents.worldbank.org/curated/en/458631468152376668/pdf/831770WP0P11260ES003000Fish0to02030.pdf.

Zertuche-González, J.A., O. Sosa-Nishizaki, J.G. Vaca Rodriguez, R. del Moral Simanek, C. Yarish, and B.A. Costa-Pierce. 2008. Marine science assessment of capture-based tuna (*Thunnus orientalis*) aquaculture in the Ensenada Region of Northern Baja California, México. Final Report of the Binational Scientific Team to the Packard Foundation. http://digitalcommons.uconn.edu/ecostam_pubs/1/.

habitat: age-specific, 200–202; changes in, 131, 133–35, 175, 223; models of, 96, 108, 272–88; nowcasts for, 272–73, 276–79, 281–87; prediction of, 96, 226, 272–88; preferences and use of, 100, 103–4, 185, 275–76, 284; spawning, 102–3; and tracers, 56, 166, 172
half-sibling pairs (HSPs), 234–36, 264
harvest-control rule (HCR), 34, 36, 61, 255–56, 263
harvest rate, historical Southern bluefin tuna, 212. *See also* catch; catchability
Hemingway, Ernest, 6
highly migratory species, 3, 59, 184, 186, 202, 248, 264, 299, 323
history, 3–6, 16, 60. *See also under* catch-per-unit-effort; *under* spawning stock biomass

ICCAT (International Commission for the Conservation of Atlantic Tunas): bluefin catch document, 27; creation of, 6, 8; GBYP, 5, 31–32, 36, 58, 60, 73; on mixing, 14–15, 29, 45, 67–68, 73, 94; recommendation objections, 33–34; reputation of, 13, 33–34; Standing Committee on Research and Statistics (SCRS) recommendations to, 8–13, 15, 20, 26, 35; stock boundary, 4, 45, 50–51, 55, 59, 67, 95, 104; stock rebuilding and recovery plan, 12, 20–21, 23, 284–85, 299; TAC levels, 7, 9–13, 28, 94; transparency, 34. *See also* management strategy evaluation; SCRS
ICCAT recommendations: Rec. 02-07, 11; Rec. 06-05, 12, 24; Rec. 06-06, 11; Rec. 08-04, 11; Rec. 13-09, 24; Rec. 14-04, 26, 33, 28; Rec. 14-05, 26, 28; Rec. 16-21, 36; Rec. 81-01, 13
index(es): abundance, 17–19, 21, 25–26, 32, 237; aerial survey, 214, 254–55, 262–63; based on POP and HSP data, 268; catch-per-unit-effort, 103, 120, 214, 237, 251–52, 255, 260, 263; combined longline abundance, 214; decline in abundance, 226; disparity between catch-per-unit-effort, 283; fishery-dependent abundance, 25, 283; fishery-independent abundance, 17, 19, 21, 227, 237–38; fishery-independent recruitment, 214, 278; Gulf of Mexico larval, 278; impact of recruitment-monitoring, 263–64; juvenile, 21, 228; longline, 25–26, 214, 251–52, 263; management procedure input, 261; recruitment, 231, 237, 251; relative abundance, 222, 231, 251, 254; spotting, 222; uncertainty in abundance, 226
Indian Ocean: chemical tracer studies in, 167–68; feeding studies in, 197; migration into, 222–23; spawning ground catch, 217; Taiwanese Southern bluefin tuna catch in, 216–17
Indian Ocean Tuna Commission (IOTC), 36
interactions, unwanted, 270
International Commission for the Conservation of Atlantic Tunas (*see* ICCAT); Atlantic-wide research program for bluefin tuna (*see* GBYP)
invasive species, 323
isothermal layer depth, 199–200
IUU (illegal, unreported, and unregulated) activity, 26–27, 185

Japan, Sea of, 134–36, 138–41, 147–48, 151–52, 156, 158, 168–70, 173, 176–77, 184
joint-venture vessels, 215
juveniles: abundance estimates of, 60, 177, 222, 228–29, 231–38; acoustic survey of, 229; aerial surveys of, 32, 214, 222, 229–30, 254; bait boat index of, 21; bycatch of, 282; catch rates of, 7, 17, 220; distribution, 147, 152, 219–22, 226–27; dives of, 222; energy intake, 137, 185, 197–200; forecasting habitat of, 273–76, 284; growth of (in cages), 307–8, 310; growth rates, 216–17, 220, 237; as indicative of spawning ground, 134–36, 140; management of, 228; migration of, 54, 74–76, 86–87, 95, 137–38, 142, 169, 176, 222–23; mixing of, 54, 59–60; movement of, 185; recruitment of, 228; ship-based survey of, 226; spatial dynamics of, 216; weaning of, 305–6

Kindai University, 316, 323
Kobe matrices, 21, 24, 34–35
Kuroshio Current, 147, 154

Kuroshio Extension, 152
Kuroshio-Oyashio Transition Region (KOTR), 147, 152, 155, 157

larvae: physiology of, 138; as prey, 136; rearing of, 299–305, 310, 315–16; separation between young-of-the-year and, 56; spatial dynamics of Southern bluefin tuna, 221, 227; and surveys, 13, 18–19, 218, 278–81, 284
Leeuwin Current, 221
length composition: of catch, 252; differences in, 141
length frequency data, 216–17, 219
life cycle: of Balearic-born bluefin tuna, 76; closing in captivity, 308, 310, 315–16, 323; of Pacific bluefin tuna, 142
life history: Atlantic bluefin tuna, 17, 67–68, 78, 95–96; and challenges for domestication, 315–16; markers for, 56–58, 166, 170; Pacific bluefin tuna, 131–33, 142, 165, 185, 200–201; Southern bluefin tuna, 215, 217, 281
light-based geolocation, 99, 220
longevity: maximum, 133; relationship to skipped spawning, 85
longline fishery/fleet, 10, 32, 141, 184, 226; catches, 7, 213; catch-per-unit-effort, 18–19, 248–52, 254–55, 263; Indonesian, 217–19, 232, 282; Japanese, 7–9, 17, 25–26, 138–39, 212, 214–15, 219, 251–52; Taiwanese, 137, 139, 201; unreported catches, 265
lunar cycle, 222
luxury goods, 326–28; technology in, 325, 327–28

macronutrients in aquaculture, 319–20
Maine, Gulf of, 50, 110, 283
management: and advice, 10, 15–16, 249, 252, 262, 264–65; and decisions, 34, 178, 281; dynamic ocean, 271, 276–77, 282–85; flexible, 272; history, 5–6; ICCAT creation for international, 8; model-based, 250; science-based, 67–68, 236, 238; spatial, 202, 226, 238, 270–71, 276–77, 284; stock, 147–48, 316. *See also* CCSBT; fisheries managers; GBYP; ICCAT; regional fisheries management organization; spatial management; *under* population; *under* subpopulation
management approaches: conventional quota-based, 270; for drift gill net fishery, 282; dynamic spatial, 276–77, 284
management procedure (MP), 34–37, 124, 211, 214–15, 218, 226–31, 236–38, 249–66; candidate, 35–36, 231, 250–60; empirical, 250, 254, 256, 264; simulation-tested, 260, 264; testing, 215, 270
management strategy evaluation (MSE): 32, 36, 61, 123–125, 231, 252–253, 265
marker: anthropogenic, 166, 172; chemical (in otoliths), 46–48, 55, 59, 166; chemical (in vertebrae), 47; microsatellite, 80–81. *See also* stable isotopes
Markov-Chain Monte Carlo (MCMC) simulation, 121
mark-recapture, 32, 109, 111; Bayesian model, 186–200; close-kin (CKMR), 219, 231–38; gene tagging, 229–31, 237–38; multicohort, 252. *See also* tag-recapture
MAST (multistock age-structured tag-integrated stock-assessment model), 15, 118–25
*Mattanza* (Maggio), 3
maturity: 50%, 83, 138, 218, 251; 95%, 158; age at first, 82–83, 87; from analysis of sex hormone profiles, 83; in aquaculture, 315; from conventional tagging, 169; effects on seasonal migration patterns, 137; from electronic tags, 95; gonadal, 134; from histological study, 83, 220; interconnectedness of, 131; length at 50%, 219; markers in ovaries, 220; ogive, 122, 218; schedules of, 30, 67, 75, 95–96, 169, 220, 238; from tracers, 58, 173
maximum sustainable yield (MSY), 11–12, 20, 22–25, 27–28, 119, 123, 201, 253
Mediterranean Sea (MED): adults in, 70–74; aquaculture in, 9, 29, 299–310, 313, 316; bluefin tuna catch, 10–11, 18–20, 27; genetic structure, 81; historical catches in, 5–8, 299; illegal, unreported, and unregulated fishing in,

26; juveniles in, 74–76; management, 11–13, 16, 26, 33, 82; migrations to/from, 54, 57–58, 70, 73, 77, 82; natal homing to, 55–56, 58, 70; as origin, 67–68, 71, 74–76, 101; population structure hypotheses, 68–70; spawning, 29, 45, 54–56, 58, 70–76, 82–86; spawning stock biomass estimates, 120; stock composition, 51–53, 68–70, 72, 77–80, 118; stock mixing, 50–54, 59, 68–70, 117; stock status, 24–26; subpopulation structure, potential, 29, 47, 67–87, 236; tag tracks, 97, 99, 107
Mexico, Baja, 184, 187, 197, 200
Mexico, Gulf of (GOM): catch-per-unit-effort fishery indicators, 17–19; chemical tracers in, 167; fishery, 10; larval index, 278; larval survey, 18–19; in MAST, 118–20; migrations/movements, 97–109, 111; natal homing to, 55–56; population structure, 80–81; residents, 57; spawning, 70, 84, 279, 281; spawning grounds, 29, 45, 48, 94, 283; stock composition, 51, 53–54
microsatellite loci, 232
Mid-Atlantic Bight, 110, 280
migration, 14, 67, 69; acoustic tracking of, 87, 96, 103, 105, 109–110, 149, 226; age-specific, 56–58, 185; annual, 98; aquaculture effects on, 323; bioenergetics of, 202; in the California Current, 286; change in, 7; chemical tracers of transpacific, 165–78; cyclical, 137, 222; dynamism of, 148, 155; ecology, 46, 61–62, 147; effect on population dynamics, 45; fixed boundaries of, 59; history, 165–66, 173; impacts of, 46; of individuals, 45, 49, 70, 222; juvenile, 74–75, 138, 150, 219; long-term, 75; Mediterranean Atlantic bluefin tuna, 49, 54, 56, 67–87, 103; ontogenetic changes in, 96, 131, 137, 271; Pacific bluefin tuna, 170; patterns in, 45, 56, 70; and predator energy intake, 202; range of, 184, 222; restricted, 77; revealed by electronic tagging, 54, 94–113, 142, 184–85, 200–202; routes of, 53, 76; seasonal/spawning, 59, 69, 71–73, 86, 82, 99, 100, 133–38, 156–58, 217, 224, 238, 271; skipped spawning, 84; and surveys, 265; and tracers, 170, 172; transatlantic, 29, 49, 57, 75, 117, 118; transpacific, 137, 147, 152–55, 158, 165–78. *See also* movement
migratory contingent, 70, 155
mixed layer, 71, 221, 282
mixed stock. *See* mixing
mixed-stock analysis. *See* stock composition analysis
mixed-stock spatial dynamics, 96, 117–25, 200, 220–27
mixing, 45–62, 69–70, 77–78, 80; in CKMR, 235; effects of, on stock assessment, 28–30, 32, 45–46, 58–59, 95–96, 103–4, 117–18; as modeling challenge, 117, 123; in Pacific bluefin tuna, 142; potential importance of, 13–16, 21, 28, 67, 96, 98; in Southern bluefin tuna, 227
modal progression analysis, 132
models: dynamic, 228, 255; integrated growth, 217–18; mark-recapture, 186–200; movement, 14, 187–96, 200; multistock age-structured tag-integrated, 15, 118–25; plausible, 265; prey, 286–87; spatial, 117–25, 200, 226. *See also* operating models; Predictive Ocean Atmosphere Model for Australia; simulation models; statistical models; *under* habitat
monitoring: electronic, 226; genetic-based, 229; long-term, 109–11, 149; of illegal, unreported, and unregulated activity, 26–27, 185; programs for, 211, 214–16, 220, 229, 237–38; recruitment, 227–31, 263–64; series, 214, 251, 253–54, 264; video, 277
Monterey Bay Aquarium, 96
mortality rates: in aquaculture, 304, 306–7, 316; fishing, 12, 15–16, 24, 27, 189, 193, 195, 200, 202, 248–49; natural, 19, 21, 30, 185, 188, 200
movement, 32, 45, 96, 117, 138, 165, 211, 271; age-specific, 56, 58, 87, 118, 221–29; climate impacts on, 111, 112; determined by chemical markers, 46, 61–62, 166–67, 169, 177; diffusion model, 14; dynamics of, 117, 124–25, 169; and ecology, 46; estimates of, with Bayesian mark-recapture model, 187–196, 200; fidelity of, 98; in MAST,

movement (*cont.*)
119–25; and matrices, 59, 61, 125; parameterized by state-space methods, 125; parameterized by telemetry data, 125; patterns of, in Pacific bluefin tuna, 152, 155–56; rates of, 118, 185, 201; revealed through tagging, 98, 104, 107, 158, 165, 176, 221, 225–26; seasonal, 61, 200; spawning, 70, 73, 83–84, 99, 102–3, 108, 219, 224; transatlantic, 46, 49–50, 54–55, 57–58, 117; transpacific, 176, 178; vertical, 70, 73, 83, 101, 108–9, 150, 156, 158. *See also* migration
MSE. *See* management strategy evaluation
multistock age-structured tag-integrated stock-assessment model, 15, 118–25

natal homing, 55–56, 58, 77, 86
natal origin: of Atlantic bluefin tuna in Gulf of St. Lawrence, 52–54; of Atlantic bluefin tuna off North Carolina, 100–101; of Atlantic bluefin tuna targeted by high seas fishery, 54–55; effects of, on management, 58–61, 67–68; indicated by chemical markers, 31, 46–48, 50; of MED Atlantic bluefin tuna, 70–76, 86–87; in modeling, 32, 118–21, 124; of Pacific bluefin tuna, 177. *See also* nursery
National Research Council (NRC), US, 14, 45
natural tracer, 61–62. *See also* marker; otolith
North Atlantic Ocean, central (CNAO), 9, 51, 53, 55, 59
North Carolina, 52–53, 74, 97–101, 104, 111
Northeastern Atlantic Ocean (NEAO), 21, 52, 54–55, 56–58, 76
Northwestern Atlantic Ocean (NWAO), 19, 49–54, 56–58, 61
nursery: age-specific ingress and egress from, 46; indicated by otoliths, 46–48, 55–60, 166; indicated by postlarval abundance, 168; phase of hatchery production, 316

observation uncertainty, 252
ocean tracking network (OTN), 109–10
ongrowing, 306–8
ontogenetic habitat: shift in, 57, 131, 133–35; utilization of, 100
ontogenetic migrations, 96, 178, 223, 271
ontogeny: and archival tags, 98; changes over, 131
oocytes, 83, 140, 218
operating models (OMs), 61; CCSBT unified model, 250–53; electronic tagging to develop, 95; in management strategy evaluation, 36–37; reconditioned, 259, 262; reference set of, 214, 252–53, 257–58, 260–62; in Southern bluefin tuna stock assessment, 218, 233, 237; spatial, 112, 118, 123–25, 200, 226. *See also* simulation models
Oppianus, 3, 5
OTN, 109–10
otolith: and age estimation, 132, 140, 142, 216, 220, 236; analysis of, to determine Pacific bluefin tuna movement, 166, 176, 178; of juvenile Atlantic bluefin tuna, 47–50; microchemistry of, 15–16, 29, 32, 45–62, 87, 103, 117, 119–20. *See also* stable isotopes
overfishing, 3, 22–25, 33, 95, 147, 185, 299; incentives for, 327

Pacific Ocean, eastern (EPO): fisheries in, 133, 282–84; fishing mortality rates in, 188, 193–95; migrations to/from, 137–38, 147–48, 152–55, 165–78, 185; northern, 186, 193–96; priors, 191; southern, 186, 193–95, 202; tag reporting rates, 196
Pacific Ocean, western (WPO): isotopic signature, 167, 173–74; migrations, 147–60, 172, 177, 184, 195–96
panmictic population, 68, 70, 77, 79
parasites, 321, 325
parent-offspring pair (POP), 232–34, 262, 264–65
pathogens, 323, 325–26
patterns, resident, 56
performance criteria, 257
performance statistics, 36, 257–58
piston line survey, 229
plausible models, 265
POAMA, 274, 277–78, 282
pole-line fishery, 138, 230

pollutant tracer, 46
population: depletion of eastern Atlantic bluefin tuna, 9; depletion of western Atlantic bluefin tuna, 8, 299; depletion of Southern bluefin tuna, 217, 221, 223; dynamics of, 45, 117, 123–25, 189–192, 200, 215–16, 227–28, 237, 255, 264; estimates of, 15; genetics of, 80, 86–87, 96, 124; genomics of, 61–62; management of, 67, 87, 95, 124, 224, 270; natal homing in, 56, 59; otolith assignment of, 46, 60; ranges of, 14; rebound of Southern bluefin tuna, 237; structure hypotheses, 68–70, 77–81, 85–86; structure of, 45–47, 61–62, 67–70, 72, 77–81, 85–87, 165–66, 220, 235. *See also* subpopulation
postovulatory follicles, 83, 140
predator interactions, 321, 323–24
Predictive Ocean Atmosphere Model for Australia (POAMA), 274, 277–78, 282
prey: diel migrations of, 222; effects of, on distribution, 284, 286; effects of, on transpacific migration, 152, 155, 170–71; models of, 286–87; species of (larval, juvenile, adult), 136–39. *See also* feeding
process uncertainty, 252
profitability, 276, 316
projections: abundance, 16, 19; catch, 250, 259; climate, 287; movement, 112; recruitment, 19, 22; of selectivity patterns, 22; spawning stock biomass, 19, 259; stock status, 22, 25–27. *See also* forecasts
purse seine fishery: Atlantic bluefin tuna, 6–10, 18, 20–21, 27, 75, 299; Pacific bluefin tuna, 136, 138–40, 188; Southern bluefin tuna, 221, 238, 248, 273–74

quota, 123, 125, 250, 265, 273, 287, 299; compliance with, 27; negotiations of, 33, 259, 288; over quota, 212; reduction of, 25, 251; sale of, 21; shares of, 284–85; unwanted capture without, 226, 271, 276–77, 285

ram ventilator, 317
range contraction, 212, 223
rate of increase, intrinsic, 228
Real Time Monitoring Program (RTMP), 215
recruitment: dynamics of, 95, 253; estimates of, 212, 214; failure of, 257–58; functions of, 59; importance of the Great Australian Bight to, 227; indices of, 214, 231, 237–38, 251; levels of, 60, 248–50, 253, 256; monitoring, 211, 227–31; monitoring index, 263–64; parameters, 252, 255; scenarios for Atlantic bluefin tuna, 11, 19–20, 22–30; spawner, 214
regional endothermy, 317, 325
regional fisheries management organization (RFMO), 33–34, 36, 248–49, 257, 264–66
reproduction, 82–86; analyses of, 131; in aquaculture, 299, 308, 315–16; biology of, 82, 85; costs of, 84; dynamics of, 227; exhaustion due to, 138; and maturity, 218; output of, 84, 131, 142, 170; parameters of, 85, 220; potential, 82, 85, 264; and season, 83–85; status of, 137, 139; strategies for, 96; traits of, 82–83. *See also* spawning
research, 31–35; analysis of sex hormone profiles, 83; aquaculture, 299–300, 315; electronic tag, 96, 98; future needs and plans, 68, 86, 138, 202, 236–38, 319, 323, 325–26; otolith chemistry, 46, 54; strategic fisheries, 211, 215, 227–29; uncertainties in, 47, 86, 168; vessel-associated studies, 134. *See also* GBYP
resident individuals, 56–57, 70, 73, 170–74, 176
resident populations, 56, 73, 79, 170–71, 177–78, 230, 323
resource efficiency, 316, 319, 327–28
resource intensity, 312, 318–20, 326
review: of candidate MPs for Southern bluefin tuna, 254; of environmental impacts of marine aquaculture, 321–26; of ICCAT Bluefin Tuna Working Group, 68; of Southern bluefin tuna spawning stock abundance, 233; of Southern bluefin tuna TAC increase, 263

robustness test, 257–60; of sublinear relationship between CPUE and abundance, 260
RTMP, 215

Safina, Carl, 3
satellite imagery, 279
scientific observers, 10, 215–16, 229, 277
SCRS (Standing Committee on Research and Statistics): advice (recommendations to ICCAT), 8–13, 23, 25, 28, 31, 35; assessment models, 67–68; catch estimates, 10, 20–21, 28; ICCAT software catalogue, 35; illegal, unreported, and unregulated catch identified by, 26; stock assessment, 15–17, 21; two-area model, 14; and uncertainty, 34; unreported catch estimates, 7
seasonal upwelling, decline in, 222
sea surface temperature (SST): and distribution of Pacific bluefin tuna, 135, 138, 152, 154, 156; and distribution of Southern bluefin tuna, 222, 224, 274–76, 278–80, 282; energy intake and, 199–200; and spawning, 109, 138, 141, 224
selective breeding, 316, 320, 323
set-net fishery, 136
sex ratio, 84–85, 217
sexual dimorphism, 132, 217
simulation models, 15, 34–36, 61, 118, 121, 124–25, 231
single nucleotide polymorphism (SNP), 80, 234
site fidelity: to feeding and foraging areas, 73, 75–76, 98, 103; postlarval, 235; to spawning grounds, 55, 71–72, 74, 98, 100
size structure, 173, 177
Slope Sea, 84, 94, 100–111
*Song for the Blue Ocean* (Safina), 3
Southern California Bight, 197, 286
spatial distribution, 77, 201–2, 213, 271, 284
spatial dynamics, 220–27; adult, 224; data on, 237–38; gravity in, 119; juvenile, 221–22; larval, 221; in management, 224–27; in MAST, 117–19; mixed-stock, 123–25
spatial management, 9, 202, 226, 238, 270–71, 276–77, 284
spatial models, 117–25, 200, 226
spatial structure, 124, 185
spawner-recruit relationship, 23, 30
spawning, 71, 73, 87, 141, 158, 281; in aquaculture, 300, 304, 310, 316, 323; area of origin identified with microchemistry, 177; areas of Atlantic bluefin tuna, 29–30, 45–46, 51–59, 68–78, 84–86, 94, 96, 101–3, 118, 281–82; conditions for, 100, 141, 224, 280, 284; duration of, 83, 104, 139–41, 218; fidelity of, 55, 72, 98, 100; frequency of, 30, 83, 85, 140, 218; grounds for, identified by electronic tags, 108–9, 138, 158; grounds for, identified by juvenile catch, 134; in Gulf of Mexico, 107; importance of archival tags to understand, 98, 100–6; indeterminate, 83, 140; and migration, 71, 82, 99, 104–5, 138, 148, 158, 173, 219; of Pacific bluefin tuna, 134–39, 141, 147–48, 168–70, 184; and populations, 95, 217, 264; rate of, 82; season for, 104, 139–40, 147, 218, 301; skipped, 83–85, 87, 227; of Southern bluefin tuna, 217, 221, 224, 227, 235, 281–82. *See also* reproduction
spawning stock: interrelations, 82; size and age distribution of, 220; status, 228
spawning stock biomass (SSB), 19–26, 95–96, 119–20, 214, 232, 253, 262, 265; decline or depletion of, 11, 22, 24, 95–96, 185, 212, 250–51, 265; estimates of, by CKMR, 216, 219; estimates of, by "knife-edge" maturity relationship, 218; estimates of, projected levels, 19, 23–24, 26, 30, 96, 120, 228, 231–33, 253, 259, 262–65, 278–79; historical, 95, 250–51, 253; proxy for, 279; rebuilding, 22, 24–25, 28, 250, 256, 260, 264
spotter planes, 221
spotter tuna, 230
stable isotope analysis (SIA), 47, 55, 138, 166–68, 170–77, 184–85
stable isotopes, 46–48; $\delta^{18}O$, 47–50, 53, 55–57, 166–68; $\delta^{13}C$, 47–50, 53, 166–68, 171, 184–85

standing stock biomass, 118; decline, 11, 95–96, 249 (*see also* stock depletion)
Stanford University, 96, 174
statistical models, 39, 41, 75, 118, 121; MAST, 15, 118–25. *See also* virtual population analysis
St. Lawrence, Gulf of (GSL): acoustic survey in, 32; electronic tagging, 76, 102–5, 109–11; indices, 11, 17–19; in MAST estimates, 120; stock composition, 51, 53
stock assessment: Atlantic bluefin tuna, 9–36, 45–46, 58–61, 95–96; best, 252, 264–65; challenges, 117–18, 234; Pacific bluefin tuna, 185, 191–93, 201–2; Southern bluefin tuna, 212, 214–15, 218–20, 227–38, 249–52, 262, 278; uncertainties in, 26–30, 34–35, 59, 67–85, 201–2, 249, 281, 284
stock composition analysis, 21, 48, 51–54
stock depletion, 8, 185, 202, 212, 235, 250, 253, 262, 265–66, 270
stock identification, 47–62, 67, 70, 96
stocking density, 303, 305, 314
stock mixing. *See* mixing
stock of origin, 31, 58, 60, 96, 118–19, 121, 124, 185
stock productivity, 59–61, 82, 118, 220, 249, 262, 281; uncertainties in, 249, 281
stock rebuilding: CCSBT plan for Southern bluefin tuna, 211, 214–15, 220, 227–28, 231, 238, 248–66; electronic tags for Pacific bluefin tuna, 202; ICCAT plan for Atlantic bluefin tuna, 11, 15, 19–20, 22, 28, 95–96
stock-recruit relationship, steepness of, 252
stock status: Atlantic bluefin tuna estimates of, 22–28; Atlantic bluefin tuna compared to historical, 95–96; and CITES, 123; effects of two-area VPA model on, 14–15; of Gulf of Mexico population, 103; improvement in, 28; in MAST model, 118; and overfishing, 185; of Southern bluefin tuna, 211, 214–15, 237, 249, 251–52, 262, 265; uncertainties in, 23, 28–30, 34–35, 61, 226, 201–2, 235, 249–52, 263, 281; of western stock, 123
stock structure, 28, 118, 158, 160, 200, 233–35
stomach content, 134, 136–37
Stone Age, 3, 5
structural uncertainty, 252
subpopulation: Atlantic bluefin tuna, 47, 67–87, 236; management of, 67–68, 70, 87; Pacific bluefin tuna, 155; structure of, 68, 70, 72, 77–78, 85, 87
substructure. *See* subpopulation
surface death, 302, 304
surfacing behavior, 221, 224
sustainable utilization objective, 266
swordfish, 270, 283, 285

Tag-A-Giant (TAG), 96, 100–101
tagging programs: Canadian electronic, 105; dedicated spawning ground, 281; genetic, 229–31, 263; insights into age-varying spatial dynamics, 237; Southern bluefin tuna conventional, 228; Tag-A-Giant, 96, 100–101; Tagging of Pacific Pelagics, 186
tag-recapture: acoustic, 109–11; Bayesian, 186–90; limits of, 165, 177–78; in MAST, 118–19, 122–23; numbers of Pacific bluefin tuna for, 151–52; pop-up satellite, 87; revealing Pacific bluefin tuna migration, 169–70
Taiwan, 137–41, 168, 173, 176, 184, 201, 212, 216–17, 249
Tasman Sea, 167–68, 217, 222–24
Tesla Motors, 327
thermoregulation, 3, 156
*Thunnus maccoyii*, 8, 94, 133, 165, 167, 197, 201, 211, 270
*Thunnus orientalis*, 5, 47, 94, 131, 147–48, 165, 167, 184, 202, 270
*Thunnus thynnus*, 3, 45, 67, 94, 133, 165, 235, 270, 299–300
Tiffany & Company, 327
TOPP (Tagging of Pacific Pelagics), 99, 105, 186
total allowable catch (TAC): Atlantic bluefin tuna catch levels compared to, 7, 10; facilitating rebuilding of Atlantic bluefin tuna, 95; global, 231; in Kobe II matrices, 24; management advice to ICCAT on, 9–13; in management procedures, 255–64; reductions in, 212, 284–85; and stock status projections, 27–28
trace elements, 46–47

tracking: with double tags, 109; Gulf of Mexico bluefin tuna, 103, 107–8; of network lines, 109–10; Pacific bluefin tuna migration, 149–50; progress of rebuilding models of, 95. *See also* electronic tagging
tracking data, 96, 97, 103
tracks: of Atlantic bluefin tuna from western releases, 97; of Atlantic bluefin tuna juveniles, 86; and Atlantic bluefin tuna spatial distribution, 77; of Atlantic bluefin tuna tagged in Bay of Biscay, 76; of Atlantic bluefin tuna visiting GOM and MED, 70, 105, 107–8; and foraging grounds, 85, 98; importance of archival, 98–100, 103; importance of multiyear, 71, 75, 86, 87, 99–100, 103, 112; to inform SCRS movement models, 32; originating in GSL, 102, 104–5; originating in MED, 86; of Pacific bluefin tuna juveniles, 158; predicted, 190, 220; of Southern bluefin tuna juveniles, 202; and spawning grounds, 71, 100–104. *See also* electronic tagging
trap fishery, 5–7, 9, 18–21, 25–26, 54, 73–74
TRCC (Tuna Research and Conservation Center), 174, 324
troll fishery, 136, 138, 229
Tsushima Island, 151–52
Tsushima Warm Current, 147, 152, 158
*Tuna: A Love Story* (Ellis), 3
tuning criteria, 256
tuning parameter, 255
uncertainty: in MAST, 120; in maturity, 218; in movement parameters, 121; in offshore aquaculture, 326; in reproductive parameters, 85; in seasonal movements and stock productivity, 61; in ship-based surveys, 226; in stock assessment, 23, 28–30, 34–36, 201–2, 235, 249, 251–52, 263, 281
United Nations Convention on the Law of the Sea (UNCLOS), 248–49

Valentino, 327
variation, interannual: in abundance, 76–77; in catch, 257–58; in distribution, 271; in fecundity, 220; oceanic, 287; in otoliths, 48; in recruitment, 23; in transpacific migrations, 155, 175–76
Veblen good, 327. *See also* luxury goods
virtual population analysis (VPA), 13, 20–21, 25, 58, 60, 118; effect of mixing on, 15, 32; with two independent stocks, 14–19
von Bertalanffy equation, 187
von Bertalanffy growth function, 132, 139
von Bertalanffy growth parameters, 321

weaning, 300, 303, 305–6, 310
*Wicked Tuna* (*National Geographic*), 3
WPO. *See* Pacific Ocean, western

Yaeyama Islands, 137–41
yellowfin tuna, 7, 168, 171–72, 226, 270, 276–77
young-of-the-year (YOY), 80–81